Food Safety and Quality Assurance

FOODS OF ANIMAL ORIGIN

SECOND EDITION

Food Safety and Quality Assurance

FOODS OF ANIMAL ORIGIN

SECOND EDITION

William T. Hubbert
DVM, MPH, PhD, Dipl ACVPM

Harry V. Hagstad
DVM, MPH, Dipl ACVPM, Fellow ACE

Elizabeth Spangler
DVM, PhD, Dipl ACVPM

Michael H. Hinton
BVSc, PhD, DSc, FRCVS, FRC Pathology

Keith L. Hughes
MVSc, PhD, Dipl Bacteriology (London), MACVSc

Blackwell
Publishing

William T. Hubbert, DVM, MPH, PhD, Dipl ACVPM, is Adjunct Professor, School of Veterinary Medicine, Louisiana State University.

Harry V. Hagstad, DVM, MPH, Dipl ACVPM, Fellow ACE, is Professor Emeritus, School of Veterinary Medicine, Louisiana State University.

Elizabeth Spangler, DVM, PhD, Dipl ACVPM, is Associate Professor, Atlantic Veterinary College, University of Prince Edward Island, Canada.

Michael H. Hinton, BVSc, PhD, DSc, FRCVS, FRC Pathology, is Professor, School of Veterinary Science, University of Bristol, England.

Keith L. Hughes, MVSc, PhD, Dipl Bacteriology (London), MACVSc, is Dean, School of Veterinary Science, University of Queensland, Australia.

© 1991, 1996 Iowa State University Press, Ames, Iowa 50014

Original © 1982 Iowa State University Press as *Food Quality Control: A Syllabus for Veterinary Students;* © 1986 Iowa State University Press as *Food Quality Control: Foods of Animal Origin*

Orders: 1-800-862-6657
Office: 1-515-292-0140
Fax: 1-515-292-3348
Web site: www.blackwellprofessional.com

Blackwell Publishing Professional
2121 State Avenue, Ames, Iowa 50014

Printed on acid-free paper in the United States of America

First edition, 1991
  Second printing, 1993
Second edition, 1996

Library of Congress Cataloging-in-Publication Data

Food safety and quality assurance: foods of animal origin / William T. Hubbert ... [et al.].—2nd ed.
          p.   cm.
     Previous ed. cataloged under Hubbert.
     Includes bibliographical references and index.
     ISBN 0-8138-0714-X
     1. Food of animal origin—Health aspects. 2. Food of animal origin—Contamination. 3. Food industry and trade—Quality control. 4. Foodborne diseases—Prevention. 5. Food adulteration and inspection. I. Hubbert, William T.
RA601.H82   1996
664′.907—dc20
                                                                                    96-1347

Contents

Preface

This text is a successor to *Food Quality Control: A Syllabus for Veterinary Students*, published in 1982, *Food Quality Control: Foods of Animal Origin*, published in 1986, and the previous edition of *Food Safety and Quality Assurance: Foods of Animal Origin*, published in 1991. This edition is approximately 50% larger than the previous edition. Major additions include a new Chapter 1, introducing veterinary medicine and food safety, as well as new topics in subsequent chapters regarding eggs, risk assessment, safe food handling at home, and the evolution of food inspection in Australia and the United Kingdom. Throughout the text, as with earlier editions, comments from the review panel have been particularly valuable in identifying areas in need of addition or revision.

Unlike texts in which attention is focused on only one part of food animal production or processing, an overall view of the food chain is presented so that the reader may better recognize potential sources of contamination.

The purpose of this text is to prepare students to

- Identify human health hazards in foods of animal origin.
- Identify the role of veterinarians in preventing introduction of health hazards into the food chain.
- Identify governmental and private sector organizations and their activities in regard to maintaining safety and wholesomeness of foods of animal origin.
- Identify principles of safe food production, processing, and handling.
- Collect and analyze data relevant to investigation of foodborne disease outbreaks.

Material presented is designed to serve as a guide to current knowledge. In-depth information may be obtained from the publications listed in the bibliography following each chapter.

Unlike the last edition and the other predecessors listed at the beginning of the preface, this edition goes beyond North America in scope. The current practice of food safety has developed a scientific structure on a base of social and political history. Therefore, we find many interesting regional differences, not the least of which are the terms used to describe the products and functions. There is a global trend toward harmonization among these markets in the interests of increasing opportunities for trade. As economic trade barriers fall, and new trading groups are forged by governments, it will become increasingly important to understand food issues, both regulatory and biological, from a global perspective.

Abbreviations

The following abbreviations taken from the International System of Units (SI) are used in this text: *millimeter (mm), centimeter (cm), meter (m), kilometer (km), milligram (mg), gram (g), kilogram (kg), tonne, metric (t), milliliter (ml), liter (l), second (s), minute (min), hour (h),* and *day (d).* In addition, the following abbreviations accepted by the Council of Biology Editors are used: *inch (in.), foot (ft), yard (yd), ounce (oz), pound (lb), pint (pt), quart (qt), gallon,* imperial or U.S. *(gal), week (wk), month (mo),* and *year (yr). Mile* and *ton* are not abbreviated.

Acknowledgments

The authors are deeply indebted to the following persons who constituted a review panel to evaluate this edition. Their comments and suggestions were invaluable in the development of this book.

Dr. A. A. Adesiyun
University of the West Indies
Champs Fleurs, Trinidad and Tobago

Dr. Eddie Andriessen
Department of Primary Industries and
  Energy
Adelaide, Australia

Dr. Tore Aune
Norwegian College of Veterinary
  Medicine
Oslo, Norway

Dr. Loinda R. Baldrias
University of the Philippines at Los
  Banos College
Laguna, Philippines

Dr. Paul C. Bartlett
Michigan State University
East Lansing, Michigan, United States

Dr. George W. Beran
Iowa State University
Ames, Iowa, United States

Dr. Asa B. Childers
Texas A&M University
College Station, Texas, United States

Dr. Peter Cowen
North Carolina State University
Raleigh, North Carolina, United States

Dr. David W. Dreesen
University of Georgia
Athens, Georgia, United States

Dr. J. R. Egerton
University of Sydney
Camden, Australia

Dr. Don A. Franco
Food Safety and Inspection Service
Washington, D.C., United States

Dr. Constantin Genigeorgis
University of California
Davis, California, United States

Dr. John C. Gordon
Ohio State University
Columbus, Ohio, United States

Dr. Michael H. Hinton
University of Bristol
Langford, England

Dr. Keith L. Hughes
University of Queensland
Brisbane, Australia

Dr. Sarah Kahn
Embassy of Australia
Washington, D.C., United States

Dr. Reuven A. Kathein
Hebrew University of Jerusalem
Jerusalem, Israel

Dr. Anne A. MacKenzie
Agriculture Canada
Moncton, New Brunswick, Canada

Dr. Ingmar Mansson
Swedish University of Agricultural
   Sciences
Uppsala, Sweden

Dr. Robert B. Marshak
University of Pennsylvania
Kennett Square, Pennsylvania,
   United States

Dr. Geoffrey Mead
University of London
Potters Bar, England

Dr. Patrick M. Morgan
Oklahoma State University
Stillwater, Oklahoma, United States

Dr. David A. A. Mossel
University of Utrecht
Utrecht, Netherlands

Dr. John C. New
University of Tennessee
Knoxville, Tennessee, United States

Dr. Paul L. Nicoletti
University of Florida
Gainesville, Florida, United States

Dr. Kevin D. Pelzer
Virginia Polytechnical Institute and
   State University
Blacksburg, Virginia, United States

Dr. Alan Royal
New Zealand Meat Research and Devel-
   opment Council
Wellington, New Zealand

Dr. Pamela L. Ruegg
University of Prince Edward Island
Charlottetown, Prince Edward Island,
   Canada

Dr. A. Mahdi Saeed
Purdue University
West Lafayette, Indiana, United States

Dr. M. D. Salman
Colorado State University
Fort Collins, Colorado, United States

Dr. C. Donald Seedle
Kansas State University
Manhattan, Kansas, United States

Dr. Peter Seneviratna
Canberra, A.C.T., Australia

Dr. Donald F. Smith
Cornell University
Ithaca, New York, United States

Dr. Richard E. Smith
Louisiana State University
Baton Rouge, Louisiana, United States

Dr. Ronald D. Smith
University of Illinois
Urbana, Illinois, United States

Dr. Albert E. Sollod
Tufts University
North Grafton, Massachusetts, United
   States

Dr. Diana M. Stone
Washington State University
Pullman, Washington, United States

Dr. David A. Stringfellow
Auburn University
Auburn, Alabama, United States

Dr. Clyde E. Taylor
Mississippi State University
Mississippi State, Mississippi, United
   States

Dr. James G. Thorne
University of Missouri
Columbia, Missouri, United States

Dr. Michael V. Thrusfield
University of Edinburgh
Easter Bush, United Kingdom

Dr. J. U. Umoh
Ahmadu Bello University
Zaria, Nigeria

Dr. C. M. Veary
University of Pretoria
Onderstepoort, Republic of South Africa

Dr. Saul T. Wilson, Jr.
Tuskegee University
Tuskegee, Alabama, United States

The authors gratefully acknowledge the expert editing advice provided by **Sandra L. Hubbert,** BAJ, MJ.

Food Safety and Quality Assurance

FOODS OF ANIMAL ORIGIN

SECOND EDITION

1

Veterinary Medicine and Food Safety

Objectives

- Describe the impact of veterinarians on food safety and quality assurance throughout the food chain and explain how they are prepared to perform these functions.

- Describe the impact of veterinarians on food safety and quality assurance throughout the food chain and explain how they are prepared to perform these functions.

- Explain how societal and technologic changes affect food safety in veterinary medicine.

Definitions
Foods of Animal Origin

Veterinarians are recognized for their expertise in maintaining the health of animals and provide leadership worldwide in protecting people from disease acquired as a result of eating foods of animal origin, such as fish, meat, milk, poultry, and eggs. Although this list includes most of the food of animal origin consumed by the world's people, it does not include all products actually consumed. Because veterinarians are often asked questions involving events affecting human and animal interaction, we begin with a brief overview of the types and sources of foods of animal origin.

Food is composed of inorganic (inorganic chemicals and water) and organic (carbon-based) nutrients. All of the organic nutrients (carbohydrates, fats, proteins, vitamins) we eat come from living organisms, either plants or animals.[1,2,24,107] Among the 15 main groups of food commodities (Table 1.1) recognized by the Food and Agriculture Organization (FAO) of the United Nations, 9 are of plant origin: cereals (wheat, rice, corn), roots and tubers (potatoes, manioc), pulses (edible seeds, usually dried, of beans, lentils, peas), nuts and oilseeds, vegetables, fruits, spices, stimulants (coffee, tea), and alcoholic beverages (beer, wine). This list does not include many minor foods of plant origin (marshmallow), edible fungi (mushrooms), or alcoholic beverages of animal origin, such as fermented milk from goats (kefir) or mares (koumiss). Two of the main groups include commodities of both plant and animal origin: sugars and honey and oils and fats (including butter, lard, and tallow). Actually, honey is plant nectar that has been concentrated and partially hydrolyzed by honeybees.[42] The remaining four groups comprise foods of animal origin: meat and offals, eggs, fish and seafood, and milk. Here, the term *offal* refers to edible parts of the animal other than muscle meat, e.g., blood and liver. In the FAO list, cheese is included under milk but is not specifically identified. These groups are widely used by nutritionists in human dietary studies because they represent the most significant nutrient sources worldwide.[17]

Besides foods of animal origin, listed in Table 1.1, people also consume flesh from members of every class of vertebrate: sharks, bony fishes, amphibia (frog legs), reptiles (alligators, snakes, turtles), wild birds, and mammals.[185,196,211,241] In addition to bird eggs, we eat eggs from turtles and fish (sturgeon [caviar], lumpfish). We drink milk from a great variety of mammals in addition to our own mothers' milk. Bird's nest soup, a Chinese delicacy, is prepared from the nest of a Southeast Asian swallow. The bird builds its nest from a gelatinous material formed by partial digestion of a marine plant. People also use a whole host of invertebrates for food, including edible insects (larvae, grasshopper legs), crustacea (crabs, crayfish, lobsters, shrimp), molluscs (clams, mussels, octopuses, scallops, slugs, snails, squid), echinoderms (sea cucumbers, sea urchins), and annelids (earthworms).[26,60,103]

Obviously, consumption of most of these animals or their products is restricted by availability and preference. Few of the species consumed are avail-

Table 1.1. FAO main food commodity groups, by origin

Plant	*Animal*
1. Cereals	10. Meat and offals
2. Roots and tubers	11. Eggs
3. Pulses	12. Fish and seafood
4. Nuts and oilseeds	13. Milk
5. Vegetables	
6. Fruits	*Plant and animal*
7. Spices	14. Sugars and honey
8. Stimulants	15. Oils and fats
9. Alcoholic beverages	

able commercially although many of them occasionally may have been involved in human foodborne illness because they were the vehicle of an infectious, parasitic, or toxic agent (see Chapter 3). It is for this reason veterinarians must be aware of them so they may recognize and reduce the risk from their associated hazard.

Since early history, a small number of bird and mammal species have been domesticated for food.[141] The origins of the birds include Africa (guinea fowl, ostrich), Asia (pheasant), Europe (duck, goose), India (chicken), the Mediterranean (pigeon), and Mexico and tropical America (muscovy duck, turkey). Domestication of food-producing mammals originated in Europe (cattle, pig, rabbit, reindeer, sheep), western Asia (goat, horse), and India (cattle, water buffalo). Domesticated camelids are found in Asia (Bactrian camel), Arabia and northern Africa (dromedary), and South America (llama). Subsequent selection and geographic dispersal have resulted in the development of diverse breeds within most of the species to provide food (meat, milk, eggs) and fiber (feathers, hides, wool). More recently, several species of cervids (deer, elk) as well as several aquatic (freshwater and marine) species (e.g., catfish, salmon, shrimp, tilapia) have been reared commercially for food. Emphasis in Chapters 2 and 4 is on the terrestrial and aquatic species and their products harvested commercially for food (Fig. 1.1).

The Food Chain

Ecologists define the food chain as (1) the transfer of food energy from plants to herbivores, who eat them, and then to carnivores, who subsequently eat the herbivores (grazing food chain) or (2) the feeding of microorganisms on dead organic matter and the organisms that subsequently eat the herbivores (detritus food chain).[165] The ultimate source of energy is radiant energy from the sun for plant growth. Because food chains are interdependent in every ecosystem, their

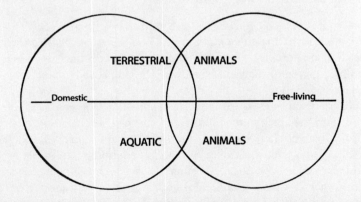

Fig. 1.1.   Sources of foods of animal origin.

overall activity is referred to as a *food web*. Unicellular marine plants (phyto-plankton) are eaten by unicellular marine animals (zooplankton) that are the principal food source of baleen whales. In addition to being of interest to zoologists, food chains are often the underlying reason for specific patterns of foodborne disease because of the movement of disease agents through them. For example, phytoplankton (diatoms and dinoflagellates) and multicellular marine algae may be sources of toxins in certain seafood (see Chapter 3). Some third-stage parasitic larvae in fish muscle are transferred sequentially to other predator fish (paratenic host) without undergoing further development (anisakids). These larvae then cause significant allergic enteric reactions when finally eaten by previously sensitized persons.

Although food chains are essential to the survival of all animals in nature, it is the human food chain that is of most importance to the well-being of people. This chain begins with all the activities of agriculture (plant and animal) and fisheries (freshwater and marine) involved in producing and harvesting food, progresses to intermediate processing and marketing functions, and ends when the food is prepared and eaten by the consumer. For foods of animal origin, the chain doesn't begin with the animals or their products we eat but with the production of the plants that are eaten by the animals at the beginning of the chain. Ruminants are herbivores and eat grass, whereas swine are omnivores and also eat some foods of animal origin; however, today the food chain may be much more complicated with, for example, recycling of animal wastes to herbivores and occasional contamination of feed with toxic industrial chemicals (e.g., polybrominated biphenyls). It is this food chain "from farm to table" that attracts public attention in regard to food safety (Fig. 1.2).[208] Every stage is essentially subject to human manipulation.[44,51,157] Veterinarians have an important role in ensuring the quality and safety of foods of animal origin throughout the food chain (see Chapter 2).

Preharvest and Postharvest Food Safety

Food safety, or *food hygiene,* has been defined by a Joint FAO/WHO (World Health Organization) Expert Committee on Food Safety as, "All conditions and measures that are necessary during the production, processing, storage, distribution, and preparation of food to ensure that it is safe, sound, wholesome, and fit for human consumption."[149]

Preharvest and *postharvest* are terms applied in food commerce to distinguish "on-farm" production functions from "in-plant" processing functions. Although *farm gate* is often used to identify the preharvest endpoint for all farm products, this is not a clear dividing line for characterizing food safety aspects of food animals and their products. The time when the physiologic activity of the live animal ends is a more useful point of demarcation between preharvest and postharvest in the evaluation of food safety risks. Before this point, exposure to pathogens and toxic chemicals can still result in systemic distribution via the bloodstream as well as the expected bodily response. The body of the live ani-

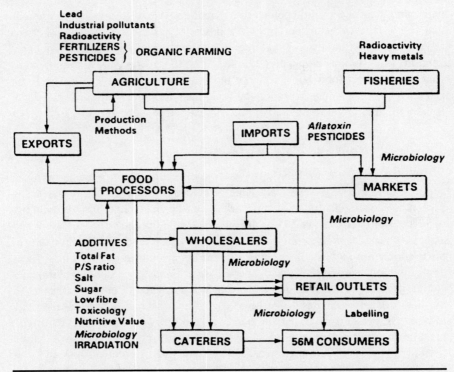

Fig. 1.2.   The food chain. Bold lowercase = general concerns; capitals = mainly consumer-led concerns; italics = mainly science-led concerns; 56 M = population of the United Kingdom. Source: (208)

mal still responds to physical trauma with hemorrhage, etc. Animals for slaughter are still alive on the farm, whereas they are carcasses of meat in the plant (abattoir). Milk, preharvest, is still in the udder, whereas postharvest it may be in a bulk tank on the farm or in a milk-processing plant. Transport of live animals to slaughter is a preharvest activity, whereas transport of milk from the farm is a postharvest activity. Food may be exposed to numerous postharvest activities before it is eaten by the consumer.

For on-farm-produced foods of animal origin, preharvest and postharvest food safety are extremely important concepts. The animal is most likely to be exposed to infectious and parasitic pathogens as well as residue-producing chemicals while still alive on the farm. This is also true of aquatic species reared in commercially maintained beds or ponds. If on-farm exposure to these foodborne hazards is prevented, they do not enter and persist in the food chain. Parasitism with foodborne parasites can only be established in the living animal host. Chemical residues can only accumulate from exposure while the animal is still alive. These residues cannot be inactivated or removed later from the meat or milk. This does not preclude possibilities for postharvest contamination of the food with pathogens or harmful chemicals. Maintaining a healthy, pathogen-free animal on the farm is the primary prevention. On the other hand, all we do later to kill pathogens in food by cooking, freezing, pasteurization, etc., is secondary prevention. Veterinarians play a pivotal role in on-farm primary prevention that enhances preharvest food safety.[8,9,67] For the benefits of primary prevention to be successful in food safety, however, the food must also be continuously protected from contamination throughout the postharvest period (see Chapter 2).

Quality Assurance and Quality Control

Today, *quality assurance* (QA) and *quality control* (QC) are terms used commonly in industry to indicate integral industry functions. QA refers to the "corporate" oversight function to clearly state the "corporate" product quality objectives and goals as well as to affirm that the QC program is functional and achieving these objectives and goals. QC, on the other hand, is an on-line or production function that establishes and administers the day-to-day policies, procedures, and programs at the "plant" level.[215] Kraft Foods, for example, designed a companywide QA program outlining what was expected to ensure safety in all its food products, whereas it was up to each plant manager to see that the daily in-plant QC activities in the production of its product were done satisfactorily to meet the overall corporate QA goal. QA and QC programs are used throughout the food-processing industry to ensure that standards of product input and output, including governmental food safety and consumer acceptance standards, are being met.

During the 1980s, food-animal producer organizations began developing QA programs, principally in response to the need to control chemical residues in their products. In the United States, for example, the National Pork Producers Council developed a QA program with the individual swine producer undertaking the daily QC functions. These food animal industry-driven programs require the establishment of an effective client-patient-veterinarian relationship on the farm. Although initiated for residue prevention, these on-farm QA programs form an excellent basis for complete herd health programs.

Veterinarians' Input into the Food Chain
Yesterday (before the 1950s)

In the late 1700s, the human population of Europe was returning to the numbers that existed before the great plague of the mid 1300s. With this growing human population, an adequate food supply was critical, and at this time, the European cattle population, of major dietary and economic importance, was threatened by rinderpest.[195] The disease had spread from Africa by way of the Middle East. The first two veterinary schools, Lyon and Alfort, were founded in the 1760s in France to seek ways of combating this scourge and to produce graduates skilled in its control. This was the beginning of veterinary medicine as the health profession we recognize today. During the next century, graduates of the French schools were instrumental in the establishment of other schools in Europe, including the United Kingdom, and North America. The graduates of these schools were in demand by governments to spearhead control of catastrophic contagious diseases of livestock. The numerous examples of eradication of diseases such as rinderpest, foot-and-mouth disease (FMD), and contagious bovine pleuropneumonia years before the causes of these diseases were elucidated testifies to their skill. In particular, rinderpest was eradicated from Europe and FMD was eradicated in North America each time after several introductions. The success of these programs, particularly from the mid 1800s through the early 1900s, to control and eradicate catastrophic diseases of livestock had a great impact on the availability of animal protein for human consumption.[251]

To prevent marketing of diseased meat, in 1853 Germany was the first country to introduce veterinary inspection of animals at slaughter at all abattoirs (including those for horses), livestock markets, and "farmers'" markets in the district of Berlin.[118,213] The state of Prussia passed a meat hygiene law in 1868, and a national law was passed in 1900. Routine microscopic examination of pork for cysts of *Trichinella spiralis* began in Berlin in 1865. This was done for two reasons: (1) trichinosis from eating uncooked sausage was recognized as a serious human health problem, and (2) the significance of trichina cysts in pork was known. Grossly observable lesions, such as the abscesses of tuberculosis and other diseases, parasite cysts, and anthrax hemorrhages, were among the principal reasons for rejection of cattle. Meat inspection was a major topic at the Third International Veterinary Congress held in Zurich in 1867. During the late 1800s, governments in several countries organized meat inspection programs based on the German model.[23] Although protection of an export market was the main driving force behind establishment of some programs, the underlying justification was always protection of human health.

Throughout the nineteenth century, veterinarians were among the leaders in promoting a safe milk supply.[21,37,65,79,117,172,174,189,205,206,232] This was particularly important for children because this valuable food was frequently implicated in outbreaks of childhood diseases such as scarlet fever. Veterinarians in many countries promoted "clean milk from healthy cows." In 1886, Bernard Bang, a Danish veterinarian, published his monumental findings on bovine tuberculosis. This work included a description of the tuberculin test, arguably the most valuable tool developed for the diagnosis of any infection in animals. By the beginning of the twentieth century, tuberculin testing of dairy cows had become a widespread requirement of municipalities in many countries for the sale of milk. It was not until the early 1900s, however, that the efforts of Bang and other veterinarians succeeded in getting recognition that *Mycobacterium bovis* was a human pathogen. In 1897, Bang demonstrated *Brucella abortus* to be the cause of bovine brucellosis, while 20 yr later another veterinarian, Charles Carpenter, proved the relationship between cows' milk and human brucellosis. He isolated *B. abortus* from students in a U.S. university dormitory who drank raw milk and developed undulant fever. He also isolated the agent from milk obtained from the dairy of a faculty member whose infected cows supplied the dormitory. Commercial pasteurizers were introduced in the 1890s. Although for several decades veterinarians encouraged pasteurization to kill milkborne pathogens, sale of pasteurized milk grew slowly until the 1920s because of public and industry resistance.

In 1885, Daniel Salmon, who was the **first** graduate of a U.S. veterinary school (Cornell, 1876), while searching for the cause of swine fever (hog cholera), published a description of the type species, *Salmonella choleraesuis,* of the bacterial genus later named in his honor. Worldwide, members of the genus *Salmonella* are the most important foodborne pathogens. Salmon was the first head of the U.S. Bureau of Animal Industry (BAI) and, in 1890, established the first federal meat inspection program in the United States. He was instrumental in ensuring the inclusion of food hygiene in early U.S. veterinary curricula because he made it a requirement that graduate veterinary inspectors were employed by the BAI. Through his contributions to food safety, Salmon served both his profession and the public well.

Canning to preserve food was introduced in the late 1700s by the Frenchman Appert. Gail Borden marketed the first canned sweetened condensed milk in 1857.[255] Commercially canned foods, however, were frequently a source of botulinum toxin until the 1920s. Then Karl F. Meyer, a Swiss veterinarian working in the United States, demonstrated the time-temperature-pressure relationship needed to kill the spores of *Clostridium botulinum,* a process that is used by the industry today.

Beginning in 1917, the United States utilized the tuberculin test in a nationwide campaign of voluntary herd testing to reduce bovine tuberculosis.[156] Many veterinarians in private practice as well as employees of state and federal agencies were involved. By 1931, the prevalence of infected herds had been reduced sufficiently so that mandatory testing was initiated, including payment of indemnity to owners of reactor cattle slaughtered. At this point, the program had evolved over nearly 2 decades with continuous dialogue among representatives of livestock producer organizations, veterinarians, and government officials. The program had been highly successful and was perceived to be well accepted by all parties as providing considerable economic and public health benefits. In spite of this, producer groups in several states objected to mandatory testing, and they organized revolts in the form of mobs of hundreds of people. The governor of one state considered activating the militia to provide armed escorts for the veterinarians doing the testing. In the meantime, it became evident that many cattle producers had joined the revolt because of misinformation distributed by individuals opposed to testing. Once they had been adequately informed, producer opposition and tensions subsided. A few of the initial agitators were arrested. This so-called cow war, or TB war, is an excellent example of the need to develop widespread informed support for any disease control activity. The biomedical skills developed by veterinarians have occasionally preceded the public's understanding of their benefit.

Today (the 1950s to 1990s)

By the 1950s, most of the major diseases that cause significant losses (e.g., by death or abortion) in flocks and herds had been effectively controlled in many countries. The success of these disease control measures led some to believe that the days of the veterinarian in private food animal practice were numbered, particularly because on-farm disease control practices were perceived by producers as an additional production cost: treatment of a sick animal was added to the money lost with decreases in weight gain or milk yield. During this time, dairy farmers still had regular contact with veterinarians because of periodic testing of their milking herds for brucellosis and tuberculosis. On the other hand, a study of U.S. swine producers indicated that less than one-third utilized the services of a veterinarian. In the past 40 yr, the role of food animal veterinarians has changed from almost exclusive emphasis on diagnosing and treating diseased animals to include preventing infection and promoting health and production efficiency. Many producers now expect veterinarians to be an integral part of their food animal quality assurance (QA) program, with a valid veterinarian-client-patient relationship as a component. The demand by producers for veterinarians able to lead in the design and management of flock and herd health strategies will grow as long as producers perceive this as a route to increased production efficiency and product quality.[183,184,229] In addition, the population health promotion skills of veterinarians are being sought by today's rapidly expanding farm-reared fish and wild game industries, production areas almost unheard of 40 yr ago.

In the past, cooperative efforts in the control of food animal disease between veterinarians, both in private and public practice, were usually limited to specific programs such as the eradication of bovine brucellosis or tuberculosis.[108] The industry-driven QA programs, such as the Pork Quality Assurance program of the National Pork Producers Council (NPPC), will force greater communication among veterinarians, who will have input throughout the food chain. This trend to devise efficient means of benefiting producers and consumers by focusing on critical points throughout the food chain will certainly continue.[104]

Veterinary Medicine as a Health Profession

Two centuries ago, the first curricula in veterinary medicine were developed in response to the public need for effective control of catastrophic livestock diseases. Today, veterinary medicine is the only health profession that includes food safety as a significant part of the instruction in its professional curriculum as well as part of its postgraduate practice.

Much of the early stimulus for population-based components of the veterinary curriculum came from the need for animal disease control by veterinarians in public practice. This was an early emphasis of the European schools. In contrast, the first English school (Royal Veterinary College, London) was established to provide veterinarians for private practice, with an emphasis on the treatment of horses. These dual themes—public versus private and prevention versus cure—have provided the fuel for hours of discussion among veterinary curriculum designers everywhere. Over the years, these debates among academics reflected the profession's evolving perception of food safety because the faculty initially formed that perception among its students.

Soon after veterinary schools were established in Germany, formal training in a German school was required for their veterinarians to practice in either the public or private sector. In countries such as the United States, on the other hand, graduation became an early requirement for employment by public agencies but not for private practice. For example, from 1892, all veterinarians employed by the BAI were required to be graduates of a recognized veterinary school. (The BAI began accrediting U.S. schools, i.e., reviewing the suitability of their curricula against BAI standards, in 1906.) In contrast, many U.S. veterinarians in private practice were nongraduates who learned their skills through apprenticeship rather than formal academic instruction. Around the beginning of the twentieth century, many state governments in the United States introduced veterinary practice acts that required the passing of an examination to qualify for licensure. The examination typically included sections on materia medica, surgery, and food hygiene. Although the early acts did not require applicants to be graduates, nongraduates failed to pass the examination. At that time, the relatively limited materia medica and surgery could be learned by apprentices, whereas instruction in food hygiene was only available at a veterinary school. In Kansas, for example, the first practice act was passed in 1909, the same year the first class of veterinarians graduated from Kansas State College. It was 50 yr later before that practice act was amended to require all applicants to be graduates. In the meantime, however, no nongraduate was licensed. It is evident that food hygiene instruction played a significant, although subtle, role in the evolution of the U.S. veterinary profession in the early 1900s. This is in spite of the fact that the number of hours devoted to the subject has never been comparable to many European schools where a much higher proportion of the graduates are employed full-time in food hygiene.

Providing a safe and adequate food supply has been a central theme in the evolution of the veterinary profession throughout its history. Food-related issues will continue to affect veterinary education and practice in the future.

Driving Forces for Change: Past, Present, and Future Issues

The Human Population

During the past 90 yr, the human population of the world has increased 3.3-fold, with the number more than doubling since 1950. Whereas a century ago most people lived on farms, today most live in towns and cities. The number of people remaining on farms to produce the food for these urban dwellers is now only a fraction of what it was a few decades ago (<2% in some countries). This imbalance between the urban and rural human population will continue with even more to feed and fewer to feed them. Today's efficient food production systems require skilled persons to ensure a sufficient quantity of safe food products. Not only has the population increased, its structure has also changed to increase the numbers of those with greatest risk of acquiring severe foodborne disease, e.g., infants, elderly, undernourished (especially war- or famine-associated), and AIDS-affected persons.[10] This includes a disproportionate increase in the number of potentially immunocompromised individuals, including many who receive immunosuppressive drug therapy. For example, the percentage of elderly (>65 yr) in the developed world has doubled in the past 30 yr to more than 12%. Particularly since the advent of antibiotics, many more children with diseases such as Down's syndrome are surviving for much longer today.

Changes in eating habits in developed countries reflect today's emphasis on speed, resulting in a greater demand for prepared foods and fewer people who are willing to take the time to prepare food safely starting with basic (raw) ingredients.[80,100,167,221,256] Because the retail food market is the principal connection to the food chain for most of today's consumers, their knowledge of preharvest production is minimal.[7,41,68,72,74,77,91,129,227,228,245,254] This lack of knowledge may lead to actions by consumers that can be detrimental to producers, particularly when they rely on inaccurate sources of information regarding the safety and quality of the foods they eat.[18,19,25,132,191,199] For example, a product such as milk from cows injected with bovine somatotropin (BST) may be avoided because it is perceived to be unsafe. Also, because of the long postharvest (processing) portion of today's food chain, producers and consumers frequently fail to identify with one another. All too often when a problem arises, as with *Escherichia coli* O157:H7 in meat, consumers are willing to place the blame on producers or processors and ignore their own role in ensuring safe food by adequate handling or cooking.

The dietary habits of people are often classified loosely into two groups: omnivores (those who eat foods of plant and animal origin) and vegetarians.[16,27,52,59,75,81,99,134,135,137,139,152,155,176,200,207] Actually, vegans are the only true vegetarians because they are the only group that excludes all foods of animal origin from their diet. Lactovegetarians consume dairy products, and lacto-ovovegetarians consume eggs as well as dairy products. There are also semivegetarians, those who do not eat red meat but may eat some fish and poultry. The underlying reasons for these dietary patterns may be physiologic (based on required nutrients or food intolerance or sensitivity), sensory (based on food taste or texture), or cognitive (based on sociocultural beliefs and knowledge).[38,54,76,106,193,237] The most widely recognized sociocultural beliefs affecting food are the religious laws that proscribe some foods and require the strict handling of others. Also, proponents of vegetarianism and "clean living" activists encourage avoidance of foods of animal origin on various nutritional, moral, spiritual, and ecological grounds.[61] Representatives of these latter groups, along with animal rightists, are frequent participants in public debates (particularly if there is impending legislation) on animal welfare and the safety of foods of animal origin.

It is always important to establish dietary needs based on valid scientific evidence and not to reject any diet offhand.[34,58,116,119,120,125,164,171,181,202] Recognizing the etiology of the classic nutritional diseases involves associating a clinical syndrome with a diet in which an essential nutrient is virtually absent—beriberi (vitamin B_1), pellagra (niacin), scurvy (vitamin C)—or, in the case of kwashiorkor, inadequate quality and quantity of dietary protein.[115] On the other hand, assessing the need for foods of animal origin is not as easy when the diet includes an adequate amount of fruits, vegetables, **and** energy. Rickets can be a problem in children on a vegetarian diet and with limited exposure to sunlight. Vegan diets can present a greater risk of osteoporosis for postmenopausal women. Iron-deficiency anemia is more frequent among vegetarian adults and macrocytic anemia is seen among infants breast-fed by vegan mothers. Slow growth and delayed maturity is associated with vegetarian diets in children.[4,45,50,62,64,66,96,112,123,138,151,170,187,190,219,231,244] Humans cannot synthesize 10 of the 20 amino acids required to form proteins. These essential amino acids must be obtained from biological sources, and the quantities of five (isoleucine, lysine, methionine, threonine, and trytophan) may be limited in foods of plant origin. In contrast, proteins of animal origin have a balanced amino acid content and are generally more digestible than plant proteins. We should not overlook, however, the positive facts that persons on vegetarian diets often have lower blood pressure and lower serum cholesterol levels. Although the value of milk as a source of protein and calcium in the diets of infants and growing children is clear, we must also accept the fact that it is possible to survive later in life on a diet lacking foods of animal origin. In fact, two fatty acids essential in the mammalian diet, linoleic and γ-linolenic, must be obtained from foods of plant origin.

Cardiovascular disease and cancer are the chief causes of death among adults in many developed countries, and both are linked to diet.[13,33,53,73,82,84,85,86,92,93,105,121,126,127,128,130,145,159,162,163,168,179,182,192,201,210,216,222,230,242] Elevated serum cholesterol from dietary sources is associated with increased risk of coronary heart disease. A greater risk of cancer of the colon and prostate is associated with diets high in fat. These dietary factors have a strong correlation with foods of animal origin, and dissemination of this information has stimulated major changes in patterns of food consumption. Nutrition (health) education and other societal pressures (e.g., lean is more attractive) will continue to affect the demands of consumers.[47,109,113,133,177,178,181,224,226] As these demands change, so will the decisions by producers, processors, and marketers regarding how to meet them (e.g., lower fat and trim cuts).

Preharvest Food Technology

Although the human population of the world has grown apace in the past century, per capita consumption of foods of animal origin has largely remained the same or increased in affluent societies. To meet this demand successfully, food production (as well as the marketing system) had to increase at least three-fold during the past 90 yr. This began with the efforts of government and industry to reduce the frequency of catastrophic diseases of livestock and poultry to levels no longer of economic significance. The early techniques used were isolation, quarantine, and slaughter of exposed or affected animals. Later, vaccines that prevented infectious disease (preferably also infection) or drugs that killed parasites were introduced for many of the diseases. Bactericidal drugs came even later. When the proportion of surviving progeny increased significantly as a result of effective disease control, the results of genetic selection for desirable economic traits (e.g., growth rate, carcass quality, feed efficiency, egg production, and milk production) began to have an effect. We finally had enough surviving animals in the breeding flock or herd to selectively retain the more productive animals and to cull those that were less productive. Until the past century, the major genetic selection pressure among domestic animals had been for resistance to endemic disease.

In areas where livestock production has not been limited to extensive range pasturage, the efficiency of production per land unit has been increased by the introduction of improved feed production, fencing, soil treatment for mineral deficiencies (copper, iodine, molybdenum, phosphorus, etc.), use of fertilizers, and other husbandry practices. Artificial insemination and machine milking of dairy cows, squeeze chutes for handling beef cattle, centralized formulation of livestock and poultry rations, and individual cages for laying hens are examples of technology that has become commonplace within the last 50 yr. Generally, there are fewer flocks and herds, but they are of greater size.[69,195,239] When herd size was small and all the feed was produced on the farm, the overall impact of production management errors was minor. The increases in herd size and other complexities of management may significantly lower unit costs of production, but any errors greatly increase the risk of affecting large numbers of animals.[146,147] As herd size and density (e.g., feedlots) increase, the risk of adverse environmental impact also increases.

Castration to produce barrows, steers, or wethers is an ancient practice considered useful to produce slaughter animals with a greater rate of gain and carcass quality. Of course, it was also used to produce work oxen and to prevent indiscriminant breeding. In recent decades, however, it has been shown that entire (intact) young ruminants and pigs gain weight faster (with no temporary setback), and the meat from them is equally acceptable to consumers. Yet some markets still penalize producers who sell entire calves or lambs. With today's consumer demand for meat with less fat, these entire animals should sell at a premium.

Products of biotechnology such as the growth hormones, BST and porcine somatotropin (PST), offer more-efficient milk production and a leaner pork carcass of potentially greater consumer appeal.[3,14,20,101,131,150,153,173,175] Public response to such products, however, has been less than favorable. Public acceptance of the products of biotechnology has been selective. For example, bioengineered insulins were readily accepted. On the other hand, use of diethylstilbestrol (DES) as a growth promoter in livestock was banned in the United States in 1979 because of public concern even though DES in meat was never shown to be a hazard to human health. It was the therapeutic use of DES in pregnant women that was the health hazard to the human fetus, not the minuscule quantity found in beef.

Bovine spongiform encephalopathy (BSE or "mad cow" disease) was first recognized in the United Kingdom in 1986.[6,12,28,31,48,78,140,148,212,218,223,234,235,243, 246,247,248,249,250,252,257] Recycling of rendered animal tissue into feed has been associated with the appearance of BSE, particularly with a trend during the 1970s away from the use of hydrocarbon solvent extraction of fat from meat and bone meal. Although shifts from batch to continuous processing also occurred that involved temperature and pressure changes, these features do not appear to be causally related to BSE. Recycling of ruminant protein to ruminants was prohibited in the United Kingdom in 1988. The infectious prions involved are similar to those that cause scrapie in sheep and Creutzfeldt-Jakob disease and kuru in people. BSE was first thought to be a variant of scrapie because scrapie is prevalent in the United Kingdom and considerable rendered sheep tissue was recycled in animal feed. This possible interspecies transfer by food and the fact that similar human diseases exist, one of which (kuru) was known to be foodborne, generated speculation in the media regarding the human health risk. The public response in the United Kingdom resulted in a dramatic decrease in beef consumption. Although the food-associated public hysteria has passed, the problem persists as a serious concern to animal health officials in the United Kingdom as well as in countries importing live cattle from the United Kingdom.

Animal welfare has been a recognized social issue since societies for the prevention of cruelty to animals were formed in many countries a century ago.[10,15,22,30,36,46,56,70,87,88,89,90,111,122,142,144,166,188,203,209,217,238,253] Veterinarians and organized veterinary medicine have been and are leaders in the movement. Currently, the World Veterinary Association promotes five "freedoms" in relation to the welfare of animals: freedom from hunger and thirst, physical discomfort and pain, injury and disease, fear and distress, and freedom to conform to essential behavior patterns. It is this last freedom that has become a major point of contention between farm animal producers and animal rights groups. Vast sums of money have been spent on campaigns to outlaw the rearing of veal calves in crates and maintenance of laying hens in individual cages. These and similar campaigns have had their successes in some countries, and they will continue. In many instances, the limits placed on "natural" behavior and habitat have resulted from a producer decision to improve disease control in the flock or herd. If we are to justify rearing of farm animals in conditions other than a so-called natural habitat, then we must be prepared to generate the science-based information needed to show that the rearing system used supports the five freedoms promoted by the World Veterinary Association. On-farm freedom from injury and disease is important to preharvest food safety.

Organic and *natural* are terms used in association with foods supposedly produced under certain defined conditions.[35,39,110,136,186] Many consumers perceive foods with these labels to be safer or more healthful. For example, some people who consider themselves to be chemically sensitive seek these products in the belief they are free of chemical residues. A considerable international movement exists promoting food that is a product of organic farming. *Organic farming* is defined as "a production system that avoids or largely excludes the use of synthetically compounded fertilizers, pesticides, growth regulators, and livestock feed additives. To the maximum extent feasible, organic farming systems rely upon crop rotations, crop residues, animal manures, legumes, green manures, off-farm organic wastes, mechanical cultivation, mineral-bearing rocks, and aspects of biological pest control to maintain soil productivity and tilth, to supply plant nutrients, and to control insects, weeds, and other pests." The term *natural food* lacks a widely held definition and seems to have fewer adherents. Foods of animal origin are being produced and marketed now under criteria established by various private and public organic certifying organizations. Because of significant consumer demand, efforts are underway to codify organic food criteria into consistent standards that can be recognized nationally and internationally. There is some overlap between what is perceived to be the production of organic food, promoted as being less detrimental to the environment than so-called conventional agriculture, and sustainable agriculture, mentioned below.

Soon, the most significant pressure for change in agriculture and fisheries may be the need to find more "sustainable" methods of food production.[55,71,97,160,198,240] The yields of marine fisheries are no longer increasing as a function of improved harvesting technology. The energy cost expended on the production of chemical fertilizers, etc., to produce a crop is, in some instances, exceeding the energy return in the harvest. One definition of *sustainable agriculture* is "one that, over the long term, enhances environmental quality and the resource base on which agriculture depends; provides for basic human food and fiber needs; is economically viable; and enhances the quality of life for farmers and society as a whole." Many lands still are best suited for grazing, and efforts at crop production in these areas have generally been destructive of the environment. Although animal agriculture has often been considered as less efficient than crop production, it may be the inputs of animal agriculture and fisheries that will be the critical factor in devising sustainable integrated means of world food production in the future. In the past, production methods have aimed for maximum yield, whereas now we need to focus on optimum yield (i.e., the greatest difference between input and output).

Postharvest Food Technology

Since early history there have been methods of preserving foods of animal origin postharvest to extend their availability beyond times of plenty and to make them more portable. Fish, meat, and poultry have been dried, salted, pickled, or smoked. Cheese has been made from milk, and it has also been salted. Among the five basic methods of food preservation (chemical, dehydration, refrigeration, canning, and irradiation), only the first two date back to primitive societies.[255] Early use of refrigeration was limited to regions where ice was available, and then it was often seasonal (unless stored in ice houses). Techniques to expand the use of refrigeration (including the introduction of mechanical refrigeration) as well as the introduction of commercial canning only evolved in the past century.[220] Extensive preservation of fresh animal products at home has developed since the 1920s, when home electrification and the electric refrigerator became available. At the same time, sale of pasteurized milk began, a process that both extended the shelf life of milk and made it safer to drink. In the 1950s, a process for making milk powder (a product easily transported worldwide) was introduced from which, for the first time, milk could be reconstituted in a liquid form that was widely acceptable to consumers. Commercial freezing of fresh food began to expand rapidly at this time, with the introduction of complete individual meals ("TV dinners"). Today, UHT milk can remain on the shelf at room temperature for months without loss of quality or becoming unsafe. Of course, uses of atomic energy in food preservation began in this century. Although the improvements in food preservation technology are evident, today's consumers may now keep stored food too long. This is associated with a reduction in quality and may also be unsafe.

As postharvest food technology continues to become more complex, we can expect continued public concern about safety, and there will be a continuing need for credible scientific effort to allay this concern.[43,57,63,83,98,158,161,169,204,214] At the beginning of this century, there was public outcry regarding the sanitary status of abattoirs. Later, pasteurization of milk was considered unnatural and a hazard because it destroyed vitamin C (even though milk is not a significant source of this nutrient). Today, we are embroiled in debate concerning the safety of food that has been irradiated.[40,94,236] Public concern must be allayed if new technology is to be accepted.

Outbreaks of foodborne infection or intoxication have always been a concern to public health agencies, and inspection of public food-handling facilities (e.g., processors, markets, and restaurants) has been a major activity of their sanitarians.[5,29,102,114,124,154,194,197,225,233] In spite of this effort, foodborne disease outbreaks have become more widespread, with steadily increasing numbers of persons being affected as postharvest food technology has become more centralized and sophisticated. Outbreaks of botulism, campylobacteriosis, salmonellosis, and other foodborne diseases continue to make news headlines. Public reaction is usually swift and negative in relation to the food involved. Sales decline and considerable effort is needed to restore public confidence in the product. Occasional breakdowns in safety along the postharvest food chain can be expected as long as human manipulation of the product is involved. Although education of industrial and domestic food handlers will help to reduce the risk, we cannot expect the zero risk some consumers demand unless all food is further processed to be "ready to eat" and no longer fresh.

Bibliography

1. Abu-Tarboush, H.M., and A.A. Dawood. 1993. Cholesterol and fat contents of animal adipose tissues. *Food Chem.* 46:89–93.

2. Ackman, R.G. 1989. Nutritional composition of fats in seafoods. *Prog. Food Nutr. Sci.* 13:161–241.

3. Acuff, G.R., R.A. Albanese, C.A. Batt, et al. 1991. Implications of biotechnology, risk assessment, and communications for the safety of foods of animal origin. *J. Am. Vet. Med. Assoc.* 199:1714–1721.

4. Allen, L.H., J.R. Backstrand, E.J. Stanek III, et al. 1992. The interactive effects of dietary quality on the growth and attained size of young Mexican children. *Am. J. Clin. Nutr.* 56:353–364.

5. Anderson, R.K., D.A. Fenderson, L.M. Schuman, et al. 1976. *A Description of the Responsibilities of Veterinarians as They Relate Directly to Human Health.* Report to Bureau of Health and Manpower, Department of Health, Education, and Welfare, Washington, D.C.

6. Anonymous. 1993. Citing "mad-cow" fears, Rifkin group petitions FDA to halt "cow cannibalism." *Nutr. Week* 23(24):1–2.

7. Ashwell, M. 1991. Consumer perception of food-related issues. *BNF Nutr. Bull.* 16:25–35.

8. American Veterinary Medical Association. 1992. Public and corporate veterinary practice symposium. *J. Am. Vet. Med. Assoc.* 200:295–336.

9. ——. 1992. AVMA food safety workshop. *J. Am. Vet. Med. Assoc.* 201:227–266.

10. ——. 1994. AVMA animal welfare forum: The veterinarian's role in farm animal welfare. *J. Am. Vet. Med. Assoc.* 204:363–395.

11. Aubert, C. (translated by C.J. Cusumano, W. Greenberg, and N. Bellotto). 1985. *Hunger and Health: Eleven Key Questions on Farming, Food, and Health in the Third World.* Emmaus, Pa.: Rodale Press.

12. Barlow, R.M., and D.J. Middleton. 1991. Oral transmission studies of BSE to mice. *Curr. Top. Vet. Med. Anim. Sci.* 55:33–39.

13. Barnard, N.D. 1991. The edge against cancer. *Veg. Times* 170(Oct.):18, 20–21.

14. Bauman, D.E., P.J. Eppard, M.J. DeGeeter, et al. 1985. Responses of high-producing dairy cows to long-term treatment with pituitary somatotropin and recombinant somatotropin. *J. Dairy Sci.* 68:1352–1362.

15. Baumgardt, B., and H.G. Gray. 1993. *Food Animal Well-being 1993—Conference Proceedings and Deliberations.* West Lafayette, Ind.: Purdue University.

16. Beardsworth, A.D., and E.T. Keil. 1993. Contemporary vegetarianism in the U.K.: Challenge and incorporation? *Appetite* 20:229–234.

17. Becker, W., and E. Helsing (eds.). 1991. *Food and Health Data: Their Use in Nutrition Policy-making.* WHO Reg. Publ., Eur. Ser., 34. Copenhagen: World Health Organization Regional Office for Europe.

18. Beggs, L., S. Hendricks, N.E. Schwartz, et al. 1993. Tracking nutrition trends: Canadians' attitudes, knowledge, and behaviours regarding fat, fibre, and cholesterol. *J. Can. Diet. Assoc.* 54:21–25.

19. Bender, M.M., and B.M. Derby. 1992. Prevalence of reading nutrition and ingredient information on food labels among adult Americans: 1982–1988. *J. Nutr. Educ.* 24:292–297.

20. van den Berg, G. 1991. A review of quality and processing suitability of milk from cows treated with bovine somatotropin. *J. Dairy Sci.* 74(Suppl. 2):2–11.

21. Billings, F.S. 1884. *The Relation of Animal Diseases to the Public Health, and Their Prevention.* New York: D. Appleton.

22. Birbeck, A.L. 1991. A European perspective on farm animal welfare. *J. Am. Vet. Med. Assoc.* 198:1377–1380.

23. Biro, G. 1987. Department of Food Hygiene: Meat hygiene, milk hygiene. *Acta Vet. Hung.* 35:147–151.

24. Blank, M.L., E.A. Cress, Z.L. Smith, et al. 1992. Meats and fish consumed in the American diet contain substantial amounts of ether-linked phospholipids. *J. Nutr.* 122:1656–1661.

25. Bloyd-Peshkin, S. 1991. Mumbling about meat. *Veg. Times* 170(Oct.):66–68, 70, 72, 75.

26. Bodenheimer, F.S. 1951. *Insects as Human Food.* The Hague: W. Junk.

27. Boonin-Vail, D. 1993. The vegetarian savage: Rousseau's critique of meat eating. *Environ. Ethics* 15:75–84.

28. Bown, W. 1992. BSE five times worse than feared. *New Sci.* 134:9.

29. Brandly, P.J., G. Migaki, and K.E. Taylor. 1966. *Meat Hygiene.* 3d ed. Philadelphia: Lea and Febiger.

30. Broom, D.M. 1987. Applications of neurobiological studies to farm animal welfare. *Curr. Top. Vet. Med. Anim. Sci.* 42:101–110.

31. Brown, P. 1991. The clinical epidemiology of Creutzfeldt-Jakob disease in the context of bovine spongiform encephalopathy. *Curr. Top. Vet. Med. Anim. Sci.* 55:195–202.

32. Brunser, O., J. Espinoza, G. Figueroa, et al. 1992. Field trial of an infant formula containing anti-rotavirus and anti-*Escherichia coli* milk antibodies from hyperimmunized cows. *J. Pediatr. Gastroenterol. Nutr.* 15:63–72.

33. Burr, M.L. 1989. Fish and the cardiovascular system. *Progr. Food Nutr. Sci.* 13:291–316.

34. Burton, B.T., and W.R. Foster. 1988. *Human Nutrition.* 4th ed. New York: McGraw-Hill.

35. Byng, J. 1993. EC organic food standards. *Br. Food J.* 95:16–17.

36. Caneff, D. 1993. Family farmers should move toward the animal welfare movement. *Am. J. Alternative Agric.* 8:4, 46.

37. Cary, C.A. 1898. *Dairy and Milk Inspection.* Alabama Agric. Exp. Stn. Bull. 97. Birmingham, Ala.: Roberts and Son.

38. Chandra, R.K., S. Puri, and A. Hamed. 1989. Influence of maternal diet during lactation and use of formula feeds on development of atopic eczema in high risk infants. *Br. Med. J.* 299:228–230.

39. Clarke, R. 1991. Standards for organic foods. *Food Aust.* 43:12–14.

40. Conley, S.T. 1992. What do consumers think about irradiated foods? *FSIS Food Safety Rev.* 2(3):11–15.

41. Conning, D.M. 1988. The public perception of food safety. *J. R. Soc. Health* 108:134–135.

42. Crane, E. 1990. *Bees and Beekeeping: Science, Practice, and World Resources.* Ithaca, N.Y.: Comstock Publishing.

43. Cuthbertson, W.F.J. 1989. What is a healthy food? *Food Chem.* 33:53–80.

44. Dabeka, R.W., A.D. McKenzie, G.M.A. Lacroix, et al. 1993. Survey of arsenic in total diet food composites and estimation of the dietary intake of arsenic by Canadian adults and children. *J. AOAC Int.* 76:14–25.

45. Dagnelie, P.C., W.A. van Staveren, S.A.J.M. Verschuren, et al. 1989. Nutritional status of infants aged 4 to 18 months on macrobiotic diets and matched omnivorous control infants: A population-based mixed-longitudinal study I. Weaning pattern, energy, and nutrient intake. *Eur. J. Clin. Nutr.* 43:311–323.

46. Daly, C.C., E. Kallweit, and F. Ellendorf. 1988. Cortical function in cattle during slaughter: Conventional captive bolt stunning followed by exsanguination compared with Shechita slaughter. *Vet. Rec.* 122:325–329.

47. Darrall, J. 1992. The response of the food chain to healthy eating. *Br. Food J.* 94(4):7–11.

48. Davies, P.T.G., S. Jahfar, I.T. Ferguson, et al. 1993. Creutzfeldt-Jakob disease in individual occupationally exposed to BSE. *Lancet* 342:680.

49. Dealler, S.F., and R.W. Lacey. 1990. Transmissible spongiform encephalopathies: The threat of BSE to man. *Food Microbiol.* 7:253–279.

50. Devlin, J., R.H.J. Stanton, and T.J. David. 1989. Calcium intake and cows' milk free diets. *Arch. Dis. Child.* 64:1183–1184.

51. Dewailly, E., A. Nantel, J.-P. Weber, et al. 1989. High levels of PCBs in breast milk of Inuit women from Arctic Quebec. *Bull. Environ. Contam. Toxicol.* 43:641–646.

52. Dingott, S., and J. Dwyer. 1991. Benefits and risks of vegetarian diets. *Nutr. Forum* 8:45–47.

53. Donnan, P.T., M. Thomson, F.G.R. Fowkes, et al. 1993. Diet as a risk factor for peripheral arterial disease in the general population: The Edinburgh artery study. *Am. J. Clin. Nutr.* 57:917–921.

54. Duchen, K., and B. Bjorksten. 1991. Sensitization via the breast milk. *Adv. Exp. Med. Biol.* 310:427–436.

55. Dunlap, R.E., C.E. Beus, R.E. Howell, et al. 1992. What is sustainable agriculture? An empirical examination of faculty and farmer definitions. *J. Sustainable Agric.* 3:5–39.

56. Dunn, C.S. 1990. Stress reactions of cattle undergoing ritual slaughter using two methods of restraint. *Vet. Rec.* 126:522–525.

57. DuPont, H.L. 1992. How safe is the food we eat? *J. Am. Med. Assoc.* 268:3240.

58. Eastwood, M., C. Edwards, and D. Parry (eds.). 1992. *Human Nutrition: A Continuing Debate.* London: Chapman and Hall.

59. Edenharder, R., M. Mitschke, and K. Jung. 1992. Urinary mutagenicity in vegetarians and people on a mixed-Western diet. *Zbl. Hyg.* 192:494–508.

60. Elzinga, R.J. 1987. *Fundamentals of Entomology.* 3d ed. Englewood Cliffs, N.J.: Prentice-Hall.

61. Engs, R.C. 1991. Resurgence of a new "clean living" movement in the United States. *J. Sch. Health* 61:155–159.

62. Erinoso, H.O., S. Hoare, S. Spencer, et al. 1992. Is cow's milk suitable for the dietary supplementation of rural Gambian children? 1. Prevalence of lactose maldigestion. *Ann. Trop. Paediatr.* 12:359–365.

63. Evers, S. 1991. Diet-disease relationships: Public health perspectives. *Prog. Food Nutr. Sci.* 15:61–83.

64. Faith-Magnusson, K. 1989. Breast milk antibodies to foods in relation to maternal diet, maternal atopy, and the development of atopic disease in the baby. *Int. Arch. Allergy Appl. Immunol.* 90:297–300.

65. Fleming, G. 1875. *A Manual of Veterinary Sanitary Science and Police.* Vols. I and II. London: Chapman and Hall.

66. Flynn, A. 1992. Minerals and trace elements in milk. *Adv. Food Nutr. Res.* 36:209–252.

67. Food Animal Production Medicine Consortium. 1992. *Providing Safe Food for the Consumer.* Proceedings of a workshop held in Washington, D.C., Nov. 19–21, 1992.

68. Forsythe, R.H. 1993. Risk: Reality versus perception. *Poult. Sci.* 72:1152–1156.

69. Fox, G., and P. Bergen. 1990. Explaining changes in dairy farm size in Ontario. Working paper 90/27. Guelph, Ontario: University of Guelph.

70. Fox, M.W. 1981. Productivity and farm animal welfare. *Int. J. Stud. Anim. Prob.* 2:283–284.

71. Francis, C.A. 1990. Sustainable agriculture: Myths and realities. *J. Sustainable Agric.* 1:97–106.

72. Francis, F.J. 1992. *Food Safety: The Interpretation of Risk.* Ames, Iowa: Council for Agricultural Science and Technology.

73. Franco, E.L., L.P. Kowalski, B.V. Oliveira, et al. 1989. Risk factors for oral can-

cer in Brazil: A case-control study. *Int. J. Cancer* 43:992–1000.

74. Freudenthal, R.I., and S.L. Freudenthal. 1991. *Food Facts and Fictions.* Green Farms, Conn.: Hill and Garnett Publishing.

75. Gallagher, C.R., and J.B. Allred. 1992. *Taking the Fear Out of Eating: A Nutritionists' Guide to Sensible Food Choices.* Cambridge, Engl.: Cambridge University Press.

76. Galler, J.R. (ed.). 1984. *Nutrition and Behavior.* New York: Plenum Press.

77. George, K.P. 1992. The use and abuse of scientific studies. *J. Agric. Environ. Ethics* 5:217–233.

78. Gibbs, C.J., Jr., L. Bolis, D.M. Asher, et al. 1992. Recommendations of the international roundtable workshop on bovine spongiform encephalopathy. *J. Am. Vet. Med. Assoc.* 200:164–167.

79. Giblin, J.C. 1986. *Milk: The Fight for Purity.* New York: Thomas Y. Crowell.

80. Gibney, M.J., M. Moloney, and E. Shelley. 1989. The Kilkenny project: Food and nutrient intakes in randomly selected healthy adults. *Br. J. Nutr.* 61:129–137.

81. Giem, P., W.L. Beeson, and G.E. Fraser. 1993. The incidence of dementia and intake of animal products: Preliminary findings from the Adventist Health Study. *Neuroepidemiol.* 12:28–36.

82. Glattre, E., T. Haldorsen, J.P. Berg, et al. 1993. Norwegian case-control study testing the hypothesis that seafood increases the risk of thyroid cancer. *Cancer Causes Control* 4:11–16.

83. Goodman-Malamuth, L. 1986. Animal drugs. *Nutr. Action Health Lett.* 13(5):1, 5–7.

84. Goodnight, S.H., Jr. 1989. The vascular effects of ω-3 fatty acids. *J. Invest. Dermatol.* 93:102S–106S.

85. Goodnight, S.H., M. Fisher, G.A. FitzGerald, et al. 1989. Assessment of the therapeutic use of dietary fish oil in atherosclerotic vascular disease and thrombosis. *Chest* 95(Suppl. 2):19S–25S.

86. Goto, Y. 1992. Changing trends in dietary habits and cardiovascular disease in Japan: An overview. *Nutr. Rev.* 50:398–401.

87. Grandin, T. 1980. Problems with kosher slaughter. *Int. J. Stud. Anim. Prob.* 1:375–390.

88. ———. 1987. New humane slaughter system installed at Utica Veal. *Off. Proc. Ann. Meet. Livest. Conserv. Inst.* 1987:13–16.

89. ———. 1990. Improving kosher slaughter. *Humane Soc. News* 34(Spring):9–10.

90. Grandjean, P., G.D. Nielsen, P.J. Jorgensen, et al. 1991. Reference intervals for trace elements in blood: Significance of risk factors. *Scand. J. Clin. Lab. Invest.* 52:321–337.

91. Gravani, R., D. Williamson, and D. Blumenthal. 1992. What do consumers know about food safety? *FSIS Food Safety Rev.* 2(1):12–14.

92. Greus, P.C., J.L.A. Sanchez, I.F. Pons, et al. 1992. Correlation between mortality trends of ischaemic cardiopathy and some nutritional factors in Spain, 1968–1986. *Eur. J. Epidemiol.* 8:770–775.

93. Grobstein, C. (chrmn.). 1982. *Diet, Nutrition, and Cancer.* Washington, D.C.: National Academy Press.

94. Gunther, J.A. 1994. The food zappers. *Popular Science,* Jan.: 72–77, 86.

95. Hady, P.J., and J.W. Lloyd. 1992. Economic issues for dairy practitioners. Part II. Impact of dairy farm size on veterinary services. *Compend. Contin. Educ. Pract. Vet.* 14:1641–1645.

96. Haffejee, I.E. 1990. Cow's milk-based formula, human milk, and soya feeds in acute infantile diarrhea: A therapeutic trial. *J. Pediatr. Gastroenterol. Nutr.* 10:193–198.

97. Hall, D.C., B.P. Baker, J. Franco, et al. 1989. Organic food and sustainable agriculture. *Contemp. Policy Issues* 7:47–72.

98. Hall, R.H. 1992. A new threat to public health: Organochlorines and food. *Nutr. Health* 8:33–43.

99. Hardinge, F., and M. Hardinge. 1992. The vegetarian perspective and the food industry. *Food Technol.* 46:114, 116, 121.

100. Hartman, A.M., C.C. Brown, J. Palmgren, et al. 1990. Variability in nutrient and food intakes among older middle-aged men. *Am. J. Epidemiol.* 132:999–1012.

101. Hecht, D.W. 1991. Bovine somatotropin safety and effectiveness: An industry perspective. *Food Technol.* 45:118, 123–124, 126.

102. Heidelbaugh, N.D., and E.L. Menning. 1993. Safety of foods of animal origin. *J. Am. Vet. Med. Assoc.* 203:199–204.

103. Henderson, I.F. (chrmn.). 1989. *Slugs and Snails in World Agriculture.* BCPC Monogr. 41. Thornton Heath, United Kingdom: British Crop Protection Council.

104. Hentschl, A. (chrmn.). 1991. Milk and dairy beef residue prevention: A quality assurance protocol. *J. Am. Vet. Med. Assoc.* 199(2 insert):1-23.

105. Herrero, R., N. Potischman, L.A. Brinton, et al. 1991. A case-control study of nutrient status and invasive cervical cancer. I. Dietary indicators. *Am. J. Epidemiol.* 134:1335–1346.

106. Hill, S.M., A.D. Phillips, M. Mearns, et al. Cows' milk sensitive enteropathy in cystic fibrosis. *Arch. Dis. Child.* 64:1251–1255.

107. Holt, G. 1991. Increasing the "pulse" rate: Opportunities for product development. *Br. Food J.* 93(5):17–23.

108. Hubbert, W.T. 1979. Perspective on veterinary preventive medicine in the United States. *J. Am. Vet. Med. Assoc.* 174:378–379.

109. Hulshof, K.F.A.M., M. Wedel, and M.R.H. Lowik. 1992. Clustering of dietary variables and other lifestyle factors (Dutch Nutritional Surveillance System). *J. Epidemiol. Community Health* 46:417–424.

110. International Federation of Organic Agriculture Movements. 1992. *Basic Standards of Organic Agriculture and Food Processing.* Tholey-Theley, Germany: IFOAM.

111. Jackson, W.T. 1988. On-farm animal welfare law in Europe—using the law. *Appl. Anim. Behav. Sci.* 20:165–173.

112. Jacobsen, B.K., and I. Stensvold. 1992. Milk—a better drink? *Scand. J. Soc. Med.* 20:204–208.

113. Jamrozik, K., R. Jamieson, and C. Fitzgerald. 1992. An oral history of changes in the Australian diet. *Med. J. Aust.* 157:759–761.

114. Johnston, A.M. 1990. Veterinary sources of foodborne illness. *Lancet* 336:856–858.

115. Jukes, T.H. 1989. The prevention and conquest of scurvy, beri-beri, and pellagra. *Prev. Med.* 18:877–883.

116. Kant, A.K., A. Schatzkin, T.B. Harris, et al. 1993. Dietary diversity and subsequent mortality in the first national health and nutrition examination survey, epidemiologic follow-up study. *Am. J. Clin. Nutr.* 57:434–440.

117. Karasszon, D. (translated by E. Farkas). 1988. *A Concise History of Veterinary Medicine.* Budapest: Akademiai Kiado.

118. Kautni, J. 1985. Historie der Deutschen Fleischbeschaugesetzgebung unter Berucksichtigung der darin geannten Tierkrankeiten. Inaugural-Dissertation. Hannover: Tierarztliche Hochschule Hannover.

119. Kent, G. 1985. Fisheries and undernutrition. *Ecol. Food Nutr.* 16:281–294.

120. Kon, S.K. 1975. *Milk and Milk Products in Human Nutrition.* 2d rev. ed. FAO Nutr. Stud. 27. Rome: Food and Agriculture Organization.

121. Kono, S., K. Imanishi, K. Shinchi, et al. 1993. Relationship of diet to small and large adenomas of the sigmoid colon. *Jap. J. Cancer Res.* 84:13–19.

122. Koorts, R. 1992. The development of a restraining system to accommodate the

Jewish method of slaughter (Shechita). *Proc. Meat Sci. Technol. Serv. Meat Ind.* 7:41–51. Irene, Republic of South Africa: Irene Animal Production Institute.

123. Kostraba, J.N., J.S. Dorman, R.E. LaPorte, et al. 1992. Early infant diet and risk of IDDM in blacks and whites. *Diabetes Care* 15:626–631.

124. Koulikovskii, A., and Z. Matyas. 1985. Past, present, and future activities in food hygiene of the Veterinary Public Health Unit (VPH) of the World Health Organization. *Int. J. Food Microbiol.* 2:197–210.

125. Krebs-Smith, S.M., F.J. Cronin, D.B. Haytowitz, et al. 1990. Contributions of food groups to intakes of energy, nutrients, cholesterol, and fiber in women's diets: Effect of method of classifying food mixtures. *J. Am. Diet. Assoc.* 90:1541–1546.

126. Kromhout, D. 1989. n-3 fatty acids and coronary heart disease: Epidemiology from Eskimos to Western populations. *J. Intern. Med.* 225(Suppl. 1):47–51.

127. Kromhout, D., A. Keys, C. Aravanis, et al. 1989. Food consumption patterns in the 1960s in seven countries. *Am. J. Clin. Nutr.* 49:889–894.

128. Kyogoku, S., T. Hirohata, Y. Nomura, et al. 1992. Diet and prognosis of breast cancer. *Nutr. Cancer* 17:271–277.

129. Lacey, R. 1992. Scares and the British food system: Problems and policies in relation to food-related health issues. *Br. Food J.* 94(7):26–30.

130. Lapre, J.A., and R. Van der Meer. 1992. Dietary modulation of colon cancer risk: The roles of fat, fiber, and calcium. *Trends Food Sci. Technol.* 3:320–324.

131. Lemieux, C.M., and M.K. Wohlgenant. 1989. *Ex ante* evaluation of the economic impact of agricultural biotechnology: The case of porcine somatotropin. *Am. J. Agric. Econ.* 71:903–914.

132. Levy, A.S., S.B. Fein, and M. Stephenson. 1993. Nutrition knowledge levels about dietary fats and cholesterol: 1983–1988. *J. Nutr. Educ.* 25:60–66.

133. Lewis, C.J., L.S. Sims, and B. Shannon. 1989. Examination of specific nutrition/health behaviors using a social cognitive model. *J. Am. Diet. Assoc.* 89:194–202.

134. Ling, W.H., and O. Hanninen. 1992. Shifting from a conventional diet to an uncooked vegan diet reversibly alters fecal hydrolytic activities in humans. *J. Nutr.* 122:924–930.

135. Lombard, K.A., and D.M. Mock. 1989. Biotin nutritional status of vegans, lactoovovegetarians, and nonvegetarians. *Am. J. Clin. Nutr.* 50:486–490.

136. Lovisolo, R. 1993. Revised draft guidelines for the production, processing, labelling, and marketing of organically/biologically produced foods. Rome: Codex Alimentarius Commission, Food and Agriculture Organization.

137. Lowik, M.R.H., J. Schriver, J. Odink, et al. 1990. Long-term effects of a vegetarian diet on the nutritional status of elderly people (Dutch Nutrition Surveillance System). *J. Am. Coll. Nutr.* 9:600–609.

138. Luke, B., and L.G. Keith. 1992. Calcium requirements and the diets of women and children. A review of dairy resources. *J. Reprod. Med.* 37:703–709.

139. Maenpaa, P., A. Karinpaa, M. Jauhiainen, et al. 1991. Effects of low-fat lactovegetarian diet on health parameters of adult subjects. *Ecol. Food Nutr.* 25:255–267.

140. Marsh, R.F. 1991. Risk assessment on the possible occurrence of bovine spongiform encephalopathy in the United States. *Curr. Top. Vet. Med. Anim. Sci.* 55:41–46.

141. Mason, I.L. (ed.). 1984. *Evolution of Domesticated Animals.* London: Longman.

142. Mayer, E. 1993. Animal welfare. *World Vet. Assoc. Bull.* 10(1):6–10.

143. Mela, D.J. (ed.). 1992. *Dietary Fats: Determinants of Preference, Selection, and Consumption.* London: Elsevier Applied Science.

144. Mench, J.A., and A. van Tienhoven. 1986. Farm animal welfare. *Am. Sci.* 74:598–603.

145. Michel, A., M. Glick, T.C. Rosenthal, et al. 1993. Cardiovascular risk factor status of an old order Mennonite community. *J. Am. Board Fam. Pract.* 6:225–231.

146. Miles, H., W. Lesser, and P. Sears. 1992. The economic implications of bio-

engineered mastitis control. *J. Dairy Sci.* 75:596–605.

147. Miller, G.Y., and P.C. Bartlett. 1991. Economic effects of mastitis prevention strategies for dairy producers. *J. Am. Vet. Med. Assoc.* 198:227–231.

148. Miller, L.D., A.J. Davis, and A.L. Jenny. 1992. Surveillance for lesions of bovine spongiform encephalopathy in U.S. cattle. *J. Vet. Diagn. Invest.* 4:338–339.

149. Miller, S.A. (chrmn.). 1984. *The Role of Food Safety in Health and Development.* WHO Tech. Rep. Ser. 705. Geneva: World Health Organization.

150. ——— (chrmn.). 1991. *Strategies for Assessing the Safety of Foods Produced by Biotechnology.* Report of a joint FAO/WHO consultation. Geneva: World Health Organization.

151. Mills, A.F. 1990. Surveillance for anaemia: Risk factors in patterns of milk intake. *Arch. Dis. Child.* 65:428–431.

152. Moreiras-Varela, O. 1989. The Mediterranean diet in Spain. *Eur. J. Clin. Nutr.* 43(Suppl. 2):83–87.

153. Morton, D., R. James, and J. Roberts. 1993. Issues arising from recent advances in biotechnology. *Vet. Rec.* 133:53–56.

154. Mossel, D.A.A., and C.B. Strujik. 1990. *Prevention of the Transmission of Infections and Intoxications by Foods: The Responsibility of the Veterinary Public Health Profession.* San Antonio, Tex.: American College of Veterinary Preventive Medicine.

155. Musaiger, A.O. 1993. Socio-cultural and economic factors affecting food consumption patterns in the Arab countries. *J. R. Soc. Health* 113:68–74.

156. Myers, J.A., and J.H. Steele. 1969. *Bovine Tuberculosis Control in Man and Animals.* St. Louis, Mo.: Warren H. Green.

157. Namihira, D., I. Saldivar, N. Pustilnik, et al. 1993. Lead in human blood and milk from nursing women living near a smelter in Mexico City. *J. Toxicol. Environ. Health* 38:225–232.

158. Nesheim, M.C. 1991. Dietary guidelines: Their implications for the food industry. *Trends Food Sci. Technol.* 2:138–140.

159. Nettleton, J.A. 1991. ω-3 fatty acids: Composition of plant and seafood sources in human nutrition. *J. Am. Diet. Assoc.* 91:331–337.

160. Newman, A. 1993. Defining sustainable agriculture for the tropics. *Environ. Sci. Technol.* 27:1004–1006.

161. Ney, D.M. 1991. Potential for enhancing the nutritional properties of milk fat. *J. Dairy Sci.* 74:4002–4012.

162. Nicklas, T.A., L.S. Webber, B. Thompson, et al. 1989. A multivariate model for assessing eating patterns and their relationship to cardiovascular risk factors: The Bogalusa Heart Study. *Am. J. Clin. Nutr.* 49:1320–1327.

163. Nicklas, T.A., L.S. Webber, S.R. Srinivasan, et al. 1993. Secular trends in dietary intakes and cardiovascular risk factors of 10-yr-old children: The Bogalusa Heart Study (1973-1988). *Am. J. Clin. Nutr.* 57:930–937.

164. Nordoy, A., L.F. Hatcher, D.L. Ullmann, et al. 1993. Individual effects of dietary saturated fatty acids and fish oil on plasma lipids and lipoproteins in normal men. *Am. J. Clin. Nutr.* 57:634–639.

165. Odum, E.P. 1971. *Fundamentals of Ecology.* 3d ed. Philadelphia: W.B. Saunders.

166. Oldham, J. 1988. Aspects of farm animal welfare. *Vet. Rec.* 122:20.

167. O'Neill, O., and A.M. Fehily. 1991. Nutrient intakes of men in South Wales: A comparison of surveys taken in 1980–1983 and 1990. *J. Hum. Nutr. Diet.* 4:413–419.

168. Ozonoff, D., and M.P. Longnecker. 1991. Epidemiologic approaches to assessing human cancer risk from consuming aquatic food resources from chemically contaminated water. *Environ. Health Perspect.* 90:141–146.

169. Paardekooper, E.J.C. 1984. *Trends Mod. Meat Technol.* Netherlands: Wageningen.

170. Parent, M.-E., M. Krondl, and R.K. Chow. 1993. Reconstruction of past calcium

intake patterns during adulthood. *J. Am. Diet. Assoc.* 93:649–652.

171. Passmore, R., and M.A. Eastwood. 1986. *Davidson and Passmore Human Nutrition and Dietetics.* 8th ed. Edinburgh: Churchill Livingstone.

172. Pegram, T.R. 1991. Public health and progressive dairying in Illinois. *Agric. Hist.* 65:36–50.

173. Pell, A.N., D.S. Tsang, B.A. Howlett, et al. 1992. Effects of a prolonged-release formulation of sometibove (n-methionyl bovine somatotropin) on Jersey cows. *J. Dairy Sci.* 75:3416–3431.

174. Peters, A. 1888. *The Value of Veterinary Science to the State.* Boston: Wright and Potter.

175. Peters, A.R. 1990. Improving food quality through new technology. *Vet. Rec.* 126:543–546.

176. Pluhar, E. Who can be morally obligated to be a vegetarian? *J. Agric. Environ. Ethics* 5:189–215.

177. Popkin, B.M., P.S. Haines, and K.C. Reidy. 1989. Food consumption trends of U.S. women: Patterns and determinants between 1977 and 1985. *Am. J. Clin. Nutr.* 49:1307–1319.

178. Popkin, B.M., P.S. Haines, and R.E. Patterson. 1992. Dietary changes in older Americans, 1977–1987. *Am. J. Clin. Nutr.* 55:823–830.

179. Potter, J.D. 1993. Colon cancer—do the nutritional epidemiology, the gut physiology, and the molecular biology tell the same story? *J. Nutr.* 123:418–423.

180. Prattala, R., M.-A. Berg, and P. Puska. 1992. Diminishing or increasing contrasts? Social class variation in Finnish food consumption patterns, 1979–1990. *Eur. J. Clin. Nutr.* 46:279–287.

181. Public Health Service. 1988. *The Surgeon General's Report on Nutrition and Health.* DHHS (PHS) Publ. 88-50210. Washington, D.C.: U.S. Department of Health and Human Services.

182. Reddy, B.S. 1992. Dietary fat and colon cancer: Animal model studies. *Lipids* 27:807–813.

183. Roberts, T., and E. van Ravenswaay. 1989. *The Economics of Safeguarding the U.S. Food Supply.* Agric. Info. Bull. 566. Washington, D.C.: U.S. Department of Agriculture.

184. ———. 1990. The cost-safety balance in protecting our food supply. *Farmline* 11(Feb.):8–9.

185. Royce, W.F. 1972. *Introduction to the Fishery Sciences.* New York: Academic Press.

186. Russell, S. 1991. Organic foods: Consumer viewpoint. *Food Aust.* 43:14–15.

187. Sabate, J., M.C. Llorca, and A. Sanchez. 1992. Lower height of lacto-ovovegetarian girls at preadolescence: An indicator of physical maturation delay? *J. Am. Diet. Assoc.* 92:1263–1266.

188. Sainsbury, D. 1986. *Farm Animal Welfare. Cattle, Pigs, and Poultry.* London: Collins.

189. Salmon, D.E. 1904. Veterinary science in its relation to public health. *Reference Handbook of the Medical Sciences.* New York: William Wood, 224–233.

190. Sanders, T.A.B., and S. Reddy. 1992. The influence of a vegetarian diet on the fatty acid composition of human milk and the essential fatty acid status of the infant. *J. Pediatr.* 120:S71–S77.

191. Sapp, S.G., and C.L. Knipe. 1990. Japanese consumer preferences for processed pork. *Agribusiness* 6:387–400.

192. Sasaki, S., M. Horacsek, and H. Kesteloot. 1993. An ecological study of the relationship between dietary fat intake and breast cancer mortality. *Prev. Med.* 22:187–202.

193. Sazawal, S., M.K. Bhan, and N. Bhandari. 1992. Type of milk feeding during

acute diarrhoea and the risk of persistent diarrhoea: A case control study. *Acta Paediatr. Suppl.* 381:93–97.

194. Schwabe, C.W. (chrmn.) 1975. *The Veterinary Contribution to Public Health Practice.* WHO Tech. Rep. Ser. 573. Geneva: World Health Organization.

195. ——. 1978. *Cattle, Priests, and Progress in Medicine.* Minneapolis: University of Minnesota Press.

196. ——. 1979. *Unmentionable Cuisine.* Charlottesville: University Press of Virginia.

197. ——. 1984. *Veterinary Medicine and Human Health.* 3d ed. Baltimore: Williams and Wilkins.

198. Senanayake, R. 1991. Sustainable agriculture: Definitions and parameters for measurement. *J. Sustainable Agric.* 1:7–28.

199. Shea, S., T.A. Melnik, A.D. Stein, et al. 1993. Age, sex, educational attainment, and race/ethnicity in relation to consumption of specific foods contributing to the atherogenic potential of diet. *Prev. Med.* 22:203–218.

200. Shickle, D., P.A. Lewis, M. Charny, et al. 1989. Differences in health, knowledge and attitudes between vegetarians and meat eaters in a random population sample. *J. R. Soc. Med.* 82:18–20.

201. Shu, X.O., W. Zheng, N. Potischman, et al. 1993. A population-based case-control study of dietary factors and endometrial cancer in Shanghai, People's Republic of China. *Am. J. Epidemiol.* 137:155–165.

202. Simopoulos, A.P. 1991. Omega-3 fatty acids in health and disease and in growth and development. *Am. J. Clin. Nutr.* 54:438–463.

203. Singer, P. 1990. *Animal Liberation.* 2d ed. New York: Random House.

204. Smith, A.M., and K.L. Baghurst. 1992. Public health implications of dietary differences between social status and occupational category groups. *J. Epidemiol. Community Health* 46:409–416.

205. Smith, F. 1933. *The Early History of Veterinary Literature and Its British Development.* London: Bailliere, Tindall and Cox.

206. Smithcors, J.F. 1963. *The American Veterinary Profession: Its Background and Development.* Ames: Iowa State University Press.

207. Southgate, D.A.T. 1991. Nature and variability of human food consumption. *Phil. Trans. R. Soc. Lond. Biol.* 334:281–288.

208. Spedding, C.R.W. (ed.). 1989. *The Human Food Chain.* London: Elsevier Applied Science.

209. —— (chrmn.). 1991. *Report on the Welfare of Laying Hens in Colony Systems.* Hook Rise South, United Kingdom: Farm Animal Welfare Council.

210. Steinmetz, K.A., and J.D. Potter. 1993. Food-group consumption and colon cancer in the Adelaide case-control study. II. Meat, poultry, seafood, dairy foods, and eggs. *Int. J. Cancer* 53:720–727.

211. Storer, T.I., R.L. Usinger, and J.W. Nybakken. 1968. *Elements of Zoology.* 3d ed. New York: McGraw-Hill.

212. Studdert, M.J. 1992. Bovine spongiform encephalopathy and Australia. *Aust. Vet. J.* 69:153–154.

213. Sturzbecher, M. 1973. Aus der Geschichte der veterinarpolizeilichen Lebensmittelaufsicht in Berlin in der zweiten Halfte des 19. Jahrhunderts. *Arch. Lebensm. Hyg.* 24:36–39.

214. Sutherland, A.K. 1991. Semi-automated slaughtering and dressing of cattle. *Aust. Vet. J.* 68:57.

215. Swientek, R.J., and D.D. Duxbury. 1985. Kraft's QC/QA program. *Food Process.* 46:54–57.

216. Talamini, R., S. Franceschi, C.L. Vecchia, et al. 1992. Diet and prostate cancer:

A case-control study in northern Italy. *Nutr. Cancer* 18:277–286.
217. Tarrant, P.V. 1984. The farm animal welfare research in the EEC. *Livest. Prod. Sci.* 11:457–460.
218. Taylor, K.C. 1991. The control of bovine spongiform encephalopathy in Great Britain. *Vet. Rec.* 129:522–526.
219. Tayter, M., and K.L. Stanek. 1989. Anthropometric and dietary assessment of omnivore and lacto-ovo-vegetarian children. *J. Am. Diet. Assoc.* 89:1661–1663.
220. Thevenot, R. (translated by J.C. Fidler). 1979. *A History of Refrigeration throughout the World.* Paris: International Institute of Refrigeration.
221. Thorogood, M., K. McPherson, and J. Mann. 1989. Relationship of body mass index, weight, and height to plasma lipid levels in people with different diets in Britain. *Community Med.* 11:230–233.
222. Thun, M.J., E.E. Calle, M.M. Namboodiri, et al. Risk factors for fatal colon cancer in a large prospective study. *J. Natl. Cancer Inst.* 84:1491–1500.
223. Tilston, C.H., R. Sear, R.J. Neale, et al. 1992. The effect of BSE: Consumer perceptions and beef purchasing behaviour. *Br. Food J.* 94(9):23–26.
224. Tucker, K.L., G.E. Dallal, and D. Rush. 1992. Dietary patterns of elderly Boston-area residents defined by cluster analysis. *J. Am. Diet. Assoc.* 92:1487–1491.
225. Turner, R.W.D. 1989. Beware of the cow. *Lancet* 2(8669):983.
226. Ursin, G., R.G. Ziegler, A.F. Subar, et al. 1993. Dietary patterns associated with a low-fat diet in the national health examination follow-up study: Identification of potential confounders for epidemiologic analyses. *Am. J. Epidemiol.* 137:916–927.
227. van Ravenswaay, E. 1990. Consumer perception of health risks in food. *In creasing Understanding of Public Problems and Policies.* Oak Brook, Ill.: Farm Foundation, 55–65.
228. van Ravenswaay, E., and M. Smith. 1986. Food contamination: Consumer reactions and producer losses. *Natl. Food Rev.* 33(Spring):14–16.
229. van Ravenswaay, E., and L.T. Wallace. 1983. Changing agricultural marketing programs. In *Federal Marketing Programs in Agriculture.* Danville, Ill.: Interstate, 305–326.
230. Vena, J.E., S. Graham, J. Freudenheim, et al. 1992. Diet in the epidemiology of bladder cancer in western New York. *Nutr. Cancer* 18:255–264.
231. Venkataraman, P.S., H. Luhar, and M.J. Neylan. 1992. Bone mineral metabolism in full-term infants fed human milk, cow milk-based, and soy-based formulas. *Am. J. Dis. Child.* 146:1302–1305.
232. Wagstaff, D.J. 1986. Public health and food safety: A historical association. *Public Health Rep.* 101:624–631.
233. Waites, W.M., and J.P. Arbuthnott. 1990. Foodborne illness: An overview. *Lancet* 336:722–725.
234. Walker, K.D., W.D. Hueston, H.S. Hurd, et al. 1991. Comparison of bovine spongiform encephalopathy risk factors in the United States and Great Britain. *J. Am. Vet. Med. Assoc.* 199:1554–1561.
235. Weaver, A.D. 1992. Bovine spongiform encephalopathy: Its clinical features and epidemiology in the United Kingdom and significance for the United States. *Compend. Contin. Educ. Pract. Vet.* 14:1647–1656.
236. Webb, T., and T. Lang. 1990. *Food Irradiation, the Myth, and the Reality.* Wellingborough, England: Thorsons Publishers.
237. Webber, S.A., R.A.C. Graham-Brown, P.E. Hutchinson, et al. 1989. Dietary manipulation in childhood atopic dermatitis. *Br. J. Dermatol.* 121:91–98.
238. Webster, A.J.F. 1982. The economics of farm animal welfare. *Int. J. Stud. Anim. Prob.* 3(4):301–306.
239. Weersink, A., and L.W. Tauer. 1991. Causality between dairy farm size and pro-

ductivity. *Am. J. Agric. Econ.* 73:1138–1145.

240. Weil, R.R. 1990. Defining and using the concept of sustainable agriculture. *J. Agron. Educ.* 19:126–130.

241. Wein, E.E., J.H. Sabry, and F.T. Evers. 1989. Food health beliefs and preferences of northern native Canadians. *Ecol. Food Nutr.* 23:177–188.

242. Weisburger, J.H., and R.C. Jones. 1990. Prevention of formation of important mutagens/carcinogens in the human food chain. *Basic Life Sci.* 52:105–118.

243. Wells, G.A.H., and I.S. McGill. 1991. Recently described scrapie-like encephalopathies of animals: Case definitions. *Curr. Top. Vet. Med. Anim. Sci.* 55:11–24.

244. Wharton, B. 1992. Which milk for normal infants? *Eur. J. Clin. Nutr.* 46(Suppl. 1):S27–S32.

245. Wheelock, V. 1988. Public perception of food safety. *J. R. Soc. Health* 108:130–131.

246. Wilesmith, J.W., and J.B.M. Ryan. 1992. Bovine spongiform encephalopathy: Recent observations on the age-specific incidences. *Vet. Rec.* 130:491–492.

247. ——. 1993. Bovine spongiform encephalopathy: Observations on the incidence during 1992. *Vet. Rec.* 132:300–301.

248. Wilesmith, J.W., L.J. Hoinville, J.B.M. Ryan, et al. 1992. Bovine spongiform encephalopathy: Aspects of the clinical picture and analyses of possible changes 1986–1990. *Vet. Rec.* 130:197–201.

249. Wilesmith, J.W., J.B.M. Ryan, and W.D. Hueston. 1992. Bovine spongiform encephalopathy: Case-control studies of calf feeding practices and meat and bonemeal inclusion in proprietary concentrates. *Res. Vet. Sci.* 52:325–331.

250. Wilesmith, J.W., J.B.M. Ryan, W.D. Hueston, et al. 1992. Bovine spongiform encephalopathy: Epidemiological features 1985 to 1990. *Vet. Rec.* 130:90–94.

251. Wilkinson, L. 1992. *Animals and Disease: An Introduction to the History of Comparative Medicine.* Cambridge, Engl.: Cambridge University Press.

252. Will, R.G. 1991. Is there a potential risk of transmission of BSE to the human population and how may this be assessed? *Curr. Top. Vet. Med. Anim. Sci.* 55:179–186.

253. Williams, C.M. (chrmn.). 1987. *Farm Animal Welfare in Canada: Issues and Priorities, 1987.* Report of the Expert Committee on Farm Animal Welfare and Behaviour, Canada. Ottawa: Agriculture Canada.

254. Willis, J.L. (ed.). 1990. *Food Risk: Perception vs. Reality.* Rockville, Md.: Food and Drug Administration, U.S. Department of Health and Human Services.

255. Wilson, C.A. (ed.). 1991. *'Waste Not, Want Not', Food Preservation from Early Times to the Present.* Edinburgh: Edinburgh University Press.

256. Winkler, G., H. Holtz, and A. Doring. 1992. Comparison of food intakes of selected populations in former East and West Germany: Results from the MONICA projects Erfurt and Augsburg. *Ann. Nutr. Metab.* 36:219–234.

257. Woodgate, S.L. 1991. Pilot plant studies in BSE/scrapie deactivation. *Curr. Top. Vet. Med. Anim. Sci.* 55:169–175.

2

Food Production Technology: The Food Chain

Objectives

- Distinguish between subsistence and commercial agriculture and between competitive and noncompetitive agricultural imports, and indicate an influence veterinary medicine may have on products imported.

- Identify the major geographic areas where the following types of animal production are concentrated: beef cows, dairy cows, water buffaloes, sheep, goats, swine, horses, rabbits, camelids, cervids, broilers, turkeys, ducks, ratites, eggs, and fish.

- Explain how the price of milk is established.

- Describe the components and functions of a milking machine system.

- Describe the proper methods of milking, storage, and transportation of fluid milk.

- Define mastitic and rancid milk and describe how such milk is detected.

- List procedures for control programs for abnormal milk.

- Explain the importance of marine and freshwater fish and shellfish species in the world food supply, and identify the major areas of production and harvest.

- Describe design features and explain the efficacy of various materials used in the

31

construction of slaughter and processing plants and equipment in relation to hygiene.

• Describe slaughter and dressing or processing procedures that minimize contamination of foods of animal origin.

• Identify the procedures used in preserving foods of animal origin.

• Describe the equipment features of LTH, HTST, and UHT methods of heat treating milk and the methods used to ensure their effectiveness.

• Differentiate among psychrotrophic, mesophilic, and thermophilic microorganisms, and indicate how each group may be involved in spoilage of foods of animal origin.

• Describe enzymatic, oxidative, and chemical actions that can cause deterioration of foods of animal origin.

• Describe physiologic conditions in the animal causing PSE, PSS, and DFD.

Production
Socioeconomic Factors[90,114,118,229,236,248,307]

Veterinarians in food animal practice normally are the only health professionals whose expertise affects both the food-animal producer and consumer. Although in a classic sense veterinary medicine is concerned with health maintenance in animal populations, veterinarians work directly with human clients (either private or public) whose primary reason for involvement in food animal production is to produce income. While the world's total human population has increased tremendously during the twentieth century, in the developed nations the proportion of that population engaged in agriculture has decreased even more dramatically. In Australia, Canada, the United Kingdom, and the United States these population changes are listed in Table 2.1.

With the decrease in the proportion of the total population engaged in farming there has been a concomitant decrease in the proportion of veterinarians engaged in food-animal practice. In the United States, for example, only 7% of all veterinarians are currently engaged in full-time farm practice, resulting in a ratio of one food-animal practitioner per 1,076 persons living on farms.

The practice of veterinary medicine affects two aspects of production: (1) it increases the harvest of animal agriculture, and (2) it adds to the value of foods of animal origin by ensuring food safety and quality. Obviously, both of these must be accomplished in a manner that will provide an income to the veterinarian while also ensuring a satisfactory cost-benefit ratio for the producer; i.e., the

Table 2.1. Changes in farm populations

	1900 Farm population (in thousands)	Percentage on farms	1950 Farm population (in thousands)	Percentage on farms	1990 Farm population (in thousands)	Percentage on farms
Australia	NA[a]	NA[a]	8,307	4.5	17,085	2.30
Canada	5,301	NA[a]	13,712	22.2	26,603	3.30
United Kingdom[b]	32,247	6.0	43,830	0.2	49,194	0.05
United States	91,885	34.9	151,132	16.6	247,800	1.80

[a]Data not available.

benefit obtained from veterinary services must be greater than the cost of those services.

Subsistence Agriculture. In large areas of the world, primitive **subsistence farming** prevails. In still other large areas, **subsistence herding** is the predominant agricultural activity. In a subsistence economy there is little or no surplus to be offered for sale. Because cash is not generated to pay for veterinary services, most veterinarians are employed by public agencies. Frequently, animal diseases are a major factor in limiting production from such herding to the subsistence level. Infection and undernutrition (mismanagement) reduce the quantity and quality of animals produced. Also, infections transmitted between animals and humans (such as brucellosis and tuberculosis) are often prevalent in the human population and reduce the efficiency of human labor. In the areas where subsistence farming or herding predominate, usually more than 75% of the active population (workforce) is employed in the primary sector (mainly agriculture).

Types of Tenure of Farm Operators.[34,40,60] Where private ownership of land occurs, the farm manager may own the farm outright, have one or more partners, or rent the land. As farms grow in size, fewer are owned by individuals, with the percentage operated by partnerships increasing as the size of the farms increase. The largest farms are frequently operated by corporations. In contrast to subsistence farmers, large operators generally have capital and qualified managers and thus employ veterinarians in developing efficient herd health programs. Worldwide, most farms are operated (managed) by the owners. The efficiency of the operation as well as the size has a direct impact upon the owner's ability (and willingness) to utilize veterinary services effectively. There are some differences among livestock enterprises. Generally, a greater percentage of dairy farmers own their own enterprises, compared with other livestock operations, many of which are operated by managers.

Imports and Exports. [8,10,13,15,16,18,50,62,69,86,93,122,126,148,152,153,157,173,184,189,199,204,224, 225,234,250,252,281,290,294,298,303,304,308] In many areas of the world, agricultural production is insufficient to meet the local demand, and it becomes necessary to import goods. In other areas, production exceeds the demand, and it becomes possible to acquire additional income by exporting the excess agricultural goods. This simple application of the principles of supply and demand forms the basis for the international trade of agricultural products. In the four countries of Australia, Canada, the United Kingdom, and the United States, production and consumption of foods of animal origin in 1990 were as listed in Table 2.2.

International trade of agricultural products is subject to restrictions. Some of these are designed to prevent disease, others to protect the public from shoddy or potentially hazardous products. Some are developed primarily to gain an economic advantage for the producers in the importing or exporting country. To better understand these restrictions it is necessary to define some terms. From the point of view of the importing country, imported goods may be **noncompetitive** (complementary) or **competitive** (supplementary). Noncompetitive agricultural products are those that are not produced domestically. The United States, for instance, imports large quantities of coffee, cocoa, natural rubber, bananas, tea, spices, and carpet wool, none of which are domestic products. Competitive imports are those that are produced domestically. In the United States, these commodities include meats, sugar and molasses, fruits and vegetables, wine, oilseeds, and dairy products and eggs, as well as cattle and calves. These foreign products compete in the U.S. domestic market (Fig. 2.1).

Frequently, other import limitations may be imposed on meat and other products of animal origin. Many governments establish quotas, by country, on the amount of meat that may be imported from that country. They may also impose variable levies and tariffs to protect domestic agricultural production and, in some cases, use export subsidies to help their own producers gain and maintain export markets. In 1985, the United States created the Export Enhancement Program (EEP) to help U.S. agricultural exporters match price competition from other subsidized exporters. It is not unusual for two countries to formulate spe-

Table 2.2.   Production and consumption of foods of animal origin

	Australia	Canada	United Kingdom[b]	United States
		Production (t)		
Beef and veal	1,695	922	997	10,464
Lamb and mutton	666	0	371	165
Pork	305	1,140	980	6,964
Total meat	2,666	2,062	2,348	17,593
Poultry	419	733	1,087	18,878
Fluid milk	6,500	7,900	15,016	67,912
		Consumption (kg per person)		
Beef and veal	39.3	37.8	7.7	44.3
Lamb and mutton	21.6	0.8	4.3	0.8
Pork	18.5	27.5	8.8	29.1
Total meat	79.4	68.1	20.8	74.2
Poultry	24.6	29.3	NA[a]	29.0
Fluid milk	102.0	107.7	NA[a]	104.5

[a]Data not available.
[b]England and Wales.

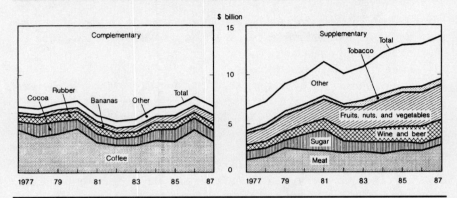

Fig. 2.1. U.S. agricultural imports by commodity, 1977–1987. Complementary imports do not compete with domestically produced products; supplementary imports do. Source: 282.

cific agreements governing the import and export of goods between them. An example is the Beef Market Access Agreement (BMAA) established in 1988 between the United States and Japan and between Australia and Japan. Japan produces about 70% of its own beef requirements, two-thirds of this from culled dairy animals. About 70% of the beef imported into Japan is grass-fed beef from Australia, and 30% is grain-fed beef from the United States. Occasionally, groups of nations will associate to facilitate the movement of goods among their members by eliminating many of these barriers and, at the same time, to create a solid front to competition from countries outside their group. Examples are the European Union (EU) and the North American Free Trade Agreement (NAFTA).

One of the earliest international bodies developed to facilitate world trade was the General Agreement on Tariffs and Trade (GATT) created in 1948. Under GATT there was excellent progress in eliminating many trade barriers. For example, it is estimated that tariffs on manufactured goods were reduced from an average level of 40% before World War II to less than 5% today. Some of the member nations, however, have bypassed the spirit of GATT by imposing non-tariff trade barriers (NTBs), such as quantitative restrictions, or by using technical standards to block imports. An example of the latter is the ban by some countries on the use of hormones in the production of meat or milk. Residues of antibiotics, hormones, and other chemicals are becoming an important issue for exporting countries. In 1995, following international approval of the Uruguay Round accords, GATT was replaced by the World Trade Organization (WTO).

A common use of technical standards is to protect member nations from importing animal disease. The greatest problem occurs between countries using an eradication policy and those using a control strategy such as vaccination. In order to assist in making decisions regarding the movement of animals or animal products between nations, the EU has classified livestock diseases into three groups:

Group 1. Diseases that could quickly devastate livestock populations over large areas, e.g., foot-and-mouth disease and hog cholera. These have been targeted by the EU for total eradication.

Group 2. Diseases whose prevalence tends to be more localized, e.g., bovine leucosis, brucellosis, and tuberculosis. Eradication is the goal for these diseases, but the steps taken to achieve this may be less stringent than for Group 1.

Group 3. Less serious diseases that the EU does not consider a serious threat to the animal population.

The United States has unilateral regulations to protect its own livestock industry. For example, countries in which foot-and-mouth disease is endemic cannot export any fresh meat into the United States (**cooked** meat may be imported). Some large exporters of fresh meat such as Argentina are thus excluded from this market. If foot-and-mouth disease is eradicated from these areas, serious pressure on the U.S. domestic fresh meat market may come from these countries.

Most of these bilateral and multinational organizations have a broad scope; they influence the international movement of nonfood manufactured goods as well as agricultural food products. Soon after World War II the World Health Organization (WHO) was created as an agency of the United Nations (UN) with responsibility for international health. At about the same time the UN created another agency, the Food and Agriculture Organization (FAO), with the objective of improving the production and distribution of food and other agricultural products. In 1963, these two agencies established the **Codex Alimentarius Commission (CAC)**, commonly referred to as "the Codex," to create internationally accepted standards for food production and safety. It has three primary functions: (1) to facilitate international trade through the removal of nontariff barriers caused by differing national food standards, (2) to protect the health of consumers and to ensure fair practices in the food trade, and (3) to promote coordination of all food standards work undertaken by international governmental and nongovernmental organizations.

The work of the Codex is carried out by Codex committees, of which there are three types: (1) general subject matter committees, such as food hygiene, pesticide residues, and analysis and sampling, (2) commodity committees, such as fish and fishery products, meat hygiene, cereals, pulses, and legumes, and (3) regional coordinating committees for areas such as Europe, Asia, North America, and the Southwest Pacific.

These committees receive technical advice for standards from the Joint FAO/WHO Expert Committee on Food Additives (JECFA) and the Joint FAO/WHO Meeting on Pesticide Residues (JMPR). Some 250 food standards, codes of practice, and related documents have been issued by the Codex. These cover a wide range of subjects: additives, compositional standards, contaminants, pesticide residues, methods of analysis, food processing, and food inspection.

The current ban by the EU on the nontherapeutic use of hormones in livestock and the importation of meat from countries that permit the use of these hormones provides an illustration of the complexities of international trade in foods of animal origin. The major exporting countries to the EU are Argentina, Australia, New Zealand, and the United States. The ban has a major impact on the United States because it produces beef from cattle receiving high-energy grain-based rations in feedlots. Among the other three major exporters, cattle are mostly range fed, and only a small percentage receive hormones because the procedure is not cost-effective under range conditions. Supported by a finding from the Codex that such a ban is "unscientific," i.e., there has not been sound evidence that consumption of meat from cattle treated with hormones is unsafe to human health, the United States has claimed the ban is a violation of GATT principles and that the EU has created a trade barrier. The EU, in turn, has denied this because it applies the same standard to domestic meat. A major difference, however, is that in the EU beef originates primarily from dairy herds, not feedlot cattle.

Livestock (Meat) Production[276]

Livestock production can be classified regionally according to the principal species or types of livestock produced. FAO data from 1988 indicate that, worldwide, ruminants represent 76% of total livestock numbers (excluding poultry). The annual production of meat, on a dressed carcass basis, however, puts swine at the top with 61.6 million tons annually. Cattle and buffalo annual production equaled 49.5, poultry 35.2, sheep and goats 8.6, and horses 0.5 million tons.

At the beginning of the 1970s, beef was produced and consumed in greater quantities than any other type of meat. Beef production continued to rise during the 1970s until, in the early 1980s, beef production had increased by 17%. During this same period, however, swine production had increased by 43% and poultry production by 87%.

Undeveloped Areas. In some areas, such as the rain forests of Africa and South America, there are practically no livestock. In these areas, hunting and fishing provide most of the animal protein consumed.

Livestock Ranching. The major livestock ranching areas of the world are those where sheep, cattle, and goats are grazed principally to produce wool, hides, or meat. In some of these areas, animals are born and reared on the range and then sold to feedlots for fattening before slaughter.

Cattle.[31,104,117,121,143,163,263,282,296] Three areas of dense commercial cattle production exist: the midwestern United States, central Europe, and the east coast of South America. Some countries, such as India, have a large cattle population, but these cattle are not part of an organized production program. The major cattle ranching countries are the United States, Brazil, Argentina, and Australia.

During the twentieth century, the cattle populations of Australia, Canada, the United Kingdom, and the United States have varied, as shown in Table 2.3.

With the exception of the United States and Canada, cattle in these countries are primarily sold for slaughter directly off the farm. In the United States and Canada, beef cattle production is a two-stage procedure. Calves are born on the range in cow-calf operations and remain there to nurse and graze until the end of the good pasture season. At this time, when the calves are 5–6 mo old, they are transferred to feedlots where they will spend up to a year before they are sent to slaughter. Modern feedlots can be huge; some can handle 1 million head of cattle at one time (Fig. 2.2). In some instances, if suitable pasture is available and a minimum of concentrate feeding is required, calves may be kept on pasture for a year and require only a few months in a feedlot before slaughter. In countries where calves are "grass fattened" rather than going to feedlots for a period of time, the animals may be 2–3 yr old when slaughtered. In spite of decreasing demand for the heavily marbled beef produced in feedlots, grass-fattened beef from Australia and New Zealand is shipped to the United States as "manufacturing grade" beef and is used primarily for hamburger, rather than expensive cuts such as steaks or roasts. Beef imported from the EU is also classified as manufacturing grade.

Sheep and Goats.[131]   The world population of sheep is approximately 1.2 billion. Most of these are in areas unsuitable for other types of agriculture. In developing countries they are also associated with subsistence economies. The major sheep-farming countries of the world are Australia, the former USSR, and China (Table 2.4).

Table 2.3.   Changes in cattle populations (in thousands)

	1900	1950	1990
Australia	8,640	15,220	25,332
Canada	7,202	8,343	12,287
United Kingdom[a]	484	8,014	8,460
United States	42,265	55,479	33,200

[a]England and Wales.

Table 2.4.   Major sheep-producing countries

Country	Number (millions)
Australia	167
USSR (1986)	142
China	134
New Zealand	95
Turkey	62
United Kingdom	42
India	41
South Africa	36
Argentina	35
Uruguay	24
United States	11

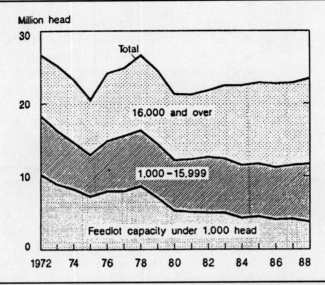

Fig. 2.2.   Fed cattle marketed in the United States by feedlot capacity, 1972–1988. Data are for 13 states. Source: 282.

During the twentieth century the sheep population of Australia, Canada, the United Kingdom, and the United States has varied, as shown in Table 2.5.

Australian sheep have been selected to produce the world's finest quality wool for garment cloth. Other areas produce wools of lesser grade (such as carpet wools) or raise meat breeds. New Zealand, for example, exports large quantities of lamb. The emphasis on export is clear from the local term for abattoirs as "freezer works."

Lambs are sold as hothouse lambs, spring lambs, or simply as lambs. Hothouse lambs are less than 3 mo old and weigh 15–30 kg (33–66 lb). Spring lambs will weigh 30–45 kg (66–100 lb). Young sheep that do not qualify as hothouse or spring lambs are sold as lambs. These are usually born in late winter or early spring and are marketed at 7–12 mo of age.

Since the beginning of World War II, sheep production in the United States has declined as a result of increasing labor costs and other production costs. The demand for wool has decreased as a result of the widespread use of synthetic fibers. And, unlike beef, lamb has never been a favorite in the average American diet.

Table 2.5.   Changes in sheep populations (in thousands)

	1900	1950	1990
Australia	70,603	108,600	166,500
Canada	2,543	1,579	759
United Kingdom[a]	16,261	12,376	41,711
United States	44,473	30,743	11,363

[a]England and Wales.

Much of the goat population in the leading goat-producing areas of the world is of the Angora type (used to produce mohair) rather than the meat and milking breeds. The United States, Turkey, and South Africa produce almost 90% of the world's mohair. In Australia, goats for meat are harvested from the feral population in the pastoral zone, and the meat is exported primarily to Singapore, the Pacific islands, and the West Indies. With good husbandry milk production can be good—680 kg (1,500 lb) per year, but frequently the average is only 200 kg (440 lb).

Swine.[37,181,185,213]   The major centers for swine production in the world are
China, Russia, the upper midwestern United States, and central Europe (Table 2.6).

During the twentieth century, the number of swine produced in Australia, Canada, the United Kingdom, and the United States varied, as shown in Table 2.7.

Most swine production units in the United States are what is termed *farrow-to-finish* operations. Each farm maintains its own breeding stock of boars and sows. Sows ready to farrow are brought to a separate facility. Usually, the farrowing unit and the nursery are in the same building, but these may be separate for more-effective disease control. Immunization, tail docking, and castration are done before the baby pigs are moved to the finishing unit, which is nearly always a separate building. The pigs will remain here until they reach market weight. This weight varies in different parts of the world and, sometimes, even in different areas of a single country. Market weight ranges from 80–110 kg (176–242 lb). Beyond 110 kg the return on money invested in feed diminishes. The increase in world swine production in the 1980s has been stimulated by improved

Table 2.6.   Major swine-producing countries

Country	Production (thousand tons)
China	344,248
Russia (1988)	79,501
United States	53,795
Germany (West and East)	37,343
Brazil	32,000
Mexico	18,662
Poland	18,546
Romania	14,711
Netherlands	14,349
Spain	14,000

Table 2.7.   Changes in swine populations (in thousands)

	1900	1950	1990
Australia	960	1,123	2,648
Canada	3,379	4,372	10,532
United Kingdom[a]	2,225	2,212	6,409
United States	52,600	60,502	54,562

[a]England and Wales.

survival of pigs to weaning age, lower feed conversion rates, access to cheaper feeds, and improved disease control.

Horses.[212]   Horse meat production depends primarily on discarded pleasure horses rather than a ranching system operated with meat production as a primary purpose. In 1980, Canada produced 16 million kg (35 million lb) of horse meat (70,000 horses), and in 1983 the United States produced 28.5 million kg (63 million lb) (125,000 horses). Most of this is exported to France, Belgium, and Japan. The price for horse meat in North America is equivalent to that paid for old cows.

Rabbits.[212]   Although accurate statistics on domestic rabbit production are scarce, many countries have a recognized rabbit industry, including Belgium, Germany, Italy, Russia, Scandinavia, Switzerland, China, India, the United Kingdom, and the United States. In addition to meat, skins and angora wool are significant sources of income.

Because rabbits reproduce and grow quickly and do not compete with humans for food, they have tremendous potential as a meat animal. The feed conversion rate is 3.0–3.5 on a high alfalfa ration (versus 8.0 for beef). An equal amount of alfalfa can produce five times as much meat as when fed to beef cattle. The fat content of rabbit meat is less than that of chicken, beef, or pork. Only tom turkey meat has a lower fat content. The primary disadvantage of commercial rabbit production is the high labor requirement and, as a consequence, most production takes place in small backyard rabbitries.

Exotic Species.[59,216,242,311,316]   In recent years, there has been increasing interest in raising several species of mammals and birds in areas where they are not normally found as domestic livestock or where they are usually harvested through hunting. In these countries they are referred to as "exotic species" although the term *exotic* is misleading because in many areas of the world they are commonly regarded as domestic animals, raised for draft, meat, or milk.

Camelids.[59,216,242,311,316]   Camelids contain two genera: *Camelus* and *Lama*. In the genus *Camelus* are two species, the dromedary, or desert, camel and the Bactrian, or two-humped, camel. The dromedary constitutes 94% of the estimated 17.5 million camels in the world. Most of these are in Africa and Asia with the majority in Somalia, the Sudan, India, Ethiopia, and Pakistan. Camels are a triple-purpose animal (draft, meat, and milk) and can survive, reproduce, and produce milk and meat in environments unsuitable for other domestic livestock. The average slaughter weight of mature dromedaries is about 450 kg (1,000 lb), dressing out at approximately 50%. Milk production, with good nutrition, ranges from 2,700 to 3,600 kg (6,000 to 8,000 lb) per lactation.

The so-called American camelids are made up of four species: the llama, alpaca, vicuna, and guanaco, all native to South America. Of these, only the llama has evoked any interest as a prospective meat animal for production outside its normal habitat. At present, activity is primarily restricted to the sale of breeding couples, much as was true for chinchillas following World War II. It remains to be seen whether or not the llama will prove to be a cost-effective domesticated meat animal.

Water Buffalo.[58,144,272] The water buffalo, another triple-use animal, has gained attention as a meat animal outside its normal habitat because of its ability to do well in low swampy areas unsuitable for domestic cattle, its resistance to many parasites, and, particularly in this day of health awareness, the nature of its meat. Buffalo meat has only 44% of the fat content of lean beef and less saturated fats (31% versus 40%). When slaughtered at 1 yr of age, the average carcass weight is approximately 115 kg (250 lb), dressing out at 44%. At 2 yr, carcass weight is between 250 and 300 kg (550 and 650 lb), dressing out at 50%. Milk production ranges from 1,500 to 1,700 kg (3,300 to 3,700 lb) per lactation.

Cervids (Red Deer, Fallow Deer, Blackbuck).[6,64,147,179,205,214,233,243,297] Beginning in the 1970s, the emphasis in several countries changed from slaughter of feral deer for venison to rearing in captivity. At slaughter, when they are usually 15–16 mo of age, live weight is approximately 45 kg (100 lb), and the carcasses dress out at 70–75%. "Velvet," the soft outer part of the growing antler, is used in the Orient in some traditional medicines; however, the cropping of velvet is inhumane, and many countries prohibit it. Problems associated with commercial deer production are numerous: low reproductive and weaning rate, expensive overwintering for young deer, long birth to slaughter interval, insufficient knowledge regarding nutritional needs, and the presence of antlers, which have to be removed (or disbudded shortly after birth) to prevent injury to other deer under confinement. Tuberculosis and brucellosis may become endemic among captive-reared wild cervids, and infected animals are difficult to detect.

Reindeer.[96,171,193] These animals have been domesticated for more than 1,000 yr. They are well adapted to arctic conditions where their main forage is lichens. Under normal herding conditions their main problems are parasites (especially warbles) and predation. Three or four wolves may kill 20 reindeer in one night. If they are confined in corrals for any length of time they develop foot rot. Another problem is a high prevalence of dark, firm, and dry meat. This is directly related to stress prior to slaughter.

Ratites.[119,135] Three members of ratites have evoked interest as potential sources of commercially produced meat: the ostrich, emu, and rhea. Ostriches, natives of Africa, are primarily grazers. Adult males will weigh 150 kg (330 lb) at 18–20 mo of age. Hens start laying eggs when 2–3 yr old. The hide represents the major return on investment, accounting for 60%. The meat represents 25%

and the feathers 15% of the profit. There are currently 12,000–15,000 ostriches on farms in the United States. About 250,000 are on farms in South Africa. The emu, a native of Australia, is also a grazer but will eat insects, grains, fruits, and vegetables. Adults weigh a little more than 50 kg (110 lb). In addition to commercial production in Australia, there are several thousand of these in the United States, primarily in Texas, Louisiana, and Arkansas. The rhea, from South America, is the smallest of the three with adults weighing a little more than 25 kg (55 lb). Rheas are grass and leaf eaters. As yet, there has not been much commercial farming of this species.

Poultry Production[10,21,22,43,111,116,196,208,218,228,257,299]

China, the United States, and Russia are the world leaders in poultry production (Table 2.8).

In 1990, the poultry production (in metric tonnes) and per capita consumption (in kilograms) in Australia, Canada, the United Kingdom, and the United States were as shown in Table 2.9.

Poultry is one of the world's fastest growing sources of meat, representing nearly one-fourth of all meat produced in 1990. The modern broiler industry started in the 1930s, when flock size was seldom greater than a few hundred birds. By the 1950s, flock size had increased to a few thousand and in the 1980s many broiler houses had a capacity of 100,000 or more. As a result of genetic selection and improved feeding practices, modern production units can produce market ready broilers in less than 6 wk. An all-out rearing system is usually followed. After a house has been emptied, it is cleaned and allowed to remain empty until the next group of birds is introduced. The time period between groups ranges from 5 to 20 d. If demand (and price) is high, a new group will be started as quickly as possible. If demand (and price) is low, or a resistant disease organism has created problems in the previous group, a longer intergroup interval may

Table 2.8. Estimated numbers of poultry (in millions)

Chickens	Ducks	Turkeys
World 10,215	World 519	World 223
China 1,849	China 325	United States 78
United States 1,540	Bangladesh 32	Russia 48
Russia 1,129	Indonesia 29	Italy 23
Brazil 550	Vietnam 27	France 20
Indonesia 410	Thailand 16	Mexico 12
Japan 334	France 11	United Kingdom 9
India 260	Mexico 7	Israel 7
Mexico 224	United States 7	Canada 6

Table 2.9. Production consumption (in metric tonnes) per capita (in kilograms)

Australia 419 t, NA[a]	United Kingdom 1,087 t, 10.9 kg
Canada 733 t, 29.3 kg	United States 18,878 t, 29.0 kg

[a]Data not available.

be used. In the latter case this provides enough time to take appropriate measures to eliminate the organism.

The modern poultry industry (**broilers** and **turkeys**) is integrated with processing plants and feed mills. The processing firm often owns the hatchery as well as the processing plant. Growing the birds is usually on a contract basis with producers who provide the facilities, utilities, and labor for payment based on the total weight of birds produced. The processor provides production supervision and retains ownership of the birds. The turkey system is more varied than that for broilers. A larger percentage of production is on processor-owned farms. Contracts may be based on a specified processing date and price formula, a minimum price contract, or a contract similar to those used in the broiler industry.

Worldwide, **duck** production expanded after World War II. Popular breeds are the Pekin and the Muscovy, which are often bred together. The resulting cross is sterile but has a faster rate of weight gain than the Pekin duck. Originally, ducks were raised near water, but experience has shown that dry-land rearing is just as efficient. In some areas, ducks are reared on fish ponds in order to utilize their feces as manure. Currently, 80% of the ducks in Taiwan are produced in this manner. In a field trial in Israel, fish (tilapia and carp) reared in ponds with ducks gained an average of 82 kg (180 lb) per hectare (2.47 acres) per day. The fish only ate natural aquatic plant life enhanced by duck feces, plus feed pellets missed by the ducks.

The increase in the population density of birds that typifies the modern poultry industry would not have been possible without effective vaccines and anticoccidial drugs. Coccidiosis has always been one of the major impediments to profitable poultry production. Sterilization-eradication of coccidial oocysts is an impractical control method because of the ubiquity of viable oocysts and their resistance to disinfectants commonly used against other pathogens of poultry. Beginning with sulfanilamide in 1940, there has been a continuing development of new, more effective anticoccidial drugs that have made possible increased production because chemotherapy is used at a price currently less than 1 cent per bird (Fig. 2.3).

A negative aspect of modern intensive poultry production is the impact upon the environment. In 1989, for example, the U.S. poultry industry generated 8.8 million tons of fecal waste (on a dry weight basis) plus 37 million dead birds from production units and condemnations at processing plants.

Transport of Livestock and Poultry[57,132,156,230,310]

After meat animals have reached a desired weight, they must be transported to a feedlot or slaughtering facility. Transportation is a stress in itself, but it may be compounded by other factors such as deprivation of food and water, crowding, and exposure to high or low temperatures when livestock are transported over long distances under harsh climatic conditions. Some of these stresses have fatal outcomes; most have deleterious effects upon meat quality. In the United

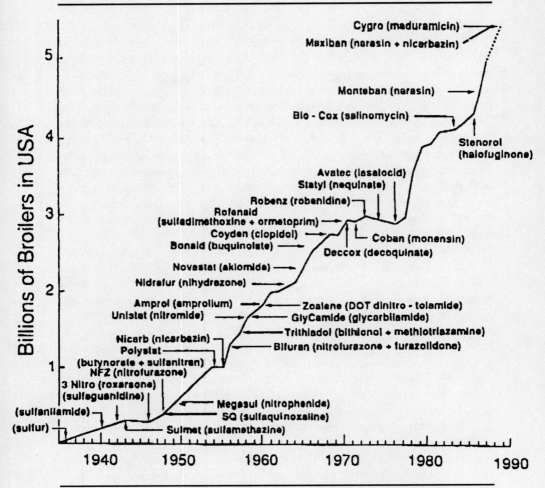

Fig. 2.3. Broiler production and year of introduction of new anticoccidial drugs. Registered names begin with a capital letter and generic names are shown lowercase. Source: 218.

States, up to 10% of feeder calves die during the first 3 wk after shipment to a feedlot. Pigs and poultry are particularly vulnerable to suffocation during transit, especially during hot weather. Among animals that survive, there may be bruising, loss of weight, increase in number of pathogenic bacteria such as *Salmonella,* or the stress-related changes that precipitate the porcine stress syndrome; pale, soft, exudative pork; or dark, firm, dry syndromes.

Under most inspection programs, **bruised** masses of muscle must be removed from a carcass before sale, producing a reduction of profit. If bruising is extensive, the entire carcass may be condemned. Bruising during transport occurs in all species but is particularly noticeable in sheep and pigs. In the United

Kingdom, the Meat and Livestock Commission reported that, in 1972 and 1974, up to 10% of lambs and 50% of pigs were damaged due to mishandling during transport. It is obvious that careless or rough handling of animals during collection, transport, or the preslaughter period is not only inhumane, it is wasteful.

Weight loss during transit is partly the result of a breakdown of fatty and muscular tissues to provide energy and heat for the fasting animal and partly the result of a loss of the water-holding capacity of muscle tissues. When fasted during transit, cattle lose weight less readily than sheep, and sheep less readily than pigs. Pigs slaughtered immediately after a short journey may provide a heavier carcass and heavier offal than pigs that have traveled for a longer time and then were rested and fed for a few days before slaughter. Even though cattle are less susceptible to weight loss during transit than other species, under extreme conditions, such as trail drives in Australia in which cattle may walk up to 1,000 miles, 12% of initial live weight may be lost.

Transport-associated stress, if severe, can precipitate clinical **salmonellosis**. Mixing consignments of pigs and calves can also lead to significant cross-infection. Animals in a stable physiologic state are generally resistant to salmonellosis because the normal floras of the gut maintain conditions unsuitable for the multiplication of salmonellae by producing volatile fatty acids (particularly butyric acid) in sufficient concentrations to be mildly bactericidal. However, holding cattle without feeding, prior to slaughter, increases the number of salmonellae in the rumen and feces. If the rumen is accidentally incised during processing, the carcass may become contaminated with salmonellae, or there may be cross-infection of other carcasses by contaminated knives. Hence it is important that livestock be held with minimal stress for only short periods of time before slaughter and that strict hygiene is observed during the dressing procedure to prevent contamination.

Transporting chickens and turkeys to processing plants can increase fecal excretion of both salmonellae and chlamydiae. This poses a potential risk to consumers if the viscera are ruptured during removal and the carcass is contaminated with intestinal material. When large turkeys are cooked, the internal temperature of stuffing may not reach the thermal death point for salmonellae; this is a potent source for human infection. Inhalation of aerosols containing chlamydiae is a significant hazard for poultry industry employees working with stressed poultry or dressing their carcasses.

Stress associated with transport is the primary reason for depleting the glycogen reserves in muscle, which, in turn, results in the porcine stress syndrome; pale, soft, exudative pork; or dark, firm, dry musculature.

Porcine Stress Syndrome; Pale, Soft, Exudative Musculature; and Dark, Firm, Dry Musculature.[32,35,46,53,108,120,166,190,264,301,318]

Initially **porcine stress syndrome (PSS)** was recognized in swine that had been transported. The syndrome is precipitated by physical stress and is manifested by dyspnea, reddening of the skin in white breeds, increased respiratory rate, cyanosis, and acidosis. As the syndrome progresses, muscle rigidity, hyperthermia, collapse, and death follow. Animals that survive frequently develop **pale, soft, exudative (PSE)** or **dark, firm, dry (DFD)** musculature. Susceptibility to PSS is related to the presence of an autosomal recessive "halothane" gene (HAL) with incomplete penetrance. (*Halothane* is a membrane-depolarizing anesthetic gas. PSS susceptible pigs develop muscular rigidity rather than relaxation when exposed to halothane in a closed system rebreathing apparatus.)

Epidemiologic Findings about PSS. Stress susceptible animals are usually very muscular with little back fat. However, not all muscular animals are stress susceptible. There also is a breed-related susceptibility. Most breeds have less than a 10% PSS incidence. The Duroc, Large White, and Minnesota No. 1 breeds show no susceptibility. The Pietrain and some strains of Landrace have a moderate to high incidence. The German and Belgian Landrace and Dutch Pietrain breeds have the highest incidence, with 70–100% showing susceptibility when exposed to halothane. Recently, it has been recognized that signs of PSS may result from mild stress such as moving from one pen to another.

PSS/PSE Lesions on Necropsy. PSS susceptible swine have a very high incidence of PSE after slaughter. Muscle changes at postmortem are especially observable in large muscle masses such as the longissimus dorsi, gluteus medius, and major muscles of the hind legs. There is an association among muscle pH, body temperature, and the occurrence of PSE. The speed with which muscle pH drops is also a factor. PSE muscle undergoes rapid postmortem anaerobic glycolysis and results in accumulation of lactic acid and a muscle pH under 5.8. The lowered muscle pH in the presence of undissipated body heat leads to denaturation of the contractile proteins such as myosin and actin. Stress in affected animals not only increases total body heat but also the rate of anaerobic glycolysis, creating the rather unfavorable situation of a more acidotic animal with more body heat that must be dissipated. After cellular proteins are denatured, they lose their water-binding ability. The meat attains a moist appearance and is termed *exudative*. Since PSE is associated with an elevated body temperature, the meat quality is lowered even more when the animals are sent through the scalding vat. An already elevated temperature is increased by this process.

Clinical signs of PSS

1. **Temperature and blood characteristics**. Both PSS and PSE have been studied extensively in recent years. In one experiment three parameters (pH, PCO_2, and PO_2) were measured in PSS susceptible hogs (Chester White) and PSS nonsusceptible pigs (Poland China). The control groups were maintained at 22°C (72°F). The experimental groups were stressed by exposure to heat of 45°C (113°F) for 3 min. Blood gas levels were then determined on all groups. The Chester White group was unable to compensate. Consequently, there was

an increase in PCO_2 and a concomitant decrease in pH, as well as a dramatic decrease in PO_2 (Table 2.10).

2. **Temperature and respiration rate**. There was also a steady increase in respiratory rate in the susceptible Chester White hogs, indicating inability to adapt to the higher ambient temperature. The Poland Chinas were able to adapt to the new environment, as indicated by a rapid decrease to normal respiratory rate after 10 min exposure to 45°C (113°F) (Fig. 2.4).

3. **Increased temperature and heart rate**. Heart rate also was recorded in this experiment. Again, the Poland Chinas exposed to 45°C (113°F) had a temporary increase in rate, which, after a short time, returned to normal. The heart rate of the susceptible Chester White hogs increased progressively to fibrillation and death.

4. **Adrenal hormone excretion**. A consistent finding in hogs afflicted with PSS is the lowered levels of adrenal hormones. Nonsusceptible Poland China pigs had consistently higher catecholamine excretion than the PSS susceptible Chester Whites.

DFD.   Dark, or high-pH, meat is a quality defect found in all meat species. It results from low muscle glycogen at slaughter preventing normal acidification of meat during rigor development. Glycogen depletion can result from stress-related increased circulating adrenaline or by strenuous muscular activity. Glycogen is also depleted during starvation.

In living muscle, the pH remains just above 7.0 except during heavy exercise, when it may decrease to 6.4. After slaughter, meat pH decreases in well-fed, rested animals. When animals are stressed before slaughter, muscle glycogen is depleted, causing reduced lactate (lactic acid) production postmortem and therefore a higher ultimate pH. Muscle fibers become swollen and more tightly packed. With the cells tightly packed, the meat then assumes a rather firm consistency. The darkening is caused by a reduction in the thickness of the bright red oxymyoglobin layer on the cut surface, revealing the underlying darker myoglobin. The darkening is also caused, to some extent, by the increased translucency of the meat, further revealing the myoglobin in the interior. Starvation will not usually affect muscle pH, but it will reduce the "buffer" of excess glycogen normally present and make the meat more susceptible to darkening after slaughter.

Table 2.10.   Influence of treatment on blood characteristics of hogs

	PCO_2		PO_2		pH	
	Before	After	Before	After	Before	After
Control (22° C)						
Poland China	51	51	36	44	7.4	7.5
Chester White	71	52	38	36	7.4	7.5
Treated (45° C)						
Poland China	54	87	38	7	7.3	6.3
Chester White	63	40	35	42	7.4	7.5

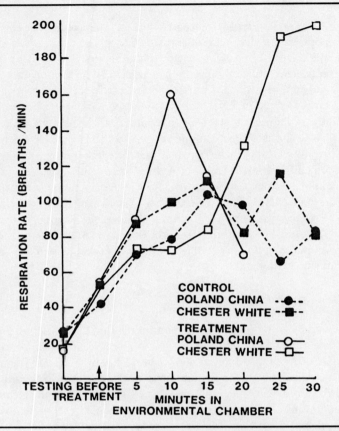

Fig. 2.4. The influence of treatment on the respiration rate of hogs.

Factors that increase the incidence of DFD include transportation, lengthy holding periods, mixing of unfamiliar animals (resulting in fighting), severe environmental conditions, and genetic predisposition. The typical history associated with this condition is stress occurring 24–48 h before slaughter, causing the animal to utilize its muscle glycogen stores. Sufficient time elapses for the excess lactic acid to be removed but not for replenishment of glycogen reserves. Consequently, anaerobic glycolysis is impaired and darker-colored myoglobin predominates. This condition is not uncommon. Fortunately, palatability is not impaired. The more expensive cuts (ribs and loins) are affected. Consumers prefer moist red meat, and DFD meat in vacuum packs tends to "green" as a result of the growth of hydrogen sulfide-producing bacteria, which occur at a high pH, not at the pH values of normal meat.

Reducing Transport Losses. The most efficient method for reducing transport-associated losses is to recognize the causes and eliminate them before they occur. Attempts to repair transport-produced damage after it has occurred usually is not cost-effective and, in some instances, can create other problems. Allowing pigs to rest for prolonged periods between transport and slaughter, for example, produces a situation in which pigs infected with salmonellae can transmit the pathogen to noninfected pigs. (The probability of this occurring is high enough to have resulted in legislation in the United Kingdom, limiting the holding time for pigs before slaughter to a maximum of 72 h.) In the United States, producers, shippers, and processors have formed the **Livestock Conservation Institute** to address and to help reduce the problem of transport-associated losses. In Canada, a federal **Health of Animals Act** came into effect in 1991.

There are numerous welfare agencies concerned with animal transportation. Among the major organizations are the **World Society for the Protection of Animals (WSPA)**, the **Eurogroup for Animal Welfare**, the **Animal Transport Association (ATA)**, and the **International Air Transport Association (IATA)**. The IATA *Live Animal Regulations*, published in 1991, is available in English, Spanish, and French.

Milk Production[49,94,159,192,284]

On a worldwide basis, centers for **commercial milk production** are limited primarily to certain parts of the United States, North Central Europe, the southeastern coast of Australia, and the north island of New Zealand. There are two main reasons for this: (1) a large part of the world does not drink milk, and (2) the areas where milk is not an important dietary item also lack the technology to handle milk in fluid form or to produce a stable powder. A large part of the production in Australia and New Zealand is used to produce butter and a high-quality milk powder for export. Because a high percentage of the milk produced is used to manufacture powder for export, the dairy industry makes maximum use of the seasonal fluctuations in pasture growth. Calving is timed to coincide with the onset of increasing growth (early spring), and cows are dry during the winter. Particularly in New Zealand, hilly terrain and high fertilizer costs restrict the production and feeding of concentrates, which, in turn, limits the potential milk yield. In 1980–1981, for example, annual production among tested pedigree cows in New Zealand ranged from a low of 2,833 kg (6,233 lb) for Jerseys to a high of 4,620 kg (10,164 lb) for Friesians. In the United States, where concentrate feeding is common, annual production among all cows involved in the Dairy Herd Improvement Association (DHIA) was 8,196 kg (18,031 lb).

In 1990, the total number of dairy cows in the world was estimated to be 219 million, and milk production totaled approximately 460,331,000 t. Of this amount, the countries of Australia, Canada, New Zealand, the United Kingdom, and the United States produced 21% of the total, distributed as shown in Table 2.11.

Milksheds.[91,125,130,146,315]   In countries with a well-developed dairy industry,

Table 2.11. Major milk-producing countries

	Number of cows	Milk (t)
Australia	1,817,000	6,500
Canada	2,389,000	7,900
New Zealand	2,260,000	7,779
United Kingdom	3,140,000	15,016
United States	11,063,000	67,260

milk production and processing constitute a major component of the total food industry. Just as the water in a river comes from a well-defined watershed, the milk supply comes from a well-defined **milkshed,** or producing area. The size of the milkshed depends largely on the size of the consuming population. The size of the producing farms (i.e., the number of milking cows per herd) determines the number of herds needed in the milkshed to serve a given consumer population. Even though milk production per cow has increased, thus requiring fewer cows, there is a trend toward larger dairy herds. Some herds supplying the Miami milkshed in the United States, for example, maintain more than 5,000 cows (Fig. 2.5). In densely populated areas of the world, each large city usually has its own milkshed. In other areas, the term may take on a different meaning. New Zealand, for example, is considered the major milkshed for much of East Asia, although many of the countries in this region are developing their own milk industry and thus providing competition.

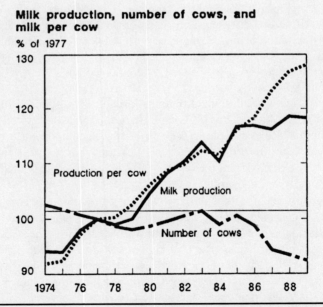

Fig. 2.5. Milk production, number of cows, and milk per cow in the United States, 1974–1988. Source: 282.

Milk Pricing.[29,36,66,79,110,133,177,194,198,268,305] Traditionally, milk producers have been
paid on the basis of quantity of fluid milk produced with a differential based
on fat content. During the past decade, alternative methods of computing the
value of milk have been introduced, such as multiple-component pricing (MCP),
which determines the price of milk on the basis of more than one component,
such as fat and protein, fat and solids not fat (SNF), or fat, protein, lactose, and
minerals. A primary reason for this change is that today a major portion of milk
produced is used for manufactured dairy products and this approach more accurately reflects the value of these products.

In 1985, the International Dairy Federation reported that 10 countries paid
on the basis of fat and protein and 6 on fat and some form of SNF. In the United
States, an increasing number of firms include a protein differential. Currently, a
minimum standard of 3.5% fat, 11.5% total solids, and 8.25% SNF is required in
most U.S. markets for Class 1 whole milk. This emphasis on the protein content
of milk is the result of the increased consumer demand for hard cheese. Much of
this approach to milk pricing originated in New Zealand, where methods for estimating the value of milk based on the market price of the end products were developed. This approach requires the inclusion of data concerning the cost of
transportation, processing, and yield formulas for butter, milk powder, and
cheese.

Milk fat is the most variable of all milk components. The breed of cow,
stage of lactation, milking interval, feed, and age contribute to this variation. In
addition, a daily variation of 1–2% in individual cows is not uncommon. There
are two principal proteins: casein (the protein of cheese) and whey. Whey proteins consist of several different proteins, all of which remain in whey when casein is transformed into curd during cheese production. Whey proteins have excellent nutritional value but, because they do not add to the yield of cheese in
most current manufacturing methods, they provide no basis for payment of premiums for milk for cheesemaking.

A second factor used in the differential pricing of bulk milk is the somatic
cell count (SCC). An elevated SCC is an indicator of udder infection and inflammation. It is also associated with the level of milk components used in MCP.
In the United States the differential for low SCC milk has not been standardized
and varies substantially from one processor to another. In the United Kingdom
there exists a standard penalty.

The retail price of milk may reflect the basic tenets of supply and demand,
with the consumer paying the full cost. Alternatively, a subsidy may be used to
determine the amount paid to producers for milk. This may be related to the consumer price index (all items). The objective of this is to ensure an adequate supply of milk at prices that are fair to producers and consumers alike. In the United
States, for example, "Milk Marketing Orders" establish minimum milk prices
within specific geographic areas that may cover parts of several states. Producers receive a federal subsidy that is based upon the relationship between production expenses and income. This is referred to as *parity,* and government support
ranges from 75–90% of parity.

Seasonal Variation in Milk Supply.[7,33,140,217]   Seasonal variation in fluid milk
production occurs because a greater proportion of cows calve in the spring
than in the fall (Fig. 2.6). This creates a problem for the dairy industry, particularly for manufacturers of nonfluid milk products, e.g., cheese and butter. It also
increases processors' costs because of the unused capacity in trucks, storage facilities, and plants during periods of low production. The variation does not create much of a problem for processors of fluid milk products because supplies normally satisfy the higher-priced fluid milk demand first, with the remainder being
sold to milk product manufacturers. Because more fluid milk is normally produced than required, fluid processors can plan their operations knowing that there
will be an adequate supply of raw milk. Manufacturers of nonfluid milk products,
however, are unable to operate their plants efficiently because, in order to process
the excess milk available in the spring, it is necessary to build a plant at a level
at which much of its capacity will remain idle in the fall.

Improved feeding practices and a more uniform calving schedule can alleviate the problem, and price differentials can motivate producers to eliminate

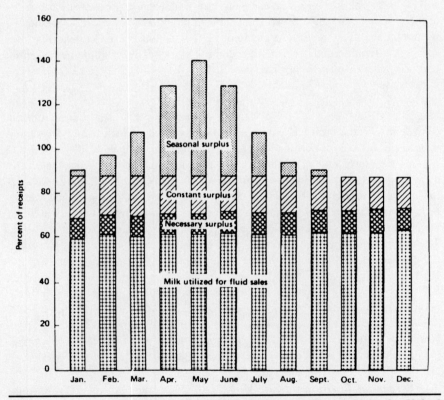

Fig. 2.6. Seasonal variations in milk supply in the United
States. The kinds of milk surplus are (1) necessary or standby,
(2) constant, (3) seasonal.

much of this seasonal variation. In the United States, for example, federal and state milk marketing orders have introduced price incentives that take the form of reduced payments per hundredweight of milk in the spring and premium payments in the fall. As a result of these programs, the average spread between high and low seasonal U.S. milk production has dropped from 54% in 1950 to 18% in 1983.

Milk Quality Control.[5,44,55,82,115,138,145,154,180,220]   The production of high-quality milk is of vital importance to the dairy industry. No other food product is subjected to the close scrutiny and regulatory control as is milk. In the United States, for example, more than 15,000 health and sanitation jurisdictions check producers, dairy plants, and stores. There are several reasons for this. One is that the structure of the dairy industry makes it is easy to monitor. This is especially true today because there are now fewer, but larger, farms producing milk. This change continues to satisfy the demand for quantity and also makes easier the monitoring required to ensure quality. Another reason is that milk, being a liquid, is easily amenable to biological, chemical, and physical evaluation. A third reason is that the almost universal acceptance of milk or milk products as a food for all age groups mandates a need for establishing and enforcing quality standards by public health agencies to protect consumers. A final reason is historical. Before pasteurization and other disease control measures were implemented, milk had been recognized as a primary vehicle for the spread of several life-threatening diseases.

Mastitic Milk.[12,14,28,51,97,235,237,238]   In the modern dairy industry, quality control begins on the farm and continues throughout the production and processing procedures. The initial critical control points include the health of the cow and all the procedures that affect the quality of the milk as it comes from the cow— sanitation, feeding, medications, and the milking process. If this milk is abnormal in any way, there is a potential threat to the health of the consumer. Definitions of abnormal milk will vary among different jurisdictions, but all are based on combinations of biologic, chemical, and physical examination of the milk. The most common evaluation is based on the number of somatic cells present in the milk and is referred to as *somatic cell count* (SCC). These cells consist of epithelial cells, sloughed off the lining of the milk-secreting alveoli in the udder, and leukocytes. The proportions of each will vary with the state of health of the udder. When a quarter (of the udder) is infected, the total number of somatic cells in the milk from that quarter will increase, and the milk is referred to as *mastitic*. Normal milk rarely has more than 1 million cells/ml, and the maximum number legally acceptable in raw milk is usually well below this.

Screening Tests for Mastitic Milk.[99,162,273,289]   There are several tests commonly used for screening raw milk samples for somatic cells. They include the California Mastitis Test (CMT), Wisconsin Mastitis Test (WMT), Modified

Whiteside Test (MWT), and Catalase Test (CT). The first three are based on viscosity, which, in turn, is dependent upon the number of cells present in the milk. The cells are lysed, releasing DNA, which then reacts with a reagent in the test solution to form a viscous material. With increasing numbers of cells, the viscosity becomes greater. The CMT also measures pH change. CT measures release of the intracellular enzyme catalase. The interpretations of test results are listed in Table 2.12. Although interpretation of the CMT is subjective, its great advantage lies in the fact that it is the only true "cowside" test; i.e., it can be performed in the barn as soon as the milk sample is collected.

Confirmatory Tests for Mastitic Milk.[141,239]   Bulk tank milk is evaluated in processing plant or regulatory agency laboratories, using either a standard microscopic plate count or an electronic somatic cell-counting technique. If the number exceeds the allowable maximum, the producer is usually sent a warning, which may list the most likely causes. Repeated high counts will result in suspension of the producer's permit to supply milk until it is once again satisfactory. In 1992, the regulatory limit for somatic cells in the EU was set at 400,000/ml. In July 1993, the limit in the United States became 750,000/ml. In Ontario, a program to lower the regulatory limit from 800,000/ml to 500,000/ml was initiated in 1989.

Misleading results may be obtained if electronic cell counters or direct microscopic counts with nonspecific stains are used for evaluating goat milk because milk secretion in the goat is of the apocrine type, rather than the merocrine secretion found in cows. This results in the secretion of cytoplasmic particles that are not cells since they do not contain DNA. Their presence is not indicative of lesions in the goat's udder. Methods for performing SCCs that are not specific for

Table 2.12.   Interpretation of screening tests for abnormal milk

Test	Score or reading	Somatic cells (leukocytes/ml milk)
California Mastitis Test (CMT)	N	0–200,000
	T	150,000–500,000
	1	400,000–1,500,000
	2	800,000–5,000,000
	3	>5,000,000
Modified Whiteside Test (MWT)	N	<500,000
	T	500,000–1,000,000
	1+	1,000,000–2,000,000
	2+	1,500,000–3,000,000
	3+	>3,000,000
Wisconsin Mastitis Test (WMT)	<10mm	<500,000
	10–20mm	500,000–900,000
	>20mm	>1,000,000
Catalase Test (CT)	<20%	<500,000
	20–30%	500,000–1,000,000
	30–40%	1,000,000–2,000,000
	>40%	>2,000,000

Note: N = negative, T = trace.

DNA are not a reliable indicator of mastitis among goats because these particles may be counted.

Procedures When a High Bacterial Count Is Reported.[39,83,172,183,203,227]
Whenever a client's milk supply is reported to have a high bacterial count, the test should be repeated to verify the count. If it is confirmed, an evaluation of the milking procedure and equipment is the first step. It is vital that proper premilking udder cleaning and postmilking teat dipping are done properly, since these two procedures form the basis of any mastitis control program. If inadequate cooling is suspected, check cooling charts on milk tanks. If poor equipment cleaning seems likely, make sure that the teat cup liners are being routinely changed and cleaned every week. High bacterial counts from equipment indicate that contamination is building up at some point on the milk-handling surface or that outside contamination is entering the system.

Sample Collection and Analysis. If the problem persists, milk samples are collected for bacteriologic analysis at the end of each pipeline at the beginning, middle, and end of the milking period. The test results of these specimens are analyzed to determine the needed corrective measures. In some instances, a further breakdown of samples may be needed, such as individual cow or milking unit samples, to pinpoint the source.

Milk samples for bacteriologic analysis must be collected aseptically. Each teat must be cleaned and then wiped with an alcohol-impregnated swab, and great care should be taken especially around the teat orifice. After discarding the first two streams, the milk should be collected by squirting it into a sterile container held to one side rather than directly below the teat so that particles do not fall into it. The milk being collected must not come in contact with the hand while manipulating the teat. Each container should be marked with the cow's ID number and udder quarter (LR, LF, RR, RF). To avoid contaminating the teat ends during the procedure, samples should first be collected from the quarters nearest the operator.

The standard plate count (SPC) is used for estimating bacterial populations in milk and milk products. (The SPC is quantitative, not qualitative. It will not differentiate between pathogens and nonpathogens.) For increased reliability, a preliminary incubation (PI) procedure may be used. In the SPC, milk is incubated at 32°C (89.6°F) for 48 h. In the PI procedure, milk samples are incubated at 13°C (55°F) for 18 h before doing the plate count. The maximum count after PI should be less than 10,000/ml. The value of the PI stems from the fact that microorganisms that make up the normal flora of the udder do not grow well at 13°C (55°F), whereas many contaminants do. The plate count of bulk tank milk that has been contaminated from the udder will not be increased because these bacteria do not normally grow well at low temperatures. A high count after PI indicates careless practices that have allowed external contamination to enter. The organisms that increase the PI count are mainly found in the cracks and crevices of equipment that routine clean-in-place (CIP) procedures do not reach. The critical areas have been found to be in plastic or rubber hoses, milking claws and gaskets, automatic takeoff sensors, and weigh jars.

Pathogens may be detected by standard culture procedures or by immunologic techniques such as counterimmunoelectrophoresis or by rapid test kits using latex agglutination or by detection of nucleases (enzymes that hydrolyze DNA). Inflammation of the udder produces changes in the composition of the milk, which can also be detected. The total protein content will not change very much, but proteins originating from the blood will increase, whereas proteins largely synthesized in the udder will decrease. Bovine serum albumin, for example, will increase from a normal of 0.082 mg/ml to as high as 4.3 mg/ml. Lactose, a sugar secreted in the udder, will decrease, but the lactose content of milk from healthy udders varies, and the changes resulting from mastitis are less reliable than SCC.

Raw and Pasteurized Count Comparison. Comparison of bacterial numbers in raw and pasteurized milk will give some clues as to the likely source of the problem. High raw and low pasteurized counts indicate contamination from the cow or milking-related equipment or environment. They also may implicate humans involved in the milking procedure. High raw counts, coupled with high pasteurized counts, indicate that contamination occurred after pasteurization. This is particularly true when Gram-negative organisms such as *Pseudomonas* spp. are found because they are very heat-sensitive. High numbers after pasteurization in a laboratory indicate the presence of thermophilic organisms.

Environmental Contamination
Water. The quality of water used on the farm is important for several reasons. Water may be a hazard to humans or animals on the farm if it is contaminated with pathogenic microorganisms or hazardous chemicals. Water for milk house and milking operations should be from a supply that is properly located, protected, and operated and should be easily accessible, adequate, and of a safe, sanitary quality. In hard water areas, a film or scale (milkstone) collects on milk utensils (including rubberware) after they have been in use for some time as a result of interaction among milk solids, detergent, and hard water. Wherever this scale is deposited in the milking system, it is a potential habitat for microbial replication and high bacterial counts in milk. A mild organic acid such as (diluted) phosphoric acid will remove the deposits.

Pavement. All areas subject to routine cow traffic should be paved for ease of general sanitation with a design that permits straight-line movement of mechanical cleaning equipment (e.g., skip loaders).

Milking Barn Arrangements. Whenever construction of a milk house, milking barn, or other building for a dairy farm is anticipated, current health code requirements should be reviewed to ensure compliance. In regions where dairies are small, barns with individual cow stanchions and bucket collection systems are used. Where large herds are the rule, barns with milking pits and various types of stall arrangement (gate, walk-through, carousel, or herringbone) are required for efficiency. Such factors as herd size, competence of available milking machine operators, and climate will influence the decision as to which design best meets individual needs. A poor choice creates management problems that will result in reduced production and quality of milk.

Equipment Cleaning.   Equipment cleaning consists of four separate steps: rinse, wash, rinse, and sanitize. The first rinse should be done as soon as possible, before the milk film dries, with a large volume of lukewarm water since hot water causes milkstone formation. Next, the equipment is washed with an approved detergent at the appropriate concentration and recommended length of time. To reduce milkstone, an organic acid detergent may be used at weekly intervals, or an acidified rinse may be used routinely. The equipment must be rinsed thoroughly to remove the detergent solution. An approved chemical sanitizer is applied when the equipment is completely clean; otherwise its bactericidal activity is reduced by any residual cleaning compound or organic matter. Each component of the milking system has specific cleaning requirements. Some parts must be disassembled completely, while others may be cleaned by circulation or CIP methods. The vacuum line, although it does not transport milk, should be cleaned periodically. Health code requirements and manufacturers' recommendations should be guides for cleaning.

Essential Components of Machine Milking Systems.[85,169,265]   The milking machine performs two basic functions: (1) it imposes a controlled vacuum on the end of the teat to open the teat orifice and provide the differential pressure (suction) needed to produce milk flow, and (2) it massages the teat intermittently to continue stimulation and prevent blood congestion in the teat end. These two functions must be performed by a properly designed and comfortable liner (inflation). A poorly fitted liner may impair letdown or place undue stress on the teat and udder tissue. Vacuum fluctuations or improper massage may lead to mastitis by irritating the sensitive tissue of the teat canal.

Following are the four components of mechanical milking (also see Fig. 2.7):

1. **Milking unit (cluster).** This is the part of the machine that is suspended from the cow, performs the milking operation, and receives the milk into a sanitary system. It includes the teat-cup assemblies, claw or suspension cup manifold and reservoir, and the connecting air and milk tubes.
2. **Pulsator.** This device controls the liner action by introducing vacuum and atmospheric pressure intermittently into the chamber between the liner and the teat-cup shell.

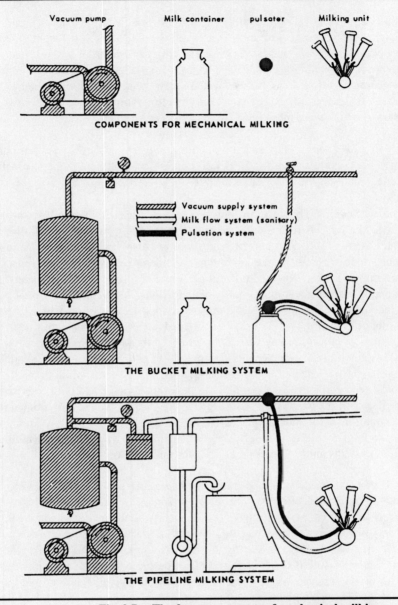

Fig. 2.7. **The four components of mechanical milking.**

3. **Vacuum pump**. This pump provides a source of vacuum to the end of the teat to cause milk flow, supplies the energy to activate the liner and massage the teat, and moves the milk through the system.
4. **Receptacle**. The milk receptacle includes all hoses, pipes, and containers for conveying and holding the product. Systems are of two basic types: (1) a

bucket system in which milk is received directly into a nearby portable bucket that is a vacuum and (2) a **pipeline** system that uses rigid stainless steel or heat-resistant glass pipe for carrying vacuum from the milk receiver to the individual milking units and for carrying milk from the units to the milk receiver. In many areas, bucket systems have been replaced by pipelines because of greater labor efficiency and the elimination of individual cow pulsators, which were more prone to malfunction.

Vacuum Lines.   The vacuum-supply piping connects the vacuum pump, reserve tank, sanitary trap just ahead of the milk receiver, and pulsator pipeline. An adequate pump is of no value if there is inadequate piping.

Milk Lines.   The **milk line** conveys milk and air from the milking units to the receiver jar, where the air is removed. Milk lines commonly installed are stainless steel tubing 50.8 mm (2 in.) or more in diameter. Most pipeline installations are now welded on the job into major assemblies. Welded milk lines look clean, improve sanitation, and prevent problems related to air leak, gaskets, and turbulence associated with union-type connections. All welded lines must have advance approval of appropriate dairy inspectors. Low-level pipeline systems, with the milk pipe at or below udder level, reduce excessive vacuum fluctuation at the claw or cup caused by vertical lift of the milk. Although it is difficult to install low lines in existing stanchion barns, all new barns and parlors should be designed for low-line systems. In general, milk pipe systems in flat barns should have no dead ends; that is, the milk pipe should be looped, and both ends should enter the milk receiver through separate inlets. Normally, the high point in the loop would be at the most remote point from the receiver. In single-string barns, a simple loop should be installed. Larger barns require double-looped (multiple-looped) systems and/or larger milk pipes for optimum performance.

Teat Cups.   The teat cup consists of an outer shell, usually of stainless steel, and an inner rubber liner for inflation. Liners are classed as molded, one-piece stretch, or ring-type stretch. Further specifications include wide (>19 mm [3/4 in.] inside diameter) or narrow bore and hardness or tensile characteristics (i.e., firm, medium, or soft). During the vacuum phase of the pulsation cycle, the liner opens and milk flows from the teat. Atmospheric air enters the outer chamber during the massage phase of the pulsation cycle, causing the liner to collapse against the teat end. The purpose of this action is to enhance circulation of blood through the tissues of the teat.

Pulsator.   The pulsator is the automatic air-vacuum valve that directs atmospheric air into the chamber between the teat-cup liner and the shell and then withdraws this air by opening a port into the vacuum systems. This intermittent air-vacuum state causes the characteristic liner action. A master pulsator is designed to operate two or more milking units simultaneously, whereas a unit pulsator operates only one. Definite and somewhat snappy pulsator action is desirable. Unit pulsators near the teat cups give a sharp action by avoiding long pipes and hoses that can cause sluggish action caused by air friction. Slave, or booster, **unit** pulsators are used either to amplify or reproduce a weak, sluggish air-vacuum signal into a sharp action near the teat cups.

Pulsation Cycle.[68,159,241] Recording fluctuations in vacuum at the teat cup during milking, with the system under full load, makes possible the most comprehensive evaluation of milking machine performance. Teat-cup liners respond to the interaction between pulsated vacuum and milking vacuum. For this reason, critical assessment of the performance of the milking machine requires information on both vacuums and an understanding of their interrelationships. Such measurements are recorded with the use of a strain gauge amplifier.

A vacuum level of 30.5 mm (12 in.) Hg at the claw is desirable. Vacuum fluctuations should be avoided. Some of the more common reasons for excessive fluctuations are inadequate vacuum pump capacity, slack V belts on the pump, flooding of milk lines, improperly functioning vacuum regulators, admission of excessive air during milking, and high-mounted milk lines (as contrasted to low lines). The vacuum with all units closed should return to normal within 5 sec after opening and then closing the milking units farthest from the pump. A pulsation rate of 45–60 per min is adequate, and a pulsator ratio (the time full vacuum is applied to the teat compared with the time the vacuum is reduced) of 60:40 is the norm for most milking machines.

Milk Storage and Transport.[24,134,267] Milk is stored on the farm and transported to the receiving station or plant in one of two ways: the can system or the bulk tank system.

Can System. Tinned metal, aluminum, or stainless steel cans, usually of 37.8 l (10 gal) capacity, are used to store milk on the farm in many parts of the world. The cans are cooled in tanks of cold water or in cabinet coolers using a cold water spray. The advantages of the can system are that it can be independent of electricity for refrigeration and enables manual collection, but cans may be roughly handled and damaged, are difficult to keep clean, and may stand in the sun awaiting collection.

Bulk Tank System. Today, in the major milk-producing areas of the world, milk is collected in mechanically refrigerated stainless steel bulk tanks for storage on the farm. Milk in such tanks is cooled rapidly; in each tank milk has contact with a refrigerated expansion plate that is an integral part of the inner liner at the bottom of the tank. Another type of bulk tank uses an ice water spray on the sides of the inner shell of the tank to cool the milk.

Because milk may be stored in bulk tanks for 3–4 d before being collected, it is important that strict attention to hygiene be observed. Partial collection should be avoided. The interior of the tank should be cleaned and sanitized after being emptied and the exterior of the tank, the plumbing, and the interior surfaces of the milk house should be kept free of dirt. Some municipalities even go so far as to require daily steam cleaning in the milk house.

Transport. Tank trucks pick up the milk from the farm and deliver it to the receiving station or dairy plant. The truck driver frequently will collect milk samples from the bulk tank on the farm and may conduct simple tests for the presence of antibiotic residues in the milk from each farm. The milk tank trucks may or may not be refrigerated since the large volume of cold milk in an insulated tank maintains a low temperature during relatively short trips from the farm to the dairy plant. Normally, no more than a 2.8°C (50°F) increase in temperature is permitted; i.e., if the milk is 4.4°C (40°F) when picked up at the farm, it must be no warmer than 7.2°C (45°F) when delivered to the receiving station. After the truck has been emptied, the interior of the tank must be cleaned and sanitized.

At the receiving station samples of milk from each farm are tested for bacterial and somatic cell content, and the milk is weighed and analyzed for fat and other components that may be included in an MCP scheme.

Sensory Abnormalities.   Milk may develop "off" flavors or odors that are primarily an issue of consumer acceptance rather than food safety. Milk often acquires an unpleasant odor when cows first graze on grass in the spring, and again in the winter when silage feeding begins. Grazing on fields containing wild onions will impart a strong odor to the milk, as will confinement in a barn freshly painted with an oil-base paint. **Rancidity** is the presence in milk of an undesirable flavor caused by certain free fatty acids. This is covered under "Deterioration of Food."

Egg Production[201,255,258,262]

Before World War II, eggs were produced commercially in small farmyard flocks, rarely exceeding 400 hens. During the war there was an increased demand, by the military, for powdered egg products. Because of a labor shortage the number of small operations decreased while the remaining farms became larger and more efficient. Today, flocks of 400,000 layers are not uncommon, and some exceed 1 million birds. At present, the principal egg-producing nations are the United States, China, Russia, and Japan.

Increased scale of production does not, of itself, equate with increased efficiency, but innovations in cage configuration, manure handling, and ventilation have caused the inflation-adjusted capital required per layer hen to decrease significantly during the past 30 yr. As evidence of the increase in efficiency, egg production per bird in 1940 was 11 dozen eggs per year. Today, 20 dozen per year is common. Efficiency that has resulted from genetic improvement and better nutrition has probably approached its peak. Currently, the feed conversion ratio is 1.8 kg (4 lb) of feed for 1 dozen eggs weighing 0.7 kg (1.5 lb).

Increased efficiency has resulted in specialization. Whereas in the past chicks were hatched and raised on the egg producer's farm, today each step of production is usually treated as a separate operation. A few breeders provide hatching eggs from genetically improved birds to pullet growers. After the pullets attain maturity, at 20–22 wk, they are transferred to laying flocks. Until now each of these steps has been usually a function of an independent operator, but the trend is toward large production and processing complexes administered by a single firm (vertical integration).

In temperate climates, where temperatures vary greatly, modern laying houses are huge, windowless, insulated, artificially ventilated buildings. Layers are in cages that range from 19 cm wide by 38 cm deep for 1 bird (8 × 16 in.) through 29 × 43 cm (12 × 18 in.) for 3 birds to "colony cages" several meters in width and length for 10–25 layers. Feeding, watering, and egg collection is mechanized so that it is now possible for one person to provide all labor needed for 100,000 laying hens.

This highly efficient egg production industry, in which the bird seems to be no more than a part of a huge machine, has led to concern regarding animal welfare. Space has decreased from about 900 cm^2 (1 ft^2) per bird when the layers were kept on the floor to about 300 cm^2 (0.3 ft^2) per bird under modern cage systems. It is difficult to predict what impact attempts to make the environment of laying hens more natural will have on efficiency. Crowding, for instance, has a detrimental effect on plumage. Feed consumption, in one study, was 27% higher at 18°C (65°F) among birds with damaged plumage than among normal birds. Therefore, in this case an improvement in environment (although it may not be more "natural" in the eyes of animal welfarists) may be cost beneficial. Putting perches in cages had no effect on total production but increased the incidence of cracked eggs and reduced egg weight. Placing an abrasive strip in cages to keep nails trimmed reduced injuries from getting nails caught in the cage and also reduced egg damage. The reduced demand for eggs that has resulted from dietary changes, coupled with these animal welfare concerns, will, in all probability, produce changes in the egg production industry in the future.

Aquatic Animal Production
Production of Fish for Food[61,168,191,221,287]

Fish Consumption Worldwide. Global per capita consumption of fishery products has increased worldwide to a level of 29 lb per year in 1989. In the United States, annual fish consumption has climbed to roughly 15 lb (6.9 kg) per capita. Fish is nutritionally more important in the diet in some developing nations than it is in the developed world. Fish supplies 6.6% of the animal source protein in the North American diet and 12% in Europe, in contrast to 29% of animal source protein in Asia and 19% of the animal protein consumed in Africa.

Sources of Fish and Seafood. Unlike meat and poultry, the majority of seafood consumed by humans is harvested from wild populations. In 1992, the most recent year for which published statistics are available, the total worldwide harvest from the sea was 90.9 million metric tonnes (MMT). This represents a decline in production from the late 1980s, at a time when per capita consumption was on the rise. Roughly two-thirds of the world fisheries' catch is used directly for food on the table. The remaining one-third goes for fish oil, solubles, and meals used in livestock and fish feeds. Of the 61 MMT of fish destined directly for human consumption, 14 MMT came from aquaculture. If per capita seafood consumption continues to grow, the worldwide demand for fish will increasingly be met by aquaculture.

Fish harvested commercially for human consumption comes from both salt and freshwater, from the wild fishery, or from aquaculture and encompasses more than 300 species of finfish and shellfish. Most of the fish consumed in developing countries comes from commercial sources, but the recreational fisheries contributes a significant share. In the United States, for example, per capita consumption of finfish, mollusks, and crustaceans outside of commercial sources is estimated at 1 to 1.5 lb (0.45 to 0.63 kg) /yr, which is 10% of the per capita consumption of commercially harvested fish products.

The market for seafood goes beyond national boundaries. In the United States far more seafood is consumed than the domestic catch can supply. The major foreign sources of seafood for the United States market are finfish from Canada and shrimp from Ecuador and Mexico. By contrast, Canada is a net exporter of fish and fish products. The supply of seafood is regionally diverse, even within a single country, and the seafood harvest industry is highly fragmented.

As world landing statistics approach a yearly total of 100 MMT, the yields from the capture fishery are beginning to peak. This is reflected in recent figures released by the FAO. Many traditional species such as herring, pilchard, cod, and tunas, which historically have formed the basis of the large international fisheries, have been overfished. Most of the recently discovered resources (Bering Sea, Falkland Islands, New Zealand, and the Antarctic) are fished at or beyond their sustainable yields. A few unconventional resources remain (e.g., Arctic krill and lantern fish), but they are of questionable commercial value for human food. Since the American invention of fish fingers in the 1950s, many unconventional fish species of rather low commercial value have made a name as raw material for various products made of minced fish meat, called *surimi*. But realistically, alternate sources of supply, such as aquaculture, and alternate sources of dietary protein will have to be substituted for wild caught fish in the near future.

Commercial Harvest in the Wild Fishery.[61,246,306]   Commercial saltwater fishing requires a general understanding of the habitat of the fish to be harvested. For example, some fish are found more frequently on the ocean bottom; catching them usually requires a net such as an otter trawl that is designed to pick up fish near the ocean floor. Other considerations regarding choice of gear include personal preference and, more important, the terrain of the ocean floor. If it is extremely rugged, line fishing may be necessary.

Line Fishing.   Hand lines may have one or several hooks attached and are used where the bottom is too rough for other types of gear or are used for large nonshoaling species like bluefin tuna. Lines are set with buoys and anchors at different depths and consist of a long, relatively horizontal line with many short vertical lines, each with a baited hook at the end. The lines are left in the water and checked periodically from a boat. Fish are harvested from long lines by a boat running parallel to the long line; the boat hauls the line up on rollers, allowing the fish to be removed and the hooks baited and returned to the water. Set lines are used primarily for catching halibut, black cod, rockfish, and other bottom dwellers. Lines with hooks may also drift from a stationary boat at any depth from the surface to the ocean bottom. Drift lines are popular for harvesting mackerel or tuna. Trolling uses a baited hook and a moving boat. Baits and lures, fishing depth, trolling speed, and type and size of hook are varied according to the species caught. Commercial trolling is used most frequently in Pacific salmon and albacore fisheries.

Gill Nets. Several different kinds of nets are used for commercial fishing. Gill nets are designed to entangle fish so that they cannot escape. A gill net is a vertical wall of net with floats on the top and anchors and drag weights on the bottom. Only fish of a fairly narrow size range will become entrapped in the net, which is strung across the path fish are expected to take. Fish swim through the mesh head first, but their bodies are too large to pass through. When they try to back out of the net, they are caught by their gills. Small fish swim through without being caught, and large fish are deflected. Fish caught in the net often die of exhaustion from struggling to free themselves. Gill damage reduces the oxygen the fish is able to extract from the water. Fish must be harvested from the gill net regularly to prevent deterioration and wastage. Careful handling of the net and the fish is necessary to prevent bruising and further carcass damage.

Seine Nets. Seine nets are used by encircling fish in a given area. The purse seine is used in deep sea fisheries for schooling fish. It accounts for a significant proportion of the world fish catch and is the most important piece of fishing gear in the United States. To use a purse seine efficiently, fish must be accurately located, and the direction and speed of movement must be tracked to determine where to let out the net. Once trapped in the purse seine, they are pumped or dipped out of the net and stored in the hold. A large purse seine can bring in up to 500 tons (450 t) of fish in one haul.

Trawlers. Trawlers are boats that haul a fishing net or trawl behind them. The otter trawl, the most common type of trawling gear, can be used for bottom-dwelling species such as cod, haddock, flounder, and hake. They may also be used for midwater species such as herring, mackerel, redfish, pollock, and shrimp. Depth of the trawl is adjusted by changing the speed of the fishing vessel or the length of the cable between the ship and the net. When used for bottom fishing, the bottom leading edge of the net is weighted and the top edge has floats attached to keep the mouth of the net open. Trawlers vary in size from small boats, 25 m (84 ft) in length, used in inshore fisheries, to factory freezer trawlers, which not only catch fish but process and freeze them on board as well.

Ancillary Gear. Modern fishing vessels are equipped with sophisticated fish-sensing devices. Sonar, radar, and other electronic gear can locate the fish, estimate the number of fish in the school, and determine their direction and speed of movement. Fleets of ships operating in waters distant from their home ports employ ships to catch the fish, a factory ship to process the catch, and still other ships to supply the fleet and carry the processed catch back home or to other markets. The quality of fish harvested depends on the care and handling of the catch, so many vessels are equipped for rapid chilling in retention tanks prior to processing, for gutting and filleting the fish, for freezing and storage at −30°C (−22°F), for canning, and for processing the offal or by-catch to fish meal and oil.

Aquaculture[45,261,288]

Aquaculture Worldwide. While fisheries' harvest overall is declining, aquaculture production is expanding. Worldwide tonnage is dominated by Asia, which accounts for just over 80% of the total weight of cultured fish and aquatic plants and nearly 80% of the total value of the world harvest. This dominance is due primarily to finfish culture, much of which is based on low-energy input systems, primarily carp. Other fish species cultured extensively in Asia include tilapia and milkfish. China dominates world aquaculture production at 47% of total tonnage, followed by Japan, India, Korea, the Philippines, and Indonesia. Average annual value of cultured fish and shellfish is increasing at a faster rate than tonnage because the market is expanding rapidly for high-end seafood products like marine shrimp and salmonids.

Aquaculture provides the opportunity for a high level of quality control over the finished product, from hatchery to harvest and processing. This is true of fish that are produced domestically, but large amounts of cultured fish and shellfish are imported into many countries, with few restrictions and little monitoring. The United States and Canada both import large amounts of shrimp from Ecuador and China, and in neither exporting country is the use of chemotherapeutic agents regulated. Chloramphenicol, banned for use in livestock in most developed countries, yet relied upon heavily in the shrimp-farming industries, is one of the compounds of concern. The Food and Drug Administration (FDA) has not examined imported seafood for drug residues. If the market for cultured fish and shellfish is to keep expanding, it will become increasingly important to regulate production and marketing to protect the public from foodborne hazards and to maintain consumer confidence in fish as a healthy source of animal protein.

Fish Farms. Many species of fish are farmed worldwide. Production systems and species farmed are locally variable and include many species of finfish and shellfish. Among shellfish, bivalve mollusks (e.g., oysters and mussels) and crustaceans (e.g., shrimp and crayfish) are farmed commercially. Fish can be farmed in extensive systems where animals live and feed in natural enclosures at low densities. Species that are farmed in this way include carp, crayfish (freshwater shrimp), and prawns (marine shrimp). In these systems fish feed on plankton naturally present in the water or plants, snails, mollusks, and other animals found on the bottom of the pond or enclosure, with little or no supplemental feed provided. Often the food supply is enhanced by fertilizing the ponds. Fish farming is also done under intensively controlled conditions. Many trout farms use concrete ponds, raceways, or fiberglass tanks and actively pump freshwater through the system. Essentially all the feed is provided by the farmer. The most intensive production systems of all are recirculatory systems, such as those used in Europe and parts of North America for eel production, where the water is used over and over again. Mechanical filters are used to remove suspended solids from the water supply, and biological filters oxidize ammonia produced by the fish. Oxygen is added to the water to achieve saturation, and very high stocking densities are achieved. This type of system is the aquaculture equivalent to modern, intensive poultry production.

For some species, such as Japanese yellowtail and blue mussels, farmed fish are a more desirable product than their wild caught counterpart. For other species like the crayfish, there is no economically viable alternative to the cultured product. For some aquatic food species, like oysters, there is a blurry distinction between farmed and wild caught, due to the conditions of management of oyster beds.

Finfish Production[123,271]

1. **Channel catfish**. The channel catfish is the major farmed fish in the United States. Catfish farming is an American success story; it accounts for over half the total fish and shellfish production from aquaculture in the United States. Most United States catfish farming and processing is in the South and Midwest. In 1988 about 295 million lb (130 million kg) of catfish were produced, representing a retail value of more than $270 million. Mississippi is currently the leading state with more than 62,000 acres in catfish ponds. Louisiana, Texas, Arkansas, Alabama, Tennessee, and Georgia are also major producers of catfish. Kansas, Florida, and California are involved to a lesser extent.

Catfish farming has become a complex operation. Initially, ponds must be constructed and filled with water. Approved chemicals may be added to the water to control microbes. Temperature control of the water is especially critical when attempting to hatch eggs. A temperature of 26.6°C (80°F) appears to be best. Lower temperatures (21°C [70°F]) retard hatching and enhance fungal growth on the eggs. Higher temperatures (32.2°C [90°F]) permit eggs to hatch so rapidly that they often deplete food supplies. Fish farms usually have separate ponds for hatching, raising fingerlings, and producing adult food fish.

Ponds intended for harvesting catfish are stocked with fingerlings at the rate of about 2000/acre (0.405 hectares) in the spring. Adult fish are harvested in the fall. The catfish farmer must be constantly aware of two potential problems. Parasitic infestation of a pond can literally wipe out a crop of fish overnight; fortunately, some of these parasites can be controlled with approved chemicals. The second problem is that of oxygen depletion of the pond. When oxygen content becomes low (less than 3 ppm), oxygen must be added to the pond. This may be done by aeration or by adding fresh water.

Catfish are harvested by combining two techniques: draining the ponds and seining. The fish are then loaded into trucks, iced, and transported to the processing plant. Crushed ice is used to minimize bruising. A disadvantage of hauling live fish is that some may be injured in the loading process or die en route and contaminate the rest of the catch with postmortem body discharges. On the other hand, iced fish are more difficult to skin, their processing cannot be delayed, and icing is an added expense.

2. **Salmon aquaculture**. Several species of salmon are farmed successfully all over the world, generally between the fortieth and seventieth degrees of latitude in both the northern and southern hemispheres, or where water temperature stays between -2° and 22°C (24° and 66°F). Ocean pen culture of salmon is important in Norway and supplies 90% of the salmon consumed in Western Europe. Salmon are also farmed in the Pacific off the coast of Chile, in the Bay of Fundy on the Atlantic coast of North America, and on the Pacific coast in both Canada and the United States. The Atlantic salmon (*Salmo salar*) is native to the North Atlantic Ocean and is farmed in Europe and eastern North America. (The Atlantic salmon is also farmed on the Pacific coast of North America.) Two species of Pacific salmon, the coho (*Onchorhynchus kisutch*) and chinook (*O. tshawytscha*) are farmed on the west coast of North America and Chile and off Japan. In contrast to catfish, which are raised in freshwater, salmon are anadromous fish, spending part of their life cycle in freshwater and part in salt water. Anadromous species in the wild migrate from the sea to spawn in freshwater. Farmed salmon are raised in freshwater hatcheries, and smolt are transferred to sea cages for the grow out stage. Atlantic Salmon take 1 to 2 yr to reach smoltification. Harvest is carried out after the second sea winter, before sexual maturity is reached.

Shellfish Production.[158,211,246] Like vertebrate fish, some species of shellfish thrive in freshwater whereas others thrive in saltwater. Shellfish of commercial importance are morphologically divided into two general categories: mollusks (e.g., clams and oysters) and crustaceans (e.g., shrimp and prawns, crayfish, lobsters, and crabs). Saltwater crustaceans and bivalve mollusks make up the majority of the shellfish harvest. Of all the commercially harvested marine animals, the crustaceans have the highest market value, although they make up less than 5% of the world catch.

1. **Shrimp**. Tropical marine shrimp are the most important to world trade and

are consumed primarily in the United States and Japan. Europeans prefer smaller cold water shrimp. The main sources of shrimp on the world market are India, Indonesia, China, the United States, and Thailand. Shrimp (*Penaeus* spp.) and prawns (*Machrobrachium* spp.) are harvested from salt or brackish water, depending on the species. Nearly a quarter of the world demand for shrimp and prawns is met by Asia and the countries of the eastern Pacific Ocean, from Mexico to Peru. The remainder is harvested from the wild.

2. **Crayfish**. Crayfish are the only freshwater shellfish of commercial importance, with the possible exception of freshwater prawns in Southeast Asia. Currently, about 500 species of crayfish exist throughout the world with more than 250 occurring in North America. The southeastern United States supplies most of the total annual production of crayfish. Louisiana accounts for most of the farmed production of crayfish and the bulk of crayfish consumption as well. It is a good example of the regional nature of fish production and consumption patterns worldwide.

Commercial production of crayfish involves aquaculture similar to that for raising catfish. In the United States most crayfish farming is done in Louisiana. In good years, the United States crayfish harvest is over 60,000 t (63,000 tons), with a value of close to $100 million (U.S.). Crayfish farming had an accidental beginning about 1950, when a farmer flooded a rice field in the fall to develop a temporary area for duck hunting. The next spring the pond was teeming with crayfish. Subsequently, the idea of crayfish farming evolved. This concept has been applied since to catfish and some saltwater shellfish. Coastal water shrimp farming (mariculture) is an example of applying fish farming to seafood production. In years to come, aquaculture will probably become extremely important in supplying the nutritional needs of the United States.

Two types of ponds are used in farming crayfish. Rice fields are popular in rice-growing areas. The general procedure involves draining the rice field 2 wk before rice harvest. As the fields dry, crayfish burrow into the soil. Later, a second growth of rice, grasses, etc., provides food for the developing crayfish. Wooded areas also are used to make ponds. This latter type of pond is less desirable as it tends to become low in oxygen because wind-aided water circulation is poor. Also, wooded areas are usually more acid, which is not conducive to crayfish production. Nevertheless, these ponds are popular because they allow use of land that otherwise would be idle.

Crayfish farming simply involves stocking the pond with brood stock. The number of crayfish stocked depends on the amount of vegetation available. Stocking usually is done in May or June. About a month later, the ponds are drained to induce burrowing of the crayfish. The ponds are flooded in the fall, which results in the release of young crayfish. The young adult crayfish then feed on vegetation until they are harvested. Depending on climatic conditions, adults may be harvested from late November through April.

Crayfish currently are harvested by hand with the use of lift nets or funnel traps. The live crayfish are kept in holding tanks and transported to processing plants where they are stored in coolers at 3.3°C (38°F).

3. **Crabs**. Crabs are harvested worldwide and market demand exceeds availability. They are marketed fresh or canned. Primary consumers of crabs are the former USSR, the United States, and Japan. Three important species of crabs are harvested in U.S. coastal waters: the Dungeness and king crabs of the Pacific and the blue crab of the Atlantic. Several types of gear are used to harvest crabs. **Trot lines** (baited lines with no hook) are used primarily for hard crabs. Wading fishermen may pass push nets or dip nets on poles in shallow waters. Crab pots, popular for harvesting blue and Dungeness crabs, are simply baited traps placed on the ocean floor.

4. **Lobsters**. Two species of clawed lobsters, the American lobster (*Homarus americanus*), found in the Northwest Atlantic from New Jersey to Labrador, and the European lobster (*H. gammarus*), in the eastern Atlantic from Norway to Morocco, are fished commercially. Clawless lobsters (spiny lobsters and rock lobsters) are also fished commercially in the Pacific Ocean.

Canada is the only lobster exporting country in the world. Canadian lobsters are marketed in the United States, Europe, and Japan. About 70,000 t of lobster with a market value in excess of $500 million are landed in North America annually. Lobster is a premium seafood, and its high price makes it the backbone of the inshore fishery industry in the Atlantic Provinces of Canada.

Lobsters are bottom dwellers that live on the continental shelf and feed on crabs, clams, mussels, starfish, sea urchins, and other lobsters. Lobsters are solitary and somewhat territorial and can migrate long distances. Little is known about the biology and the population dynamics of the lobster, but the lobster catch, unlike that of other wild fisheries, is currently on the upswing.

The lobster fishery is primarily an inshore, small-boat industry. Lobsters are taken in baited traps and usually marketed live. A small percentage of lobster is processed and sold canned or frozen, either shucked or in the shell. The live market is the most lucrative, but demand is highly seasonal and does not coincide with peak catch. Storage of live lobsters enables wholesalers to better meet market demand. Live lobsters can be held for up to 6 mo in dry land pounds, where they are stored individually in separate trays. Cold seawater is circulated through the trays under carefully controlled conditions of temperature, salinity, and oxygen concentration. Other systems of impoundment include "cars" or floating crates, wharfside ponds or tanks, and tidal ponds near the shore.

5. **Mollusks**. There are over 130,000 species of mollusks, but only a small group of them are commercially important as food. Edible mollusks fall into three groups: those with a single shell, such as snails, winkles, whelks, and abalone; bivalve mollusks, made up of clams, oysters, mussels, and scallops; and cephalopods, made up of squid, octopuses, and cuttlefish. Mollusks represent about 7% of the world's fisheries' harvest. Bivalve mollusks may come from

farmed sources or be harvested from the wild. Farming systems are generally extensive, with culture being carried out in open water. Spat (young stock) is collected in the wild and seeded in grow out areas selected for abundance of natural food sources.

a. **Oysters**. Oysters and clams are the primary commercially marketed mollusks. Oysters go to high-priced luxury markets in many developed countries, including the United States. Most oysters harvested in the United States are from the Atlantic Ocean and the Gulf of Mexico. Although oysters are present in the Pacific, overharvesting in the past has made the West Coast a less important source.

Oysters prefer a habitat of brackish water at depths down to 12–15 m (40–50 ft). Although they grow in deeper waters, there are no important oyster beds harvested commercially at depths greater than 12 m (40 ft). Oysters require a fairly hard surface on which to thrive but are little affected by fluctuations in water temperature and salinity. Because the floor of the Gulf of Mexico tends to become muddy from river drainage, shells from processed oysters are used to provide a hard surface on which to grow newly seeded oysters.

Oysters are harvested commercially from either of two sites: (1) public grounds that are regulated by strict conservation rules that control the methods and amounts harvested and (2) private grounds managed by individuals who reseed the beds after harvesting.

Mollusks may be harvested by several methods, depending on the species, the conditions of growth (bottom culture versus growth on rafts or long lines), and the local preference. Often, only hand-harvesting methods are allowed in public waters because mechanical dredgers tend to overharvest, but the most common harvesting method is dredging. A suction dredge uses a centrifugal pump to move shellfish, soil, and water onto the deck of a fishing boat, where the shellfish are separated out and the rest of the material is returned overboard. Occasionally, sand and silt may enter the mantle cavity of oysters harvested this way, and cleansing in a holding basin may be required. Product quality is excellent with hand-harvesting methods, and very good with suction dredge harvest, provided it is done carefully.

b. **Clams**. Several species of clam are harvested commercially. Procedures used to harvest and process clams are essentially the same as those for oysters. More hand harvesting of clams is done than of oysters because many species inhabit shallow waters. Hard-shell and ocean quahog clams prefer shallow oceanic shores, whereas soft-shell clams are found primarily in estuarine waters. For the latter, hand harvesting with a pail and shovel when the tide is out is the principal method used. Clams that inhabit deeper waters require dredging rigs for harvesting. Harvested clams remain alive for fairly long periods by closing their shells tightly to conserve moisture.

Large clams, such as the surf clam, are marketed as a canned product either whole or minced; smaller sizes are usually sold fresh for steaming or serving on the half shell.

c. **Scallops**. Scallops are another commercially important mollusk. Two types of scallops are harvested: sea and bay scallops. Sea scallops reside in deep ocean water and must be retrieved by dredging or with an otter trawl. Bay scallops may be harvested by hand because they are in shallow waters. Scallops must be handled carefully when harvested because, unlike other mollusks, they cannot close their shells tightly after removal from the water and consequently dry out and die quickly. Most sea scallops are shucked on the ship. Scallop processing and marketing is essentially the same as for other mollusks.

d. **Shellfish and public health**.[4,109,158,164,206] The consumption of bivalve mollusks for human food poses special public health risks. The risk of illness due to eating shellfish exceeds the risk of illness traceable to any other foods of animal origin, including poultry. This high risk comes primarily from the predilection of many people for eating shellfish raw and from the environmental conditions where these animals are produced. Mollusks are sedentary animals that filter their food from coastal and estuarine waters where they are often subject to pollution by sewage effluents and rainfall runoff from agricultural lands. The same environmental conditions that make the waters fertile and productive shellfish grounds also make the shellfish more hazardous to eat.

The usual syndrome associated with shellfish-produced illness is a gastroenteritis suggestive of Norwalk and Norwalklike viral agents with relatively mild diarrhea and no mortality. Hepatitis A and enteric bacterial infections are more serious problems that are also associated with raw shellfish consumption. The viral agents are host specific to humans, and there is no other source or reservoir of significance. Certain naturally occurring vibrios, such as *Vibrio vulnificus*, are associated with high incidence and case fatality rates among immunocompromised individuals and people with underlying liver disease. These are not associated with pollution per se but rather with warm water temperatures and postharvest problems related to extended holding times and high temperatures.

Consumer education is essential to reduce the risk of foodborne illness from shellfish. Thorough cooking of shellfish will eliminate most of the problem. Another procedure to reduce risk is environmental monitoring of water quality. Several countries have water quality standards for shellfish grow out areas. The National Shellfish Sanitation Program (NSSP) of the FDA establishes fecal coliform standards in shellfish-harvesting areas in coastal waters. Other countries have set standards based on microbial sampling of fish and shellfish after harvest. Most use fecal coliform indicators to establish the safety of shellfish or the purity of water in the grow out areas. Fecal coliform indicators do not always correlate well with the presence of human pathogens, particularly viral pathogens, which are the most common cause of illness associated with seafood. For this reason monitoring programs are not 100% effective.

Some pathogens can be removed from seafood by natural processes, such as relaying the live animal to clean water, and depuration. Relaying involves the removal of the living animal from the area of harvest to a clean area for a minimum time period, commonly 30 d. Depuration involves holding the animals in basins or tanks of clean water for several days. Seawater used in depuration is often chlorinated or treated with ozone or ultraviolet light to remove fecal pathogens. The same processes can be used to cleanse mollusks of algal toxins of risk to public health. The self-cleansing of marine vibrio and viral contamination may not proceed as quickly as for enteric bacteria and other indicator organisms; hepatitis A virus persists longer in oysters and clams than does coliform bacteria or poliovirus. Cleansing times depend upon the quality and temperature of the water in shellfish-harvesting areas, the species of shellfish, and the nature and concentration of the microbial and toxic hazards in the grow out area. Environmental control is important, but public education about the risks associated with eating raw shellfish is essential.

Processing
Plant Construction, Equipment, and Sanitation[105,124,155,165,174,210,283]

Although acceptable construction materials will vary from one jurisdiction to another (nation, state, or province), there are three basic characteristics necessary for building materials or equipment in a food-handling facility: **imperviousness to chemicals and microorganisms**, **resistance to wear and corrosion**, and **ease of cleaning**.

Floors.   The floor is perhaps the most significant item to consider in constructing a facility in which sanitation is important. All matter falls to the floor, and unless a floor is properly constructed and maintained, it can create a serious cleaning problem. No other surface is exposed to such intense wear, so durability is a prime factor. Also, floors are cleaned and disinfected, so the construction must be free from depressions, cracks, and separations.

Concrete is an excellent flooring material when properly installed. However, it may crack, allowing organic material and microorganisms to collect even with routine cleaning. The texture of the concrete is important. Fat, grease, or even water can make a smooth floor slippery and a hazard to employees. A floor that is too coarse is difficult to clean adequately.

Brick or tile can also make good flooring when properly installed. Bricks must be hard and dense to reduce breakage and absorption and must be laid over a concrete base and bonded with an acid-resistant and waterproof mortar. Tiles are very easy to clean but crack easily and, when wet, produce a very slippery surface.

Wood floors, although occasionally found in processing rooms and coolers of some older plants, are not preferred. Wood is unacceptable where water is used. If the wood floor is free of cracks and can be kept clean, it may be used in selected areas. Nevertheless, wood is one of the least satisfactory materials for flooring because it absorbs moisture and splinters may get into food products. In some countries, government regulations prohibit the use of brick or wood floors in food-handling establishments.

Synthetic poured floors are especially good because they have no seams. They can even be installed to include walls and ceiling as well as the floor, thus forming a completely seamless room. The materials used most commonly are epoxy or fiberglass plastics. These materials are quite expensive. They may be used to upgrade an unacceptable building because they can be installed over old surfaces as long as they have been adequately prepared. This method is more economical than removing the old surface and installing a base of another type.

Walls.   A smooth, flat, impervious material is most satisfactory for interior wall construction. Several materials, such as brick, tile, or plaster, are used. Unacceptable wall materials absorb moisture or are hard to clean, such as wood, plasterboard, or porous acoustical board. Concrete block is less desirable because it is difficult to clean. Unfortunately, it is commonly used. However, latex or rubberized paints can be applied to the block wall to seal it and thus make it acceptable.

Coving (curved base plate) at the junction between the wall and floor is a desirable construction feature for any building that requires frequent cleaning because it makes sanitation easier.

Bumper guards prevent damage to the wall by carts, trucks, etc. Chipped or damaged walls are difficult to clean. Aesthetically appealing guards can be used in areas where public traffic is heavy.

Ceilings.   Although ceilings are not usually exposed to direct physical wear and corrosion, they may serve as a source of contamination. Moisture (condensation) dripping on food products will carry detritus from the ceiling. Consequently, they should be constructed of a material that does not chip, peel, deteriorate, or retain dust or condensation. For the same reason, water pipes running along the ceiling should be enclosed.

Equipment.[107,167]   Equipment design and materials are important to minimize contamination. Rounded corners and lack of seams help eliminate places for growth of microorganisms. Stainless steel is used in food production plants because it is cleaned easily and is resistant to wear and corrosion. Galvanized metal may be used in certain areas of the plant. It is not as resistant to corrosion as stainless steel, however, so it must be used judiciously.

Plastics of an approved type may be used in several places in processing plants. Ease of cleaning is a primary criterion. Some plastics deteriorate rapidly when exposed to disinfectants.

Cutting boards are necessary in meat-processing plants. A board may be constructed of single-piece hardwood or an approved plastic or rubber-plastic combination. Wooden boards may be unacceptable in some countries and laminated cutting boards or chopping blocks are never satisfactory because the laminations tend to separate, especially when hot water is used in cleaning, permitting juices and small pieces of meat to lodge in the openings and become sources of contamination.

A **small-tool sanitizer** is usually required in slaughtering plants to minimize cross-contamination with infectious agents. The sanitizer maintains water at a temperature high enough to kill vegetative pathogenic microorganisms. In the United States, the required minimum temperature is 82.2°C (180°F). Between use on animals or any time when there is obvious contamination, the knives, hooks, and saws are dipped into the sanitizer. The sanitizer temperature exceeds the thermal death point of 76.7°C (170°F) for most mesophilic organisms. Most pathogenic bacteria are mesophilic. If a knife or other equipment to be sanitized is contaminated grossly with purulent exudate, dirt, or hair, it should be washed in water before sanitizing.

Water Potability.[56] All sources of water used on or around food in a slaughter plant should be examined periodically for absence of contamination by pathogenic microorganisms or toxic chemicals to ensure compliance with established standards. City water as well as water from each well used by the plant must be included in this evaluation. If nonpotable water is used in areas where there is no danger of contamination of food, the pipes carrying this water must be identified clearly to avoid cross-connections with the potable supply. Samples of water from each source are obtained at representative points throughout the system. When sampling from a tap, the water should be allowed to run freely for 2–3 min to allow sufficient time to clear the line. Water quality is assayed using a fecal coliform count as an indicator of contamination. In the United States, less than 2.2 coliforms/100 ml, based on five 10-ml samples, is generally considered acceptable.

Whereas a public health practitioner employs a standard routing for submission of water samples, there also are facilities available for a veterinarian in private practice who desires a detailed analysis of a water source. A state animal diagnostic laboratory, a department concerned with environmental quality, or a unit in a state university usually will be able to analyze water samples for microorganisms and normal mineral constituents as well as for the presence of some potentially hazardous chemicals.

Pest Control.[41,48,81,106,200,251,277,278,279,300] It is impossible to eliminate permanently all vertebrate and invertebrate pests in facilities that prepare or store food. Control is the best that can be achieved and depends on an integrated program that combines employee education, environmental sanitation, and chemical and physical control.

Employees in food-handling facilities and abattoirs are frequently unaware of the potential human health threat associated with the presence of these pests, and inasmuch as they are the individuals who will be responsible for conducting the control program, explanation of the rationale behind the procedures utilized is imperative to ensure compliance.

Environmental sanitation includes proper refuse and garbage storage and removal and daily sanitation inspections.

Effective physical and chemical control involves the use of barriers, such as screening, to prevent entry of pests into the establishment, as well as the proper use of chemicals and physical traps to control pests that have gained entry. This requires efficient utilization of approved pesticides and a familiarity with the life cycle and behavior patterns of the various species. As in environmental sanitation, frequent inspection is necessary to maintain effective control. Chemicals used in pest and rodent control programs in food-processing facilities must be restricted to those compounds approved for such use.

The effectiveness of a pest control program is evaluated by searching for physical evidence of infestation. For most insect species, this is achieved by noting their presence in the food-handling establishment. Flies are usually obvious and cockroaches less so, whereas beetles and moths usually require an examination of the food to detect their presence. Rodents usually are detected by searching for various signs that can indicate their presence, e.g.:

Droppings are one of the best indications of infestation. Rat droppings are large, up to 2 cm (3/4 in.), with blunt ends.

Burrows can be rat holes approximately 7.5 cm (3 in.) in diameter or mouse holes about 2.5 cm (1 in.) in diameter. An active burrow is free of cobwebs and dust.

Tracks, which are more clearly discerned by side illumination from a flashlight, may be observed anywhere along rat or mouse runs. Dust and soft mud are especially good places to observe tracks.

Grease streaks on walls indicate runs.

Meat Processing
Sanitary Dressing Procedures[75,215,259,260,276]

Cattle. Cattle are killed by severing the major arteries, cranial to the heart. Prior to being killed, cattle are rendered unconscious by a blow to the head or by a 120-volt pulse of electricity. In modern large-volume slaughtering plants, after stunning and bleeding, carcasses are suspended by the hind legs by inserting a hook, proximal to the coronate, under the Achilles tendon. The hook is attached to a moving overhead chain powered by electric motors. In some smaller-volume plants, the dressing procedure starts on the floor with the carcass positioned on its back by cradles or pritch poles. Regardless of the method used, great care must be taken to prevent contamination during dressing.

The first step is removal of the hide. This is usually accomplished in most large establishments with the aid of pneumatic knives and some type of mechanical hide-puller. Fecal material on the skin may be a source of carcass contamination, and washing cattle before slaughter will not reduce this problem. Recently, to remove contamination, the intact carcass has been washed and sanitized immediately after hide removal with chlorine or quaternary ammonium compounds.

After the head and hooves are removed and the animal is skinned, the head is cleaned with a high-pressure stream of cold water, marked with a tag bearing the same number as the carcass, and set aside for inspection. The carcass is positioned next to a long moving metal table and eviscerated. The viscera are placed adjacent to the carcass so that the inspector can relate one to the other. This is the second point in the dressing procedure with very high potential for contamination. During evisceration, great care must be exercised to prevent soiling the carcass with material from the gastrointestinal tract. The rectum is dissected free and tied off with a piece of string (a procedure called "dropping [or ringing] the bung") before the abdominal cavity is opened. Seepage of material from the anterior end of the gastrointestinal tract is prevented by separating the esophagus from the trachea and tying it off (a process referred to as "rodding the weasand"). The moving table surface (which is also chain driven) is automatically steam cleaned before reuse. It should be noted that the preceding description is typical of large slaughter plants, operating under U.S. regulations. Sanitary dressing procedures in other countries, and in smaller U.S. plants, will vary. For example, in small U.S. plants the viscera are placed into wheeled vehicles designed to hold the complete viscera from one animal.

After the viscera and carcass have been inspected and passed, the edible portions of the viscera are separated from the inedible and the carcass split into two halves along the spinal column, washed, and placed in a cooler to be chilled until ready to be cut up. In the United States, each half may be covered with a muslin sheet (called "shrouding"). After chilling, these clothes are removed, giving a smooth and bleached appearance to the surface of the carcass.

Veal calves may be dressed with the hides removed or still attached to the carcass. If veal carcasses are to spend several days in transit, the skin will reduce shrinkage.

Sheep and Goats. Dressing procedures for these species are similar to those for cattle except skinning is usually done by hand and requires more care to prevent contamination. Because there is little surface fat, as compared with cattle, shrouding is of little value.

Swine. Pigs are stunned, most commonly by electricity, before being killed by incising the major arteries, cranial to the heart. The first step following slaughter is immersion in a scald vat to facilitate hair removal. Two things may occur at this point that will result in carcass condemnation: (1) if a pig is not completely bled out, asphyxiation occurs, and (2) some carcasses may drop off the moving chain and become cooked in the scald vat, making the skin so friable that it will tear, which results in gross contamination of underlying tissues.

After the carcasses leave the scald vat, they pass through a dehairing machine that consists of a large revolving metal cylinder lined with rubber-surfaced paddles. These paddles, moving at high speed, rub the hair off the softened skin. The roughness of this procedure may express fecal material from some carcasses, resulting in contamination. Use of chlorinated wash water to flush the hair and debris from these machines during use helps to alleviate this problem.

After dehairing, the dressing procedure is similar to that of ruminants. Swine carcasses are not shrouded before being placed in the cooler.

Further Processing[161,175,178,202,253,309]

Offal.   In red meat abattoirs, *offal* refers to edible parts of the animal **other than** muscle meat. Sometimes, when loosely applied, edible offal is used solely in reference to edible viscera. In U.S. poultry slaughter plants, *offal* refers to inedible parts of the bird.

The offal-processing area is an excellent environment for the spread of contaminating microorganisms. When offal is retrieved from the animal, there is considerable opportunity for contamination from the intestinal tract. Water, used freely in offal processing, further distributes microorganisms. Because the animals were killed only minutes before, the temperature of the offal is still conducive to microbial multiplication. Also, animal tissue is an excellent source of nutrients for microbial growth.

It is especially important to process offal products without delay. Rapid processing ensures minimal microbial multiplication.

Processing of some of the more commonly used edible offal is reviewed below. For description of other offal products, refer to meat hygiene texts.

Liver.   Liver is one of the more commonly processed offal products. Normal processing involves removal of the gall bladder, hepatic and portal lymph nodes, and excess fat. Normally, the liver has been inspected before it reaches the offal area.

Beef Hearts.   Beef hearts also are saved as edible offal. Initial inspection is done during postmortem examination on the kill floor. At that time, the ventricle walls are incised and inspected visually. It is not until offal inspection, however, that the organ is examined by palpation.

Pork Hearts.   Pork hearts are handled somewhat differently. The organ is examined only externally at postmortem. Before offal inspection, pork hearts are sent through a slasher. It is not until offal inspection that the heart is incised and inspected internally. In dealing with either beef or pork hearts, it is essential to remove clotted blood. Failure to do so shortens shelf life remarkably. In some countries, because of possible *Erysiplothrix rhusiopathiae* infection and vegetative endocarditis, pork hearts are incised during the primary inspection.

Beef and Pork Tongues. Beef and pork tongues are saved for food by many processors. Before offal inspection, tonsillar tissue should be inspected and removed. Once the tongue and oropharyngeal tissue have been removed from the head, the tonsils are difficult to identify. Inspection of the tongue includes visual observation for ulcers and lacerations as well as palpation to detect abnormalities. Tongue worms and abscesses are important considerations when examining pork tongues. After tongues of either species are examined, they are scalded.

Head Muscle. Head muscle is valuable to the large processor because many products can be made from this tissue. In cattle, muscle from the cheek and poll areas often is salvaged. Offal inspection allows a second opportunity to examine for lesions indicating conditions commonly associated with this tissue (e.g., cysticercosis and eosinophilic myositis).

Beef Tails. Beef tails (commonly called "ox tails") are saved routinely by most plants. Contamination during hair removal is a major problem. Bruising is usually a sequela of vertebral fractures. Consequently, much trimming may be necessary before this product can be marketed.

Kidneys. Kidneys usually are saved for food. If a lesion is noted, however, the entire organ is discarded.

Lungs. Lungs are not consumed in many parts of the world. When they are used, it is mainly as sausage stuffing. The trachea, aorta, and pulmonary artery and vein are removed.

The Spleen, Pancreas, and Thymus. The spleen, pancreas, and thymus are trimmed to remove any surface fat and are primarily used in sausage manufacture.

Testes. Testes from young animals are preferred and are usually suitable for saute or frying. The only trimming required is to remove the epididymis.

Udders. Udders are sliced and washed to remove any milk and used in manufactured meat products in some countries.

Uteri. Uteri from nongravid pigs are collected and are usually poached or boiled, or they may be dried and smoked.

Blood. Blood is used to produce blood sausage, which, in certain areas of the world, is a much desired product. Sanitary collection is imperative if blood is to be salvaged. Collection usually involves the use of anticoagulants, such as citrate, and a defibrinating machine. (Anticoagulants are prohibited in some countries.) Sanitary collection is more important than the amount of citrate used because this anticoagulant is essentially nontoxic to the consumer. Blood is an excellent medium for bacterial growth, so it is a significant potential health hazard. Hog blood is not used.

Tripe.   Tripe is another tissue classed as edible offal. It includes the rumen ("blanket") and reticulum ("honeycomb"). In some areas the omasum ("bible") and abomasum ("reed" or "vell") are also popular. The first step in processing tripe is to open the rumen and to discard its contents. After the contents have been removed, the rumen and reticulum are hung, in an umbrella fashion, over a large conical structure. Subsequently, contaminated areas on the serosal surface are trimmed and the remainder washed. After washing, tripe is placed in a large tumbler machine. The tumbling and turning, in conjunction with scalding, help to remove the mucosa. After the mucosa is removed, the tripe is bleached and rinsed. Scalding, tumbling, bleaching, and rinsing all take place in the same tumbler. Finally, the product may be cooked.

Chitterlings.   Chitterlings, or swine large intestine, is another tissue classed as edible offal. Processing involves removal of fecal contents (stripping), splitting, washing, cleaning, and chilling. It is virtually impossible to remove all fecal contamination from these products.

Beef and Pork Brains.   Beef and pork brains may be saved for food, depending on the stunning procedure employed. Sometimes, even with an efficacious method, subdural hemorrhage occurs, and the brain must be condemned.

Pork Stomachs.   Pork stomachs, when used for food, are processed similarly to other intestinal viscera. The contents are first removed and then the stomach is washed. Later, it is heated in a vat at about 48.9°C (120°F). This enables easy removal of the mucosa, which before heating has the appearance of slime. Next, it is rinsed and scalded. Finally, after inspection, the stomach is chilled.

Sausage.   Sausage is a comminuted (all ingredients ground together) product composed of meat or meat food products that are seasoned with condimental substances and that may contain certain additives in permitted amounts. The product may be marketed fresh, cooked, cooked and smoked, semidry, or dry. Several compounds are added to sausage as preservatives, curing agents, flavor enhancers, color fixatives, or expanders. Their use must be controlled to ensure that the final product does not contain excessive amounts.

In the United States, 85% of the sausage is marketed in the cooked or smoked form; of this, about 55% of retail sales are frankfurters or wieners. Sausage production may vary with the type of sausage produced and the size of the operation. Following are the steps of sausage production:

1. **Batching**. This is the initial step in sausage production. All of the ingredients are set out in the correct proportions before actual processing.

2. **Grinding**. Next, the meat is ground to the desired texture (course or fine).
3. **Mixing**. This step can be done by hand or with the aid of a mechanical mixer. All the ingredients included in the final product are mixed at this time.

   The binding power of the mixture must be considered so that the final emulsion does not fall apart. The primary consideration is getting the fatty portion of the mixture to bind together. Fat globules are bound by myosin, a protein that is effective in holding the mixture together. Myosin will lose its ability to tie up fat globules if the mixture is overheated. Generally, binding power is lost at temperatures above 21.1°C (70°F). This is why ice often is added to the mixture before curing.
4. **Curing**. After the ingredients have been mixed, the product is placed in a container to cure. (Curing is described in detail in "Meat and Poultry Preservation.")
5. **Stuffing**. After the product is cured, it is ready for stuffing into the casing. Usually, casings are filled with the emulsion under pressure and tied at regular lengths. Today, many casings are made of synthetics such as cellulose. Although less common today, natural casings (intestines) are used to give a sausage a homemade appearance.
6. **Cooking and smoking**. The product is now ready for cooking or smoking. Most often, smoking is preferred because of the flavor imparted. The procedure is normally performed in large walk-in ovens with precise time and temperature controls.
7. **Showering**. During the final step in sausage production, water is sprayed on the product to cool it and to prevent bacterial growth. This rapid cooling process also prevents excessive water loss from the product and therefore helps to minimize shrinkage.

Poultry Processing[38,52,188,270]
Essentials in Prevention of Contamination during Processing

1. Only healthy poultry should be processed.
2. There should be an abundant supply (numerous taps and large volume) of potable water.
3. Birds with their feathers and feet still attached should be handled in areas separated from areas where evisceration is done.
4. Great care should be taken in removing the digestive organs, in particular the lower gut.
5. Eviscerated carcasses should be chilled as soon as possible to slow bacterial growth.
6. Personal cleanliness of employees and constant cleaning of equipment should be emphasized.

Processing Procedures
Shipping Broilers. At the end of the growing period the birds are caught by crews and placed in crates. This process is a potential cause of bruising and

broken bones. In warm climates, shipping must be done at night or very early in the morning to avoid piling and overheating. The crates are stacked on trucks for the trip to the abattoir. Cold, very hot, or rainy weather and rough roads may cause excessive losses en route.

Trucks may have to wait at the abattoir to be unloaded. Protective sheds with large ventilating fans are used to prevent losses from overheating or from drenching rain.

While birds are still on the trucks or while the crates are on the unloading dock, the shipment is checked for unusual respiratory sounds or other evidence of sickness. Shipments with a high percentage of sick birds may be rejected and the birds sent back to the farm for treatment.

The birds are taken out of the crates and hung by their legs on a moving line of shackles. This handling is another potential cause of bruising and broken bones.

Birds dead on arrival (DOA) should not be hung on the processing line, but through human error dead birds sometimes do go through processing.

Slaughtering. Humane treatment of animals is increasingly a matter of public concern. Electric stunning of birds is effected as their heads touch a brine solution to complete an electric circuit. Electric stunning causes an increase in blood pressure, so the throat must be cut immediately thereafter to prevent hemorrhages in various parts of the body. This is usually accomplished using a mechanical throat-cutting device, with a backup worker hand-cutting any birds missed by the machine. Any bird that escapes both must be condemned because it still has all its vessels engorged with blood after drowning in the scalding vat. Its respiratory tract will be full of water and contaminating bacteria.

Defeathering. Scalding (mild scald at 48.9°C [120°F], hard scald at 57.2°C [135°F]) softens the skin and facilitates removing the feathers. The hard scald also removes the outer layer of skin and produces a white-skin carcass.

After scalding, carcasses enter the rough-picking machine that removes most of the feathers. The setting of the drums (bearing rubber fingers) is very important because maladjustment may cause skin tears, broken bones, and other damage.

After rough picking, the birds enter the finishing-picking machine. Feathers missed by the machines are picked by hand. Fine hairlike feathers may be removed by passing the birds past banks of flames, a procedure referred to as "singeing."

Processing. The heads of older hens and turkeys are left attached to the carcass as an aid in detecting certain disease conditions, such as infectious laryngotracheitis, infectious coryza, and pox. Under modern methods of production, these diseases are rare among young birds; therefore heads of broilers are removed before further processing. This is near the end of the dirty side of processing.

After these steps have been completed, the legs are cut off. It is important that the carcass not be washed after this point (unless the hocks are protected from water) until the carcass has been examined to determine if joint exudates are present.

After hock cutting, carcasses pass along a chute and through rubber curtains from the dirty side to the evisceration side of the processing plant. These areas are separated by a wall. Piling up at this point can lead to contamination from the floor, when carcasses spill over, and cross-contamination from other carcasses.

After the giblets (heart, liver, and gizzard) are removed, the lungs are removed with a vacuum device, the **lung gun**. The same device is used to remove the ovary from mature hens and the testicles from males. The lungs are likely to contain bacteria that lead to spoilage. Likewise, the yolk material in a mature hen's ovary is a good culture medium, and bacteria could multiply there if the ovary is not removed.

After processing, the carcass is immersed in a **chiller** (a large vat in which carcasses are interlayed with crushed ice or a tank with very cold, overflowing water). Rapid chilling is necessary to inhibit bacterial growth. Psychrotrophic bacterial populations tend to predominate in such chilled carcasses.

After chilling, the birds are graded and packed. Whole birds are weighed automatically into weight-graded bins, and giblets, wrapped tightly in paper or plastic, are stuffed into the body cavity. Any carelessness in personal hygiene or in wrapping and placing of the giblets could introduce contaminating bacteria.

Whole birds may be shipped packed in ice or placed in acceptable (plastic or waxed paper) bags and frozen.

Trimmed birds often are cut up into parts. This operation has great potential for buildup of contaminating bacteria, so frequent cleanup of equipment is essential. The various parts are packed in plastic trays, covered with plastic or waxed paper, and then either frozen or refrigerated. In some operations, the air inside the wrapper is replaced with an inert gas to further retard spoilage.

A modern poultry-processing plant generates a tremendous quantity of inedible material. Methods for disposing of this material include burial, incineration, composting, and rendering. Burial and incineration impose environmental concerns that are becoming unacceptable. Composting is environmentally sound and relatively inexpensive. Rendering has the obvious advantage that a valuable byproduct is produced, which can be recycled as a protein supplement for broilers and other livestock. Because contaminated livestock feed has been incriminated as the cause of salmonellosis, it is critical that this material be heat treated, handled, and stored so as to prevent contamination.

Meat Preservation[23,73,98,102,129,186,269,293]

Meat preservation is primarily concerned with controlling microbial contamination and autolytic changes caused by enzyme action at the cellular level. Five general methods can be used to preserve a product and increase its shelf life: heat, cold, dehydration, irradiation, or addition of chemicals. Any of these tech-

Pork products cannot be maintained frozen as long as beef because of the relatively higher amounts of unsaturated fats in pork. With time, the unsaturated fats are oxidized.

Dehydration. Dehydration, a third general method of meat preservation, is effective for two reasons: (1) enzymatic processes within cells are slowed, thereby retarding autolytic changes, and (2) reduced moisture in the meat is unfavorable for bacterial growth. Dehydration can be accomplished by either hot-air drying or freeze-drying.

Hot-air drying is an old method used by American Indians to produce venison or buffalo meat jerky. Residual moisture content of hot-air-dried meat is about 5% whereas moisture in freeze-dried meat is reduced to 1%.

Hot-air drying preserves meats up to 12 mo before signs of deterioration appear. A disadvantage of this method is that fatty meats, such as pork, tend to become rancid unless an antioxidant is added during cooking and the product is well wrapped after final processing to prevent exposure to air. Another disadvantage is that raw meat cannot be dehydrated by this means.

Freeze-drying, on the other hand, effectively preserves both cooked and raw meat. Shelf life varies depending on the ambient temperature. Special precautions are needed with meats high in fat to avoid rancidity.

Irradiation.[89,139,160,182,247,256,266,274,280,291,292] Irradiation destroys foodborne pathogens found in meat. *Campylobacter jejuni* and *Escherichia coli* O157:H7 are most sensitive to irradiation, followed by *Staphylococcus aureus, Listeria monocytogenes,* and *Salmonella* spp. Recent studies have demonstrated that irradiation can be effective even after a meat product has been frozen and boxed. The susceptibility to irradiation of some of these organisms can vary, depending upon the temperature and fat content (which serves as an insulator) of the product.

In this method of preservation, food is subjected to ionizing radiation from a radioactive isotope of cobalt or from devices that generate a stream of electrons or X rays. The unit of measure is the kilogray (kGy). One kGy equals 100 kilorads (krad). One rad equals 100 ergs of energy absorbed per gram of absorber. High doses kill by fragmenting cellular DNA. Low doses alter biochemical reactions, thereby interfering with cell division. Irradiation also maintains the freshness of food by inhibiting the production of microbial enzymes that break down cellular integrity. By varying the quantity of irradiation, food can be treated to achieve different objectives, as follows:

1. **Radappertization**. Treatment at doses of 20–70 kGy to destroy all organisms, i.e., sterilization.
2. **Radicidation**. Treatment at doses of 0.1–10 kGy to inactivate non-spore-forming pathogens, i.e., pasteurization.
3. **Radurization**. Treatment at doses of 1–5 kGy to destroy spoilage organisms.

4. **Disinfestation**. Treatment at doses of 0.1–2 kGy to destroy insects.

History of Irradiation. As early as 1921, a U.S. patent was issued for using X rays to destroy trichinae in pork. By 1947, advances in equipment design had made it possible to preserve a variety of foods, but commercial applications were limited because of poor penetration of tissue. In 1963, the U.S. Army established a laboratory at Natick, Massachusetts, which was equipped with a cobalt-60 irradiator and a high-energy electron generator. Since then, similar facilities have been established throughout the world.

Legislation of Radiation. In the United States, radiation is considered to be an "additive" to food, rather than a process and is subject to federal regulations governing the types of food that may be irradiated and the quantity of radiation allowed. Irradiation produces substances called *radiolytic products* in foods. These products are not radioactive and are not harmful in the quantities consumed. At present, irradiation is approved for use in poultry, pork, fruits, vegetables, spices, and grains. It has not yet been approved for use in beef, and it is unlikely that it will be practical for milk or eggs because undesirable flavor changes occur. Seafood irradiation is being considered. The four major importers of United States poultry—Japan, Hong Kong, Mexico, and Canada—have not approved irradiation of poultry. Other countries have been quicker to utilize radiation as a method of preserving food. As of 1992, a total of 37 countries had approved the procedure. The World Health Organization (WHO) and the Codex Alimentarius Commission (CAC) support food irradiation and agree that food irradiated at doses below an overall average of 10 kGy are safe to eat and require no further toxicologic testing.

Procedures for Irradiation. Radiation may be applied as X rays, as electron bombardment, or as gamma rays. X rays and electrons can be generated from electromechanical sources, and gamma rays from sealed units containing isotopes of cobalt-60.

The primary advantage of X rays and electron beams over isotopes is that, being machine-generated, they can be switched on and off as the need arises. Isotope irradiators emit and decay constantly and, unless heavily used, have reduced economic return. The disadvantages of X-ray and electron beam generators are high cost and low penetrating power. Effective irradiation is limited to small low-density items (less than 0.2 g/cm^3) and sliced meat or fish up to 5 cm (2 in.) thick. Fresh or frozen uncooked whole chicken carcasses and parts can be successfully irradiated at a dose of 1.5–3.0 kGy but must be packaged for sale prior to irradiation to prevent recontamination. Costs of irradiation range from 1–1.5 cents/lb for poultry, depending on the size of the unit.

Chemicals.[9,54,72,127,231,232,245,249,285,287,313] Adding chemicals, called *curing agents,* is another method for preservation of meat products. Potential pathogenic or spoilage organisms are destroyed because an unfavorable environment is created, usually by altering the pH or osmotic pressure of the food. Caution must be exercised in the choice of chemicals because many bactericidal compounds are too toxic to be used as food preservatives.

Water.   Although water is not a curing agent, it is the largest single ingredient of any curing preparation. Because the curing preparation often is injected into the product, only potable water should be utilized. Water may increase the yield of the final product, but it does not enhance the flavor, tenderness, etc. The total amount added is usually controlled by legislation.

Salt.   Sodium chloride is an effective bacteriostatic agent, inhibiting microbial growth by increasing osmotic pressure. There is considerable variability among microorganisms in sensitivity to salt. Salt is also used for its flavor-enhancing effect.

Nitrite and Nitrate Salts.   These are used in curing as color fixatives as well as preservatives. The two salts often have been used concurrently. They have been particularly effective in providing protection against *Clostridium botulinum.* Nitrite provides a rapid initial cure whereas nitrate provides a source of nitrite during storage. Because nitrate exerts little bacteriostatic or bacteriocidal effect in the concentrations used in curing, nitrates must be converted to nitrites to be effective. Bacterial reduction of nitrate to nitrite is pH dependent and is less at pH 6.0 than at pH 7.0. Nitrite is the salt responsible for subsequent reactions such as the oxidation of the iron in hemoglobin and its conversion to methemoglobin. These agents are highly toxic if ingested in excessive quantities. Under certain conditions, nitrates may interact with amines present in meat to form nitrosopyrolidines, which can act as precursors of potentially carcinogenic nitrosamines. Because of the hazards associated with nitrates, the use of this salt in curing mixes has been greatly restricted.

Sugar.   The main purpose of adding sugar is to help reduce the harsh flavor that salt adds to a product. Sugars are useful in providing a favorable medium for growth of desirable flavor-producing bacteria. Product yield is increased somewhat by added sugar.

Phosphates.   Phosphates are included as an adjunct to curing agents to improve the water-holding capability of the product. Although the mechanism has not been established, intracellular proteins are known to increase water-binding capacity with a concomitant increase in pH. Phosphate produces a shift of the meat pH to the alkaline side. Addition of phosphates is self limiting because, in high concentrations, they precipitate out of the brine solution and crystalize on the meat. In excess, they can be toxic, so their use is restricted.

Ascorbates. Ascorbates are used as adjuncts to curing agents. These compounds aid in fixing the color of products and thus prevent fading. Ascorbates have only minimal bacteriostatic effect on spoilage and mold-producing organisms. Therefore, amounts necessary for effective microbial control are excessive and would constitute adulteration.

Acids. Certain acids may be used as bacteriostatic agents. Lactic and acetic acid are used commonly for this purpose as well as for flavor enhancement. Lactic acid bacilli are used routinely as starter cultures in certain foods or are added directly. It is this acid that gives such foods a tangy taste. As the pH of the acid decreases, a dramatic increase in bacteriostasis occurs because bacteria grow in a relatively restricted pH range. A decrease of 1 pH unit increases the bacteriostasis 10-fold.

Gases. Carbon dioxide and ozone are effective meat preservatives. Carbon dioxide is used sometimes in the holds of large ships that are transporting whole carcasses. The gas acts to retard growth of surface contaminants by reducing the amount of available oxygen.

Ozone, emitted from ozone lamps, has a bactericidal action on airborne microorganisms by oxidizing organic molecules by breaking carbon bonds, resulting in simpler organic molecules. This preservation technique has had limited use and must be used with discretion because of several disadvantages: (1) Ozone has an odor that masks any abnormal odors associated with meat aging. Consequently, substantial financial loss is possible. (2) At excessively high concentrations, ozone is hazardous to human health. (3) The gas also tends to accelerate development of fat rancidity.

A recent use of ozone is for disinfection of recycled chiller water in poultry slaughter plants. Ozone gas is produced by passing an electric current through air, changing oxygen molecules to ozone molecules. The ozone is introduced into chiller water, oxidizing organic molecules.

Control of Curing Ingredients. The amounts of sodium chloride or sugars that can be used in curing are self-limiting because of taste. There is some variability in the use of sugars. For example, maltose and lactose have low sweetening ability, so they can be added in greater quantities before the product becomes too sweet and, therefore, often are used as fillers.

Applying Curing Agents. External application involves applying either a dry or liquid form of an agent to the surface. Although this method is still used, it has been generally discarded because of poor penetration. Curing agents are more effective if they are injected into a major vessel or into the tissue (called "stitch pumping"). Modern plants have multineedled pumpers. Although greater penetration is achieved by injection, the processor still is required to retain the product while it cures. Surface application may require up to 6 wk to effect a "cure." Injection procedures, although more effective, require 7–10 d. Several alternative methods have been attempted. One such method allows the curing reaction to occur within a sealed package. For example, individual bacon slices are passed through a warm (10°–15°C [50°–77°F]) curing solution, drained, heated, and then packaged. The product is packaged in a hermetically sealed container and allowed to cure while on its way to the retailer.

Trichina Control.[65]   In the United States, where meat inspection does not include microscopic examination for trichinae in pork, and many cured products from swine are processed as ready-to-eat, a potential health hazard exists because encysted trichina larvae are transmitted readily in products that are not prepared properly. Trichina cysts are killed when an internal temperature of 58.3°C (137°F) is achieved. Pork may be certified as trichina-free if it is frozen at -40°C (-40°F). Cysts of *Trichinella spiralis* are killed instantly at this temperature. For products such as dried sausage, salting and drying will destroy trichinae, but the time required varies with the amount of salt used and how fast the drying procedure is accomplished. For example, using hams dry cured for 28 d or more, trichinae were killed at:

Temperature	Time
52°C (125°F)	4 d
37°C (98°F)	6 d
21°C (70°F)	35 d
5°C (40°F)	190 d

Control of Inedible and Condemned Meat and Poultry Products[183]
Inedible Products

Categories.   There are two types of material that may be rendered: inedible and condemned. These materials are never combined for rendering. **Inedible** materials are those that are not normally used for human food (hides, feathers, etc.). Rendering will not convert inedible materials to edible (for human consumption) products. The end products of inedible rendering are primarily used for animal feed. **Condemned** materials are those that would normally have been used for human food but have been rejected for human consumption. Many of these may be used in animal feeds.

Rendering.   Rendering in food processing commonly means separating fat from its connective tissue stroma, or the final treatment of inedible offal (called "tanking"). Low-temperature rendering (49°C [120°F]) permits survival of *Clostridium perfringens,* which can multiply at 50°C (122°F), and possibly *Staphyloccocus aureus* and salmonellae. High-temperature rendering (115°–150°C [239°–270°F]) of whole animals (including bone, feathers, blood, and viscera) is used to produce inedible fats and meals. It has always been assumed that at these temperatures all pathogens are killed. (A major problem has been recontamination of the meal during further handling.) However, in the United Kingdom an epidemic of bovine spongiform encephalopathy (BSE) has been associated with the feeding of protein supplements produced from the rendered carcasses of ruminants, especially sheep afflicted with scrapie, another encephalopathy. It is possible that the problem developed as a result of changing from batch rendering to continuous rendering, in which one infected carcass can contaminate a large quantity of material. There is concern that this problem may go beyond animal health economics and pose a human health hazard. The various spongiform encephalopathies have been considered as relatively species specific, but a 1985 outbreak of transmissible mink encephalopathy (TME) in the United States was epidemiologically linked to feeding "downer" cows to mink. A major obstacle to identifying possible cause and effect relationships between BSE and human encephalopathies, is the very long incubation periods involved. Creutzfeldt-Jakob disease (CJD), for example, has an incubation period of 10–20 yr.

There are several methods of rendering meat (Fig. 2.8).

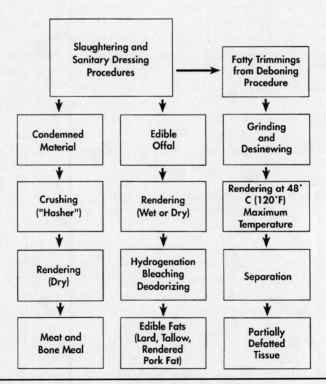

Fig. 2.8. Rendering flow diagram.

Dry rendering, which is generally used for condemned and inedible products destined for animal feed, is accomplished with the use of a large horizontal tank surrounded by a steam jacket. The tank contains an agitator that moves the meat about as the fat is separated from other tissue. If rendering takes place in a partial vacuum, high-temperature rendering is accomplished at a temperature of about 93.3°C (200°F). As the meat is processed, it is discharged from the tank into percolating vats where it separates into two layers. The lower layer is the connective tissue residue, usually referred to as *cracklings*.

Wet rendering is usually employed for removing liquid fat from edible meat. With this method, water is added to the meat after it is placed in the tank. Subsequently, steam is injected into the tank to mix water vapor with the product as it cooks. Finally, the mixture settles out in three layers: the heavy cracklings on the bottom, oil on top, and water between. As the liquid fat is drawn off during cooking, water is injected into the system, causing the remaining fat to rise to the level of the draw-off valves.

The **centrifuge method** is a more complex rendering process than the wet or dry method. It involves continual addition of meat to the system rather than rendering batches. In this method, centrifugal force rather than increased temperature is relied on to separate out liquid fat. Consequently, the process can be accomplished at relatively low temperatures. The meat is ground before centrifuging, enabling easier fat separation during centrifugation. Because of the continuous nature of this process, higher-quality meat must be used. Lower-quality meat requires varying amounts of time to be defatted, which decreases the efficiency of this process. Also, bacterial contamination is a serious problem with this method. By adding a piece of tissue that is heavily contaminated, considerable amounts of fat can be spoiled. This method is used primarily in operations in which large volumes of meat must be processed rapidly.

Edible Products. Five products can be retrieved from any of the rendering methods described: lard, rendered pork fat, tallow, oleo stock, and partially defatted tissues. Lard and pork fat are obtained from hogs, tallow and oleo stock from cattle or sheep. Partially defatted tissues are derived from beef and pork fatty trimmings.

Lard. This high-quality product is obtained from rendering fresh, clean, normal pork fatty tissue. In other words, fat associated with structures such as bone, skin, or organs is not considered edible lard. Lard is produced from "killing" or "cutting" fats. **Killing fats** are those that are removed during the sanitary dressing procedure. Fat surrounding the kidneys is an example of killing fat. **Cutting fats** are those obtained while cutting the carcass into various wholesale or retail cuts. Examples of cutting fats are back fat, belly fat, ham fat, etc.

Rendered Pork Fat. This edible product can be produced from most tissues but is of somewhat lower quality than lard. A few structures, such as stomachs and head bones, cannot be used. If a whole carcass is passed for cooking, it can be used to produce rendered pork fat. Cured products also can be used for this purpose.

Tallow.   This is fat rendered from the fatty tissues removed from beef or mutton during processing.

Oleo Stock.   This product is a high-grade tallow derived from beef or mutton but differs in that the meat is chilled and hashed before rendering. Rendering in this instance is accomplished in an open tank at about 65.5°C (150°F) as salt and water are added. Oleo stock is the fat that separates out.

Partially Defatted Tissue.   This is a lean, edible product derived primarily from beef and pork fatty trimmings, removed during the boning operation (removal of meat scraps from bone, following sanitary dressing). The raw material is high in fat but contains visible lean meat. After grinding and desinewing, the material is rendered at 40°–48°C (105°–120°F), centrifuged, and chilled or frozen before packaging. Partially defatted meat products are used in the manufacture of beef patties, corned beef hash, chili, luncheon meat, etc.

Environmental Controls.   Several environmental considerations must be reckoned with when rendering inedible products. The objective is to prevent cross-contamination of edible products. It is important that separate equipment be utilized. Also, equipment should be well segregated from other processing areas. Personnel who work in this area should have separate clothing for this purpose or do this processing last. In larger operations, different employees process edible and inedible products. Odors are controlled most effectively with use of water vapor condensers.

Product Controls.   Maintaining the identity of material containing portions of condemned animals is essential. Tanks where inedible products are placed must be sealed to ensure that an inedible product is not retrieved and subsequently processed as human food. In addition, many pathogens can be transmitted via the food chain. Precautions taken to ensure destruction of pathogens minimize the likelihood of their transmission by rendered inedible products used in animal feed.

Milk Processing[1,47,77,84,88,95,113,149,195,222,288]
Pasteurization Process.   The purposes of pasteurization are twofold: (1) to destroy any pathogenic microorganisms that might be present in the milk and (2) to enhance the shelf life of milk and milk products. Although factors such as healthy cows and milk handlers combined with adequate environmental sanitation are important in the production of safe, wholesome milk, pasteurization is still the only effective means of ensuring that pathogenic microorganisms are

eliminated from milk without significantly affecting its food value. Pasteurization does destroy some of the vitamin C in milk, but because milk is not an important dietary source of vitamin C, this minor reduction is of no practical importance.

Pasteurization extends shelf life by destroying most of the microorganisms that cause milk to spoil. The process also inactivates enzymes in milk that cause deterioration, especially lipase, which breaks down fat to fatty acids resulting in rancid flavor. Pasteurization is not sterilization. Its purpose is to kill vegetative pathogens, not all organisms. Therefore, spoilage is delayed rather than prevented.

The key to success in pasteurization is heating every particle of milk or milk product for a specified temperature and time. The main purpose of regulations dealing with inspection of pasteurization equipment is to ensure that this is achieved.

Time-Temperature Relationships.   Current recommendations for pasteurization are based on the long-time holding (LTH) time-temperature requirement of 62.8°C (145°F) to kill *Coxiella burnetii,* the causative agent of Q fever in man. The previous standard of 61.7°C (143°F) was based on time-temperature requirements to kill *Mycobacterium bovis,* considered earlier to be the most heat-resistant pathogen associated with milk. It was not necessary to change the pasteurization procedure in the high-temperature short-time (HTST) process as the temperature was sufficient to kill *Co. burnetii* and *M. bovis* (Fig. 2.9). As time-temperature relationships exceed the pasteurization line, the quality of nutritive constituents and the flavor of milk are decreased. Modern equipment has expanded the range of time-temperature relationships available. Any of the combinations in Table 2.13 will ensure destruction of the above pathogens.

In the United States, the term *ultrapasteurized,* when used to describe a dairy product, means that such product has been thermally processed at or above 138°C (280°F) for at least 2 s, either before or after packaging so as to produce a product that has extended shelf life under refrigerated conditions.

In the United Kingdom the designations for heat treatment of milk are as

Table 2.13. Time-temperature requirements for pasteurization

Temperature	Time	Common name
63°C (145°F)[a]	30 min	LTH
72°C (161°F)[a]	15 s	HTST
89°C (191°F)	1 s	HTST
90°C (194°F)	0.5 s	HTST
94°C (201°F)	0.1 s	HTST
96°C (204°F)	0.05 s	HTST
100°C (212°F)	0.01 s	HTST
138°C (280°F)	2 s	"Ultrapasteurized"

Note: LTH = long time holding. HTST = high temperature, short time.
[a]If the fat content of the milk product is 10% or more, or if it contains added sweeteners, the temperature must be increased by 3°C (5°F).

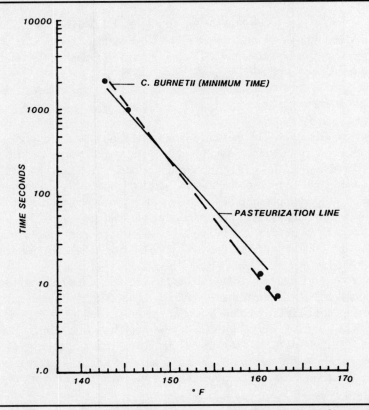

Fig. 2.9.   The time-temperature requirements for pasteurization.

follows: thermized (15 s at 57°-68°C [135°-154°F]), pasteurized (15 s at 71.7°C
[161°F]), and ultra-heat-treated (1 s at 135°C [245°F]).

During the last 2 decades, two organisms, *Listeria monocytogenes* and
Campylobacter jejuni, have been increasingly associated with foodborne disease
outbreaks, including, on occasion, diseases from consumption of milk. In nearly
all instances, the vehicle has been raw milk because *Ca. jejuni* is a fragile organism and succumbs readily to pasteurization temperatures. Some strains of *L.
monocytogenes,* however, are both thermoduric and psychrotrophic, and sub-
lethally injured organisms are capable of repair and total restoration of their pathogenicity, even under normal refrigeration temperatures. This fact has created
some doubts concerning the reliability of current pasteurization time-temperature
relationships in preventing milkborne listeriosis, but recent work has confirmed
that, when the organism is a primary source contaminant (in the raw milk), there
is no need to alter the existing standard.

If unacceptable levels of potential pathogens are present in properly pasteurized milk, it is almost certain that contamination occurred after pasteuriza-

tion. The source may be environmental, i.e., floors, bottling equipment, etc., or human. Microbiologic examination of all surfaces contacting milk after pasteurization as well as aerosols and employees is required to identify the source and prevent further contamination.

Equipment

1. **Pasteurization vat**. The pasteurization vat is rarely used today in modern dairies but may be seen in small plants or for special purposes such as for goats' milk, which must be pasteurized separately from cows' milk. A pasteurization vat or tank is used in the LTH method. Adequate agitation is needed so the temperature of the milk will not differ by more than 0.56°C (1°F) throughout the vat at any time during the holding period.

   A milk temperature thermometer is one of three thermometers required with each holding vat. This thermometer also is required with each HTST unit.

   To be sure that milk is held in the holding vat for the required time and temperature, a recording thermometer is used. The completed recording charts must be kept for review by the inspector.

   The air above the milk in the holding tank must be heated to at least 2.8°C (5°F) above the temperature of the milk to ensure that any milk particles adhering to the side or lid of the tank are pasteurized. This area is heated with an air space heater. The third thermometer measures the temperature in this air space.

2. **Clarifier**. Clarification is one of the steps done in the plant to prepare milk for the consumer. Milk is pumped through a centrifugal device, the clarifier, to remove any solid foreign material, particularly blood or somatic cells, straw, or manure. It may be clarified when cold before pasteurization or when hot during the pasteurizing process. The centrifugation isn't sufficient to remove bacteria and has little effect on bacterial count. The clarifier also may be used to separate milk into cream and skimmed milk fractions.

3. **HTST unit**. In the HTST unit, a **flow diversion valve**, which is a milk flow stop valve, is installed downstream from the holding tube to prevent forward flow of milk from the holding tube, whenever the milk temperature falls below 71.7°C (161°F). At such time the valve diverts flow back to the raw milk balance tank. The valve is critical to the HTST process and must be inspected regularly for proper function.

   Milk is maintained at a specified temperature in the **holding tube** portion of the HTST unit for a specific period. The tube must have an upward slope of at least 21 mm/cm (¼ in./ft) from the inlet to the exit to prevent formation of air pockets, which can effectively reduce the holding time.

   Cold raw milk enters the **regenerator** section of the HTST pasteurizer from the balance tank. The raw milk is heated by transfer of heat from the already pasteurized milk passing on the other side of the regenerator plate, thereby saving heat. Raw milk in the balance tank must be located so that overflow will be below the lowest level of milk in the regenerator.

The **timing pump** must be located at a point upstream from the regenerator and before the heating section. Milk will thereby be sucked through the regenerator, creating a lower pressure on the raw milk side. If a leak were to occur in the regenerator, the pasteurized milk would flow into the raw milk, not vice versa.

Location of the balance tank below the regenerator also protects against the flow of raw milk past the regenerator, should there be a leak in the plate.

If **steam** is utilized in the pasteurization process, it must be produced from potable water or water supplies acceptable to the regulatory agency involved (Fig. 2.10).

4. **Homogenizer**. Milk is homogenized by pumping it through a small orifice or other device under high pressure. The fat globules break up to such an extent that after 48 h of quiescent storage no visible cream separation occurs in the milk. In addition, the fat percentage in the top 100 ml of a quart (0.95 l) differs by no more than 10% from the fat percentage of the remainder after the latter is mixed thoroughly. When homogenization is part of the HTST system, it usually is done between the regenerator section and the heating section or at the end of the holding tube after the flow diversion valve. In the LTH system, it is done after preheating to approximately 60°C (140°F) or after pasteurization.

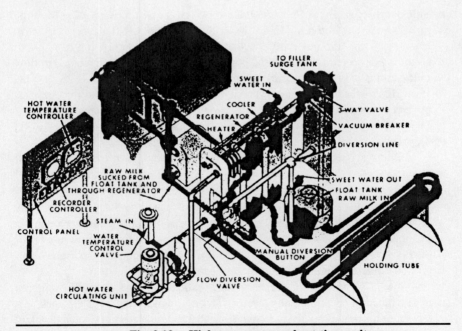

Fig. 2.10. High-temperature, short-time unit.

The temperature at the time of homogenization must be sufficient to have inactivated the enzyme lipase to avoid development of hydrolytic rancidity. Homogenization of raw or inadequately heated milk will usually lead to development of this undesirable flavor.

If the homogenizer is clean and sanitary, no significant increase in bacterial content should occur as a result of homogenization. An increase in bacterial **count** may result, however, because clumps of bacteria are broken up.

After homogenization, the milk is cooled in the same unit in the HTST system, or it is pumped to a balance or storage tank (sometimes called a "surge tank"), from which it flows to a bottling or packaging machine. The milk must be cooled to 3.3°C (38°F) or lower before bottling because cooling in the final container is very slow as it depends on air conductivity.

5. **UHT pasteurization**. In the United States, the time-temperature relationship for UHT (ultra-heat-treated) milk is 138°C (280°F) for 2 s. This is achieved either by injecting steam into the milk (which must later be evaporated) or by passing the milk over a plate heat exchanger. This destroys all vegetative forms of microorganisms and most spores, thereby producing a much longer shelf life than can be achieved by normal pasteurization. It has the added advantages of producing minimal flavor change in the milk and enabling the final product to be stored at ambient temperatures.

Evaluating Effectiveness of Pasteurization

1. **Salt conductivity test**. A salt conductivity test is used to determine the time milk is exposed in the holding tube of the HTST unit. After electrodes connected to a galvanometer are attached to each end of the holding tube, a salt solution is injected into the initial portion of the tube. The time needed to flow to the end is measured by change of conductivity.

2. **Phosphatase inactivation curve**. Alkaline phosphatase is one of the enzymes present in milk. The time-temperature relationship needed to inactivate alkaline phosphatase parallels the pasteurization line. Properly pasteurized milk, therefore, will not contain active alkaline phosphatase.

Two tests are in general use to measure inactivation of alkaline phosphatase: the rapid field method and the slower but more sensitive photometric method. Either test has the sensitivity to detect 1 part of raw milk in 2,000 parts of pasteurized milk. There are two reasons why the phosphatase test may be misinterpreted: (1) The test depends on release of free phenol by enzymatic action. Any free phenol present on laboratory equipment will result in a false-positive test. (2) Chocolate milk and flavoring substances in ice cream contain phenolic compounds that may produce a false-positive test.

The UHT process may result in reactivation of phosphatase. Phosphatase is not inactivated completely by this process, and as the product ages, detectable phosphatase activity may return. Reactivation also may be a problem with products with high butterfat, such as butter or cream.

3. **Mesophile test**. In the United Kingdom there is a microbiological standard for UHT milk. After 15 d storage at 30°C (86°F), or 7 d at 55°C (131°F), the mesophile count must be less than 100/ml.

Coliform Test.[285] The presence of coliform organisms in pasteurized milk is not an indication of inadequate pasteurization. Instead, it results from contamination of the milk from some source after pasteurization.

Milk Products.[19,20,74,78,92,112,136,142,151,176,197,295,302] Milk products that have a higher milk fat content than milk and/or contain added sweeteners have to be heated to at least 65.6°C (150°F) for a minimum of 30 min in the LTH pasteurization procedure or to at least 74.4°C (166°F) for a minimum of 15 s in the HTST method. This increase in pasteurization temperature is needed because of the slower rate of heat penetration through the more viscous product.

1. **Cheese**. Cheesemaking is basically a fermentation process in which microorganisms convert lactose or milk sugar to lactic acid. Antibiotics present in milk inhibit fermentation and cause serious economic loss to cheese manufacturers. Antibiotics appear in milk as a result of treating cows orally, parentally, or by intramammary infusion. As little as 0.05 IU/ml penicillin, 0.05 mg/ml chlortetracycline, or 0.04 g/ml dihydrostreptomycin in milk will inhibit some organisms used in cheesemaking. If a cow is treated with antibiotics by any route, her milk must not be used until the recommended withdrawal time for the antibiotic has elapsed. If an intramammary infusion product has been used, the stated withdrawal period must be adhered to for the milk from all four quarters because detectable amounts will appear in the nontreated quarters. Regulations in many jurisdictions now require that each tanker truckload of milk be certified as antibiotic-free.

    Milk to be used in cheesemaking may be pasteurized or heated to subpasteurization temperatures to destroy specific microorganisms. If a higher temperature is used, there may be texture and moisture problems and a reduction of milk protein, which will reduce total yield. In some instances, raw milk is used. Soft cheese made from raw milk is a major potential source of *Brucella* or *Salmonella* infection because the pH is not low enough to kill these pathogens.

    Cheese made from pasteurized milk ripens more slowly and may not have as strong a flavor as cheese made from raw milk. However, the quality of cheese made from pasteurized milk is consistently better than that of cheese made from raw milk because flavor defects produced by undesirable bacteria are reduced. If psychrotrophic bacteria are present in raw milk, off flavors, gassiness, and generally poor quality cheese can result; these may occur even if the milk is pasteurized because the bacteria may have produced destructive enzymes not destroyed by pasteurization temperatures.

    Milk used in cheesemaking may be contaminated from exposure to unclean equipment or infected personnel. Four organisms—*Salmonella* spp., *Listeria monocytogenes,* enteropathogenic *Escherichia coli,* and *Staphylococcus aureus*—present a serious potential hazard in cheese production. Risk of contamination by staphylococci of human origin is especially great in the cheese vat where the staphylococci can multiply in large numbers to produce enough enterotoxin to be a hazard. Once the enterotoxin has been produced, it cannot be removed because it is heat stable. Two points during cheese manufacture that are especially vulnerable to enterotoxin production are (1) holding the milk in vat pasteurizers at a temperature insufficient for pasteurization, but adequate for microbial growth, and (2) heating the curd to normal body temperature. Today, *St. aureus* is considered a lesser risk because growth and toxin production is suppressed by modern lactic cultures and acidity control in cheesemaking. All of these organisms can be eliminated from raw milk by pasteurization or heat treatment at 64.4°C (148°F) for 16 s, but the milk may become recontaminated during the cheesemaking process, with subsequent buildup of organisms or enterotoxins.

The lactoperoxidase system, which consists of treating milk with the enzyme lactoperoxidase, thiocyanate, and hydrogen peroxide, has broad antimicrobial activity. When used before heating raw milk, it creates a dramatic increase in destruction of *L. monocytogenes* and *St. aureus.*

Soft cheeses such as brie, Jalisco, or queso blanco do not undergo the ripening processes that hard cheeses do, and the pH is not as low, initially. If made from raw milk, they should be heat treated to avoid the hazard of foodborne disease, including brucellosis, listeriosis, and staphylococcal intoxication.

In the EU, cheeses made from ewe and goat milk are becoming popular. Because these cheeses are more expensive than cheeses made from cows' milk, there is a concern that adulteration of these products by the addition of cows' milk may occur. A reference test has been developed to identify any adulteration of this sort.

2. **Nonfat dried milk (NFDM).** The milk that remains after the cream has been separated is used to manufacture NFDM. To make 1 kg NFDM, 11.2 kg (25.5 lb) skim milk is required. NFDM contains 94–96% solids, 4–5% moisture, and 1.25–1.5% fat.

If the skim milk used to produce NFDM is of poor quality, the NFDM also will be of poor quality. Any enterotoxin present before the milk is dried will maintain its toxicity although the organisms are killed. An extensive outbreak of food poisoning has been associated with NFDM that had been condensed in vacuum pans maintained at a temperature that encouraged staphylococcal growth before drying. The milk should be pasteurized before processing.

3. **Ice cream.** Ice cream can be a source of foodborne infection or intoxication, primarily from use of raw or inadequately pasteurized ingredients. Outbreaks of salmonellosis have been traced to raw eggs used in the manufacture of ice cream.

Egg Processing[128,254]

After eggs are collected, they are washed in water that is maintained at a minimum temperature of 37°C (90°F) and then rinsed in slightly hotter water. The wash water contains a detergent and a sanitizer. As the wash water becomes dirty, the alkaline detergents approach a neutral pH. Unless the detergent is maintained at pH 10.0 or higher, the eggs may have a higher surface bacterial load after washing than before. Following washing, eggs are sprayed with a mist of high-grade mineral oil, candled, and sized. Candling is done by passing eggs over a bright light while being mechanically rotated on their long axis so that the albumen, yolk, air space, and shell can be checked. Oiling seals the pores in the shell and reduces loss of moisture and carbon dioxide, thereby preventing a rise in pH and ensuring a better quality egg. Carbon dioxide loss leads to thin watery whites and flat enlarged yolks. Washing and oiling of eggs for human consumption is not permitted in the United Kingdom. Eggs are usually kept for the first few days at 15.5°C (60°F) and 70% relative humidity. A colder temperature would be more effective in restricting the growth of microorganisms, but the slight advantage is not cost-effective when the added cost of refrigeration is considered. Also, maintaining eggs at the higher temperature reduces the problem of condensation ("sweating"), which occurs when eggs are removed from the cooler. In the United States, these "shell" eggs are sold on the basis of quality grades (AA, A, B) and weight, ranging from "Peewees," with a minimum weight of 206 mg (15 oz) per dozen, to "Jumboes," with a minimum weight of 412 mg (30 oz) per dozen.

In addition to being sold as shell eggs, eggs may be sold as broken-out egg products. These include liquid, dried, or frozen whole eggs, yolks, or whites. Egg products are more likely to have microbial contamination than are shell eggs. Although egg breaking previously was performed by hand, this is now done by mechanical equipment. To minimize contamination during egg breaking suitable precautions are required: (1) only edible quality eggs should be used, (2) the eggs should be candled to remove undesirables, (3) eggs should be spray washed and sanitized before breaking, (4) there should be separate washing and breaking rooms in order to reduce the probability of contamination via aerosols, and (5) there should be a separate draw-off room with filtered positive-pressure air flow. (Draw-off rooms are where liquid egg products are packaged and frozen.)

Whole eggs and yolks, before freezing, are usually treated with additives (sodium chloride or sucrose at a concentration of 10% are the most common) to prevent gelation of the yolk. If albumen is to be dried, it is desugarized; i.e., it has the glucose removed to prevent off flavors, color changes, and loss of baking quality. Desugarization is accomplished by fermentation of the glucose by lactobacilli. Yolk-containing egg products are desugarized before drying by the use of a glucose oxidase-catalase enzyme mixture obtained from fungi or by the use of yeast fermentation.

Internal contamination of duck eggs by salmonellae has always been recognized as a problem, coming primarily from the environment. It was not until the increased utilization of egg products during World War II that the same problem was recognized in chicken eggs. The source in this case was contaminated feed. In the United States, the Egg Products Inspection Act of 1970, designed to cope with this problem, required all whole eggs for use in egg products to be pasteurized at 60°C (140°F) for 3.5 min. In the United Kingdom, eggs are pasteurized at 64.5°C (148°F) for 2.5 min. Higher temperatures are required for separated yolks and for products having added salt or sugar. Pasteurization is accomplished in the same type of equipment used for milk, usually an HTST unit.

Drying of eggs was originally done by heating in pans. Later, spray drying and freeze drying techniques were developed. Today, spray drying is the most common method and may be used for all products. Pan drying is primarily utilized for albumen products, and freeze-drying is used for whole egg products.

Aquatic Animal Processing[42]

In general, the quality and safety of fish and shellfish are determined by two factors: (1) the quality of the environment they came from before harvest and (2) the standards of sanitation during harvest, handling, storage, and processing. Fish are extremely perishable. As soon as they are caught or harvested, a series of natural events begins that eventually results in spoilage. Careful handling after harvest can slow the process so that a fresh, wholesome, and safe product reaches the consumer.

The goals of processing are to reduce the temperature of fish and to keep the level of microbial contamination to a minimum. In Canada, the Department of Fisheries and Oceans has developed a series of guidelines for processing fish to ensure optimum quality. The guidelines cover bleeding, gutting, washing, icing, and storage. Not all steps are appropriate for all species harvested, but they provide guidelines for understanding the basics of proper seafood processing.

Bleeding.   Bleeding, if done, is initiated within a very short time after harvest of the live fish. Bleeding of white fish results in a fish of superior appearance, that is less easily bruised, and that has a longer frozen storage life than fish that are not bled. Bleeding of finfish is accomplished by slitting arteries behind the gills or cutting off the tail so that the caudal artery is severed.

Gutting.   Gutting is done to remove digestive enzymes and bacteria that initiate spoilage in freshly caught fish. Digestive enzymes in the gastrointestinal tract cause rapid autolysis and invasion of the flesh by spoilage bacteria. Gutting may also be helpful to prevent the migration of certain parasites from the gut into the flesh of fish. Some of these parasites, such as the anisakid nematodes, are pathogenic to humans.

Washing.   After bleeding and gutting, fish are washed with clean cold water to remove surface slime and spoilage bacteria. Washing is often done whether or not fish have been bled and gutted. To be effective, there must be a sanitary supply of washing water. Seawater is satisfactory for this provided it is not polluted.

Icing.   Temperature control is the most important element in the preservation and processing of fish. The sooner the temperature is reduced to a holding temperature of 0°–4°C (32°–39°F), the longer the quality and wholesomeness of the fish can be maintained. Ice is often used, or in colder climates chilled or refrigerated seawater may be used. An additional benefit of seawater as a chilling medium is that, because of its higher specific gravity, it provides some buoyancy to the fish, making them less likely to be damaged by crushing. In tropical climates fish spoil so quickly that, unless chilled immediately, they will be unfit for market within a few hours of death. Chilling or icing is even more effective in retarding microbial spoilage in fish from tropical waters than in cold water fish because the microbial flora that contaminate these fish are not cold adapted.

Storage.   Fish are best stored in shallow trays or boxes to prevent crushing. Although less efficient in terms of space utilization than bulk storage (as in the hold of a fishing vessel), several processing companies have adopted box storage techniques, even on the ship where space is at a premium.

Finfish Processing
Farm-raised Fish.   A practical advantage of farm-raised fish is the greater control provided at all levels of production, including harvesting and processing. In theory, this increased control should reduce contamination at all stages of processing and result in a more consistently clean, safe, and fresh food product than is possible from the capture fishery. Under these conditions there should be little difference between a fish-processing plant and any other kind of meat-processing plant. A modern catfish-processing plant, for example, closely resembles a poultry-processing plant. After the fish have been harvested, they are quickly transported to the plant and unloaded into large vats where they are kept alive until processing. Enough vats are available that the harvest from individual farmers can be kept separate for weighing and inspection. Immediately before processing the fish are stunned by an electric current. As each fish is processed, it passes from station to station where trained personnel remove fins and heads and eviscerate, skin, and apportion the meat into selected cuts and then wrap. Immediately after wrapping, the fish are frozen.

Wild (Capture) Fisheries.   Processing of wild caught ocean fish is extremely variable. The species caught, condition of the fish at harvest, size of the harvest, the distance of the fishery from shore, and local market conditions influence how fish are handled. Even the method of harvest will have an impact on product quality and subsequent processing. Trawling with a net on the sea bottom stirs up sediments and may increase the concentration of skin surface bacteria by a factor of 100 or more. Fish landed in a large net are subjected to the weight of other fish above them, and expulsion of intestinal contents may cause further surface contamination. Fish hauled up rapidly from great depths may also expel intestinal contents because of the rapid expansion of the swim bladder. With gill nets and some other commercial harvest methods fish are killed in the water and may remain there for long periods. In warm water, spoilage may have begun before they have been removed from the sea. In some inshore fisheries, fish are stored in ice on the ship and transported to a central processing facility where the processing is similar to that in a catfish-processing plant. In some deep water fisheries, fishing vessels may be out of port for up to 90 d, and the whole operation of processing and freezing is accomplished at sea.

Crustacean and Mollusk Processing

Crustaceans.   Crustaceans are extremely prone to rapid postmortem deterioration. Because of this, lobsters, crabs, and crayfish are iced and shipped alive to restaurants and retail outlets. Because of the external skeleton, which must be removed for preparing canned or frozen fillet meat, a great deal of handling and cleaning is necessary in additional processing plants. This not only increases the price of the finished products but also increases the opportunity for contamination.

Mollusks.   As is true of crustaceans, most mollusks (such as clams and oysters) are iced and shipped directly to restaurants and retail outlets. Consumption of mollusks carries with it an added risk because, in addition to the extreme amount of handling required during processing, a large percentage are consumed raw.

Deterioration of Food
Types of Microorganisms[27,67,79,188,244]

Bacteria that cause food to spoil are classified in three groups on the basis of temperatures for optimal replication: **psychrotrophic**, **mesophilic**, and **thermophilic**. Microbiologists are fond of creating tables with maximum, minimum, and optimal temperatures for each of the groups, but the proposed temperature ranges vary. The important thing to recognize is that, as regards bacteria in food, those that do well at low temperatures (refrigeration) are **psychrotrophes**, those that do well in the intermediate temperature range, and include most of the pathogens, are **mesophiles**, and those that prefer high temperatures are **thermophiles**. Two other terms related to temperature and bacterial survival often create confusion. *Psychrophiles* are psychrotrophes capable of replication at extremely low temperatures (0°–10°C [32°–45°F]). The term *thermoduric* is used to describe organisms that can "endure" high temperatures but do not necessarily replicate at these temperatures. The term is misleading because some "thermodurics" can grow below 7°C (44°F) and, therefore, are actually psychrothrophic.

Psychrotrophic bacteria, because of their ability to multiply at low temperatures, are primarily spoilage bacteria, such as *Alteromonas (Pseudomonas) putrefaciens,* with optimal replication between −1° and 20°C (30° and 68°F) and capable of relatively rapid growth at refrigeration temperatures between 2° and 10°C (35° and 50°F). The mesophilic group contains pathogens, such as *Staphylococcus aureus,* with optimal growth between 20° and 45°C (68° and 113°F). Thermophilic bacteria are primarily spoilage bacteria, such as *Lactobacillus thermophilus,* with optimal growth above 45°C (113°F). Some species can grow at 55°C (131°F) or higher.

Not all pathogens are mesophiles. *Listeria monocytogenes* and *Yersinia enterocolitica,* two recently "emergent" foodborne pathogens, are psychrotrophic. *Li. monocytogenes,* a ubiquitous organism that presents a hazard primarily to immunosuppressed individuals, multiplies slowly at 2°C (34°F), and *Y. enterocolitica* in milk can survive −20°C (−3°F) for 30 d. Both organisms are destroyed by pasteurization temperatures but have created problems in milk and milk products, indicating postpasteurization contamination followed by replication during refrigerated storage.

Deterioration of Meat[137,150,209,240,317]

Meat Quality.   The term *meat quality* refers to the physical or chemical properties that relate to its processing and palatability characteristics. Five factors are of primary importance in determining the quality of any meat product: water-holding capacity, color, texture, tenderness, and marbling.

Water-holding Capacity.   The natural moisture content of muscle is approximately 68–78%. There are inherent differences in the water-holding capacity of meats, but generally these are less important than moisture loss during processing and cooking. Excessive moisture lost in processing or through improper cooking will result in a less tender product that is perceived to be poorer in quality.

Color.   Color can influence the consumer's psychologic perception of quality of a meat product more than actual palatability. Appealing color secondarily influences the retail price of various cuts. There is, of course, a relationship between color and quality; it has been well established that, as animals age, the meat becomes darker.

Texture.   Texture of the meat product is determined by feeling and handling the tissue in the uncooked state. Coarse-textured meat is regarded as less tender. Texture is determined by the size of individual muscle fibers and the amount of connective tissue present.

Tenderness.   Two structural components of meat have been shown to determine tenderness: the collagen of connective tissue and the contractile apparatus of myofibrillar proteins. The pH of the meat and postmortem muscle shortening are primary causes of variation in tenderness. Meat with a high pH may undergo more proteolytic enzyme activity than meat with a more acid pH. Postmortem electrical stimulation of meat has been utilized to retard the extent of muscle shortening.

Marbling.   Marbling (the intramuscular fat) is an important palatability characteristic of meat and is a factor used in quality grading of meat. This factor, in addition to moisture-holding ability, has an important influence on flavor among different species. Carbonyl compounds, found in the fat-soluble part of meat, are the major contributors to flavor. The present consumer trend is for leaner, less-fatty meat. This is a result of increased health consciousness among consumers. Epidemiologic evidence of a causal association between increased cholesterol in the diet and the prevalence of cardiovascular diseases has made marbling a disadvantageous criterion for judging meat quality from the public health point of view.

Causes of Meat Deterioration.[219]   When efforts to preserve a meat product fail, it deteriorates from one or more of three causes: microbiologic, enzymatic, or oxidative.

Microbial Changes.   Microbial contamination resulting in deterioration of a meat product usually occurs as the product is handled in the plant. Before death, normal healthy tissue is free of microorganisms. The growth of contaminating organisms present on the carcass depends on environmental conditions such as the holding temperature, relative humidity, and pH of the product, as well as the inherent characteristics of the meat itself. Under different conditions, we may see seven different microbial changes in meat: acid production, gas production, slime formation, mold growth, bacterial greening, the formation of green rings, or the development of green cores.

Acid production by certain microorganisms often is desirable as a flavor enhancer. In excessive amounts, acids are undesirable.

Many microorganisms produce gas. This may be manifested as sausage casings that burst, meat with a spongy texture, or canned products with swollen lids. **Gas production** is associated most often with undercooked products or meat products that were not held at a temperature sufficient to destroy the microorganisms. Gas production indicates the possible risk of serious intoxication if the product is ingested.

Slime formation is the result of mass accumulation of microbes on the meat surface. Lactobacilli, micrococci, and yeast most commonly are responsible for this deteriorative change. Contamination of the product results from exposure to personnel or contaminated equipment. These microorganisms grow well at reduced temperatures. Good sanitation is the best means of controlling the problem.

Vacuum packaging, which renders the product devoid of oxygen, is a very effective method of combating deterioration by **mold growth**. The growth occasionally seen on the surface of products such as dried sausage is not caused by mold but rather is the result of micrococci or yeast that can survive at a lower humidity.

Greening of sausages and other cured meats is another microbial change that occurs occasionally in the production of meat. **Bacterial greening** results from surface contamination after processing. The microorganisms that cause this change grow readily under aerobic conditions and produce hydrogen peroxide. It is this metabolite that causes greening of the meat. This problem, like other changes that result from microbial contamination, can be prevented by strict hygiene and holding the final product at cold temperatures.

Green rings resembles bacterial greening but differs in that the rings are associated with a heavy population of bacteria in the sausage or other meat before cooking and processing. Strict hygienic measures must be used in processing to prevent the problem.

Another microbial change in processed meat products, **green cores**, depends on several events occurring in a series. First, the sausage emulsion must be heavily contaminated with bacteria. Subsequently, the processing temperature must be insufficient to kill the microorganisms in the center of the product. Finally, the finished product must be held at a temperature that enables growth of these organisms. This problem is prevented by increasing the processing temperature to a minimum of 68.3°C (155°F) as well as by employing good sanitation.

Enzymatic Changes. Two types of enzymes cause changes in meat products: proteinases and lipases. These enzymatic changes are not always undesirable (e.g., enzymes are often used in meat tenderization). If the process is excessive, it is undesirable.

Plants are the source of the most common exogenous **proteinases**, which are used for tenderizing meat: papain, bromelin, and ficin. Fungal proteolytic enzymes, although used in tenderizing, only give a superficial tenderness to a meat product; in other words, they tend to have less penetrability. Inherent proteolytic enzymes, the **cathepsins**, are very effective in tenderizing meat. An alternative method for tenderizing meat has been patented, but it is illegal in some countries. In this procedure, just before stunning, selected animals are injected intravenously with a tenderizing enzyme. The animal must be killed no earlier than 2 min or later than 30 min after this injection. Occasionally, an animal will have an allergic-type response to the injection, in which case it must be held for observation for 24 h and then, if normal, slaughtered in the usual way. Some people are allergic to this enzyme.

Oxidative rancidity is another, less desirable, enzymatic change that results from **lipases** acting on the fat producing "free" fatty acids causing a bitter-rancid flavor (e.g., butyric acid). Enzymatic reactions are temperature- and pH-dependent. Each enzyme has an optimal range of activity.

Oxidative Changes. Oxidation is a third process that may result in deterioration of meat products. Fat is especially susceptible to this change. Odor, flavor, and color alterations in a product result from its exposure to oxygen. Oxygen absorption in the product is accelerated by light and heat. Unsaturated fats react with oxygen to form double-bonded peroxides. In the presence of light or heat and oxygen, the reaction may be self-perpetuated until the product is fully oxidized. It follows that, if absorption of oxygen can be prevented, meat products can be prevented from deteriorating by oxidation. First, antioxidants or inert gases can be added to retard or prevent oxygen absorption. Second, storing the product away from heat and out of the light is effective. Third, using a wrapper that prevents product exposure to air is helpful. Wrappers that prevent freezer burn do not necessarily prevent oxidation. Therefore, both considerations must be kept in mind when preserving products.

Deterioration of Milk
Microbial Contamination of Raw Milk. The source of microbial contamination of milk may be within the udder, on the surface of the cow or milker's hands, or in the environment. Spoilage organisms, especially psychrotrophic bacteria, are the main cause of high bacterial counts in raw milk.

Farm Equipment. Improperly cleaned milking equipment or milk storage equipment is the usual source of spoilage organisms. If storage equipment is not properly sealed, milk may be contaminated from aerosols created during feeding, sweeping, or other activities. Adequate daily cleaning and sanitizing of equipment usually prevents high bacterial counts resulting from contaminated equipment.

Mastitis. A healthy lactating udder is not a major source of bacterial contamination. However, the udder may be a source of bacterial pathogens if there is a disease process present. Furthermore, the composition of milk from a mastitic udder is altered. Fat, casein, and lactose decrease, whereas serum proteins and minerals increase.

Hygiene. Clipping the flanks of cows significantly reduces contamination from hair and adhering debris. Careful washing of the teats with a warm sanitizer and drying before each milking reduces the bacterial count.

Storage Temperature. Storage of milk at a temperature of 4.4°C (40°F) or less is a major factor in retarding bacterial multiplication. The storage life of milk from healthy udders, produced under sanitary conditions, is significantly longer at lower temperatures. If milk is heavily contaminated, shelf life is shortened, even at low temperatures, because many of the contaminants are psychrotrophic bacteria that are capable of rapid multiplication at refrigeration temperatures.

Bacteria in Pasteurized Dairy Products. Bacteria in pasteurized dairy products may be from one or more of four sources: raw milk, plant personnel, processing equipment, and the environment within the plant. Thermophilic bacteria present in raw milk can survive pasteurization, and some psychrotrophic bacteria can also survive pasteurization if plentiful and in clumps. It may be impossible to sanitize equipment adequately as a result of flaws in design or deterioration from prolonged use.

Personnel. Dairy-processing plant personnel, either directly or indirectly, are responsible for most contamination present in finished dairy products. Employees may be the source of human pathogens because of poor personal hygiene. Of equal importance are employee work habits that determine how effectively equipment is maintained and sanitized, the cleanliness of the plant environment, and the effort put forth to prevent contamination from any environmental source.

Control. Whenever contamination occurs in a product, corrective measures can be instituted once its source has been identified. The type of bacterial contamination may be an important clue, if the organism is known to have any animal (including human) reservoir or to reside primarily in some other habitat. Systematic sampling at each critical control point in processing and handling, retracing every step back to the cow if necessary, will reveal the source.

Rancidity.[25,26,30,70,76,80,87,223,312] Rancidity is a flavor defect in milk caused by the enzymatic hydrolysis of milk lipids to free fatty acids (FFA). The enzymes are from two sources: **endogenous** enzymes, which are present in normal milk, and those of **microbial** origin.

Endogenous Enzymes. Milk of all mammals contains a lipoprotein lipase (LPL) that is synthesized in the secretory cells of the mammary gland. Cows and pigs have the highest levels. Most of the LPL is in the fat-free components of milk. LPL does not normally create rancidity because the fat present in milk is in globules protected by a membrane. Disruption of this membrane will expose the lipids to enzymatic hydrolysis resulting in the creation of FFA and imparting a rancid flavor.

Microbial Enzymes. The microbial lipases are produced by psychrotrophic bacteria, primarily the pseudomonads. One of the most important properties of these lipases is their heat stability. Many are capable of retaining some activity after high-temperature, short time (HTST) or ultra-heat-treated (UHT) treatment and may cause rancidity in stored dairy products.

Causes of Rancidity. FFA may be produced as the result of "spontaneous" or "induced" lipolysis. **Induced** lipolysis results when the milk lipase system is activated by physical or chemical means. **Spontaneous** lipolysis occurs in milk that has had no treatment other than cooling soon after milking. Microbial contamination can also contribute to hydrolytic rancidity, but most of the rancidity in milk and cream is from the action of milk lipase. Microbial lipases are of greater significance in stored milk products.

1. **Induced lipolysis**. Induced lipolysis can be precipitated by procedures that produce agitation of the milk, thereby disrupting the fat globule membrane. The agitation can be caused by faulty milking equipment admitting excessive air and producing turbulence or by homogenization of raw milk. (Lipolysis will also occur if homogenized pasteurized milk is mixed with raw, nonhomogenized milk.) Besides disrupting the fat globule membrane, agitation also hastens rancidity because it redistributes the lipase from the fat-free to the cream portion of the milk.

   Lipolysis may also be induced when fresh milk is subjected to a specific sequence of temperature changes, specifically, cooling to 5°C (40°F) or less, then warming to 25°–35°C (77°–95°F), followed by recooling to less than 10°C (50°F). This can occur if a large amount of warm milk is added to a small amount of cooled milk and then refrigerated.

2. **Spontaneous lipolysis**. Spontaneous lipolysis has been associated most frequently with milk from cows in late lactation or with a poor level of nutrition. If there is not enough protein available, the fat globule membrane is less stable, and lipase activity is more apt to occur. Milk from low-producing cows is more susceptible to lipolysis than milk from high producers.

3. **Microbial lipolysis**. Hydrolytic rancidity may result from mastitis or from contamination during processing. The tendency to rancidity increases with increasing somatic cell count (SCC). The leucocytes in milk contain a lipase that may be responsible for some of the rancidity in mastitic milk. Storage of milk in bulk tanks on the farm and at the factory may extend to several days and has led to the emergence of psychrotrophic bacteria as the dominant organisms in raw milk. Postpasteurization contamination is a common cause of lipolysis in milk products.

   The acid degree value (ADV) is used as a measure of rancidity. Normal milk has an ADV of 0.4. An increase of ADV to 1.2 is associated with a slightly rancid flavor and an ADV greater than 1.5 with a pronounced rancid flavor.

Deterioration of Eggs[83]
Egg Quality.   Interior egg quality is mostly determined by the length of time

and other conditions of storage since the egg was laid.

Albumen. The normal newly laid egg has a large amount of thick albumen, rich in mucin. Some viral infections, chemicals, and the increasing age of the hen will decrease mucin content in the egg, causing the albumen to gradually become thinner. In a newly laid egg, the yolk stands up with the thick albumen closely surrounding it. As the egg ages, the thick albumen deteriorates progressively until, finally, there is no discernible thick albumen. The **Haugh unit** is a means of expressing the quality of an egg's albumen; the height of the albumen is measured with a special instrument and integrated with the weight of the egg. The higher the albumen, the greater the number of Haugh units. Tables, special slide rules, or direct reading dials on micrometers are used to obtain Haugh unit values.

Yolk. Like the albumen, the yolk becomes less firm with time after the egg is laid. It absorbs water from the albumen, which stretches and weakens the surrounding vitelline membrane. The yolk may also have various abnormalities such as mottling, caused by feed ingredients, or blood spots from aberrant vessels crossing the stigma at the time of ovulation. Blood spots may also be in the albumen. There may occasionally be yolkless eggs when the oviduct is stimulated to form an egg by some material that enters it, such as a bit of ovarian tissue.

Egg quality is routinely monitored by "flash candling," a procedure wherein eggs are passed over a bright source of light while being mechanically rotated along their long axes, to measure the air space at the large end of the eggs and to check the shells for cracks. In an older egg with thin albumen, the yolk will be seen to move close to the shell when the egg is spun during candling.

Spoilage. The main cause of spoilage is microbial decomposition of the egg.

The major spoilage bacteria are species of *Pseudomonas, Proteus,* and *Alcaligenes.* Some bacteria grow very rapidly in broken-out egg products if the temperature in the products rises to 15.6°C (60°F) for several hours. Iron compounds in wash water may foster *Pseudomonas* growth. Molds are less of a problem than bacteria if shell eggs are stored properly.

The egg is protected from bacterial invasion, to some extent, by its shell and the underlying shell membranes. The shell, however, contains pores that are filled with a proteinlike substance that can be digested by the enzymes present in some bacteria. There are also present, within the albumen, inhibitors to bacterial growth such as lysozyme, which lyses the cell walls of gram-positive bacteria, and other compounds that alter the environment of the egg so as to interfere with microbial growth. The pH of newly laid eggs is 7.6–7.9. As carbon dioxide is lost in storage, the pH may reach 9.7, which will tend to prevent the growth of spoilage bacteria.

A current concern related to egg quality is the increase in reported human cases of foodborne disease caused by *Salmonella enteritidis.* During 1986–1990, there were 9,584 cases reported in the United States, 45 resulting in death. Of the 68 foodborne disease outbreaks reported in 1990, in which the causative organism was identified as *S. enteritidis,* 18 (26%) were traced to consumption of eggs or egg products. Recontamination of rendered feed was the primary source of infection for the laying hens, which then transmitted the organism vertically into eggs. Control of this problem requires that inedible eggs, destined for animal feed, must be denatured to prevent human consumption and then pasteurized to prevent spread of *S. enteritidis* in poultry. Flocks that have been epidemiologically linked to human outbreaks should be tested, and if found infected, the birds must be sent to slaughter and the eggs to a breaker. Liquid whole eggs must be heated to at least 60°C (140°F) for no less than 3 min. If sugar or salt is added, the temperature is raised to 63.3°C (146°F) for 3.5 min to kill *S. enteritidis.* Because the egg white has a higher pH and fewer ingredients to protect the bacterial cells, pasteurization is effective at 56.7°C (134°F). Proper washing destroys all *S. enteritidis* organisms on the exterior of the shell. A heat treatment at 54.4°C (130°F) for 15 min has been effective in further protecting shell eggs from bacteria on the eggs' surfaces. This procedure stabilizes the albumen so that thinning is slowed and pasteurizes the shell, killing many of the spoilage organisms.

Deterioration of Fish and Shellfish[246,314]

Deterioration of fish and shellfish is time- and temperature-dependent. There are three main types of deterioration: autolytic spoilage, bacterial spoilage, and rancidity or oxidative spoilage.

Autolytic Spoilage.   Autolytic spoilage begins immediately after the death of the fish and is extremely rapid. It is caused by enzymes in the gut and muscles of fish. Digestive enzymes are particularly active in the guts of fish that have been feeding heavily immediately prior to harvest. The action of gut enzymes results in autolysis of internal organs and the abdominal wall of fish and causes rapid invasion of the flesh by spoilage bacteria. This produces a visible lesion called "bellyburn" in cod or "bellyburst" in capelin and herring.

Bacterial Spoilage.   Rapid spoilage of fish results from activity of the bacteria from the gills (surface slime) and in the intestinal tract of normal, healthy fish. The surface bacteria of cold water fish are mainly psychrotrophic. These organisms survive at 0°C (32°F) and multiply rapidly at temperatures normally used for holding meat and dairy products. Most of the undesirable organoleptic changes that occur in fish are the result of bacterial growth.

Much regulatory emphasis has been placed on bacteria of public health significance in the marine environment. These are a problem primarily of polluted waters close to shore. Most fish are harvested from unpolluted waters and harbor few pathogenic bacteria. Contamination with organisms such as salmonellae and staphylococci is generally the result of unsanitary conditions during handling and processing. Certain vibrio species pathogenic to humans are present in the unpolluted marine environment and in healthy fish, as is *Clostridium botulinum* type E, which is present in small numbers in bottom sediments. Growth of clostridia and the consequent toxin production in fish is almost always due to errors in processing and handling. Foodborne illness caused by pathogens of marine origin is usually associated with the consumption of raw shellfish or uncooked, smoked, fermented, or salted fishery products.

A direct human health hazard related to bacterial spoilage and improper storage temperature is the development of dangerously high levels of histamine in the flesh of certain scombroid fish (e.g., tuna, bonito, and mackerel) and some nonscombroid fish (e.g., mahimahi, sardines, anchovies, and herring). The histamine is produced by the action of bacterial decarboxylation of the amino acid histidine found in these fish.

Shellfish (crustaceans and mollusks) contain far greater amounts of free amino acids than finfish do. This facilitates bacterial growth and spoilage. Since shrimp die very soon after catching, deteriorative changes take place earlier than in crustaceans, which can be kept alive. At first, the spoilage of mollusks is primarily caused by the presence of *Pseudomonas, Proteus, Clostridium, Aerobacter,* and *Escherichia* and, later, streptococci, lactobacilli, and yeasts.

Rancidity. The development of rancidity in fish is the result of oxidation of the oils present in the tissues. Icing, freezing, and salt curing will retard spoilage during storage, but the oil present in the flesh will eventually undergo deterioration to some extent. In herring stored at 2°–4°C (35.6°–39.2°F) the acid value (an objective measure of rancidity) doubles in 2 d and quadruples in 6 d. Oily fish such as herring, mackerel, and salmon are especially prone to rancidity. Rancidity imparts characteristic "fishy" flavors and odors to seafoods.

Bibliography

1. 3-A Sanitary Standards Committee. Periodic updates. *3-A Sanitary Standards.* Des Moines, Iowa: International Association of Milk, Food, and Environmental Sanitarians.

2. Ackman, R.G. 1989. Appendix 1, the fish list. *Prog. Food Nutr. Sci.* 13:243–289.

3. Adam, C.L. 1985. Recent developments for low ground red deer farming. *Ann. Rep. Stud. Anim. Nutr. Allied Sci.* 40:34–50.

4. Ahmed, F.E. (ed.). 1991. *Seafood Safety, a Report of the Committee on the Evaluation of the Safety of Fishery Products.* Washington, D.C.: National Academy Press.

5. Akam, D.N., F.H. Dodd, and A.J. Quick. 1989. *Milking, Milk Production Hygiene, and Udder Health.* FAO Anim. Prod. Health Pap. 78. Rome: Food and Agriculture Organization.

6. Alexander, T.L. 1988. Slaughter of deer. *Curr. Top. Vet. Med. Anim. Sci.* 48:79–84.

7. Alexander, W.H. 1973. Trends and projections in milk cow numbers and average milk production per cow, United States and Louisiana, 1925–1972. *La. Rural Econ.* 35:10–13.

8. Allen, R.J.L. 1981. Joint FAO/WHO food standards program Codex Alimentarius Commission Fourteenth Session, Geneva, 29 June–10 July 1981. Nutrition and the work of the Codex Alimentarius Commission. *Food Nutr.* 38:131–141.

9. Alonso, A., B. Etxaniz, and M.D. Martinez. 1992. The determination of nitrate in cured meat products. A comparison of the HPLC UV/VIS and Cd/spectrophotometric methods. *Food Addit. Contam.* 9:111–117.

10. Alston, J.M., and G.M. Scobie. 1987. A differentiated goods model of the effects of European policies in international poultry markets. *South. J. Agric. Econ.* 19:59–68.

11. Alston, J.M., C.A. Carter, and L.S. Jarvis. 1990. Discriminatory trade: The case of Japanese beef and wheat imports. *Can. J. Agric. Econ.* 38:197–214.

12. American Veterinary Medical Association Milk Quality Committee. 1992. Recommended minimal standards of performance for practicing veterinarians who offer milk quality control programs. *1993 AVMA Dir.* 42:71–73.

13. Anderson, K. 1983. The peculiar rationality of beef import quotas in Japan. *Am. J. Agric. Econ.* 65:108–112.

14. Anderson, K.L. (ed.). 1993. *Update on Bovine Mastitis.* Vet. Clin. North Am. Food Anim. Pract. Vol. 9, no. 3. Philadelphia: W.B. Saunders.

15. Anonymous. 1990. Global harmonization vs. democratic traditions. *Nutr. Week* 20(May 10):4–5.

16. ———. 1991. Codex Commission foils U.S. effort to open markets to beef with hormones. *Nutr. Week* 21(27):1–2.

17. Arbuckle, W.S. 1986. *Ice Cream.* 4th ed. Westport, Conn.: AVI Publishing.

18. Arbuthnott, J.P. (ed.). 1987. *Animal Health: The Control of Infection.* Dublin: Royal Irish Academy.

19. Arispe, I., and D. Westhoff. 1984. Venezuelan white cheese: Composition and quality. *J. Food Prot.* 47:27–35.

20. Ashton, W.L.G. 1984. The risks and problems associated with the import and export of domestic poultry. *Br. Vet. J.* 140:221–228.

21. Asperger, H., and E. Brandl. 1982. The significance of coliforms as indicator organisms in various types of cheese. *Antonie van Leeuwenhoek* 48:635–639.

22. Austic, R.E., and M.C. Nesheim. 1990. *Poultry Production.* 13th ed. Philadelphia: Lea and Febiger.

23. Ayres, J.C., J.O. Mundt, and W.E. Sandine. 1980. *Microbiology of Foods.* San Francisco: W.H. Freeman.

24. Bachmann, M.R. (chrmn.). 1990. *Handbook on Milk Collection in Warm Developing Countries.* IDF Special Issue 9002. Brussels: International Dairy Federation.

25. Baker, L.D. 1990. Investigating the cause of chronic milk rancidity in a dairy herd. *Vet. Med.* 85:901–902, 904–905.

26. Bandler, D.K. 1982. Rancidity: An increasingly common milk flavor problem. *Dairy Food Sanit.* 2:312–315.

27. Banwart, G.J. 1989. *Basic Food Microbiology.* New York: Van Nostrand Reinhold.

28. Barbano, D.M. 1989. Impact of mastitis on dairy product quality and yield: Research update. *Annu. Meet. Natl. Mastitis Counc.* 28:44–48.

29. Barbano, D.M., R.R. Rasmussen, and J.M. Lynch. 1991. Influence of milk somatic cell count and milk age on cheese yield. *J. Dairy Sci.* 74:369–388.

30. Barnard, S.E. 1981. Causes of rancid flavor in retail milk samples. *Dairy Food Sanit.* 1:372–374.

31. Barton, R.A. (ed.). 1984. *A Century of Achievement: A Commemoration of the First 100 Years of the New Zealand Meat Industry.* Palmerston North, New Zealand: Dunmore Press.

32. Bartos, L., C. Franc, G. Albiston, et al. 1988. Prevention of dark-cuttng (DFD) beef in penned bulls at the abattoir. *Meat Sci.* 22:213–220.

33. Bath, D.L., F.N. Dickinson, H.A. Turner, et al. 1985. *Dairy Cattle: Principles,*

Practices, Problems, Profits. Philadelphia: Lea and Febiger.

34. Beaujeu-Garnier, J. 1978. *Geography of Populations.* 3d ed. London: Longmans, Green.

35. Becker, B.A., H.F. Mayes, G.L. Hahn, et al. 1989. Effect of fasting and transportation on various physiological parameters and meat quality of slaughter hogs. *J. Anim. Sci.* 67:334–341.

36. Bessey, M.E. 1979. Can the industry in the EEC meet consumers' requirements? *J. Soc. Dairy Technol.* 32:119–124.

37. Blair, R. 1990. International swine production—capabilities and possibilities. *Outlook Agric.* 19:263–268.

38. Blake, J.P., and J.O. Donald. 1992. Alternatives for the disposal of poultry carcasses. *Poult. Sci.* 71:1130–1135.

39. Bloomfield, G. (ed.). 1987. *Bovine Mastitis.* Surrey, United Kingdom: V and O Publications.

40. Bogue, D.J. 1969. *Principles of Demography.* New York: John Wiley and Sons.

41. Bond, E.J. 1984. *Manual of Fumigation for Insect Control.* Rome: Food and Agriculture Organization.

42. Bonnell, A.D. 1994. *Quality Assurance in Seafood Processing: A Practical Guide.* New York: Chapman and Hall.

43. Brake, J., and B.W. Sheldon. 1990. Effect of a quaternary ammonium sanitizer for hatching eggs on their contamination, permeability, water loss, and hatchability. *Poult. Sci.* 69:517–525.

44. Bramley, A.J., F.H. Dodd, and T.K. Quick. 1981. *Mastitis Control and Herd Management.* Tech. Bull. 4. Reading, England: National Institute for Research in Dairying.

45. Brown, L. (ed.). 1993. *Aquaculture for Veterinarians: Fish Husbandry and Medicine.* Oxford, England: Pergammon Press.

46. Bruce, H.L., and R.O. Ball. 1990. Postmortem interactions of muscle temperature, pH, and extension on beef quality. *J. Anim. Sci.* 68:4167–4175.

47. Burton, H., J. Pien, and G. Thieulin. 1969. *Milk Sterilization.* FAO Agric. Stud. 65. Rome: Food and Agriculture Organization.

48. Campbell Soup Co. 1975. *Manual of Warehouse Sanitation Guidelines and Sanitation Procedures.* Camden, N.J.: Technical Administration Department, Campbell Soup Co.

49. Chadee, D., and H. Mori. 1992. An assessment of the 1988 Japanese beef market access agreement on beef and feed-grain markets: Comment. *Rev. Agric. Econ.* 14:327–332.

50. Chambers, R.G., R.E. Just, L.J. Moffitt, et al. 1981. Estimating the impact of beef import restrictions in the U.S. import market. *Aust. J. Agric. Econ.* 25:123–133.

51. Chamings, R.J. 1984. Effect of different methods of udder preparation on the somatic cell count and bacterial count of herd bulk milk. *J. Soc. Dairy Technol.* 37:130–132.

52. Chang, S.Y., R.T. Toledo, and H.S. Lillard. 1989. Clarification and decontamination of poultry chiller water for recycling. *Poult. Sci.* 68:1100–1108.

53. Christian, L.L., and J.W. Mabry. 1989. Stress susceptibility of swine. In *Genetics of Swine,* ed. L.D. Young. Ames: Iowa State University Press, 49–68.

54. Cingi, M.I., C. Cingi, and E. Cingi. 1992. Influence of dietary nitrate on nitrite level of human saliva. *Bull. Environ. Contam. Toxicol.* 48:83–88.

55. Claassen, M., and H.T. Lawless. 1992. Comparison of descriptive terminology systems for sensory evaluation of fluid milk. *J. Food Sci.* 57:596–600, 621.

56. Clesceri, L.S. (ed.). 1989. *Standard Methods for the Examination of Water and Wastewater.* 17th ed. Washington, D.C.: American Public Health Association.

57. Cockram, M.S., and K.T.T. Corley. 1991. Effect of pre–slaughter handling on the

behaviour and blood composition of beef cattle. *Br. Vet. J.* 147:444–454.

58. Cockrill, W.R. (ed.). 1974. *The Husbandry and Health of the Domestic Buffalo.* Rome: Food and Agriculture Organization.

59. ——. 1979. *The Camelid, an All-purpose Animal.* Vol. 1. Uppsala, Sweden: Scandinavian Institute of African Studies.

60. Cole, H.H., and W.H. Garrett (eds.). 1980. *Animal Agriculture.* 2d ed. San Francisco: W.H. Freeman and Co.

61. Committee on Assessment of Technology and Opportunities for Marine Aquaculture in the United States. 1992. *Marine Aquaculture: Opportunities for Growth.* Washington, D.C.: National Academy Press.

62. Condon, L.W. 1993. NAFTA, GATT will strongly influence poultry markets. *Feedstuffs* 65:22–23.

63. Connell, J.J. 1980. *Control of fish quality.* Farnham, Surrey, England: Fishing News Books.

64. Conroy, A.M., and I.G. Gaigher. 1982. Venison, aquaculture and ostrich meat production: Action 2003. *S. Afr. J. Anim. Sci.* 12:219–233.

65. Cowen, P., L. Shugen, and T. McGinn. 1990. Survey of trichinosis in breeding and cull swine, using an enzyme-linked immunosorbent assay. *Am. J. Vet. Res.* 51:924–928.

66. Cragle, R.G., M.R. Murphy, S.W. Williams, et al. 1986. Effects of altering milk production and composition by feeding on multiple component milk pricing systems. *J. Dairy Sci.* 69:282–289.

67. Cunningham, F.E., and N.A. Cox. 1987. *The Microbiology of Poultry Meat Products.* Orlando, Fla.: Academic Press.

68. Dairy Equipment Testing Co. 1976. *Analyzing Milking Machine Performance with the DETCO Dual Vacuum Recorder.* Whittier, Calif.

69. Dawson, R.J. 1993. The role of the Codex Alimentarius Commission in food control, safety, quality and trade. *Proc. 11th Int. Symp. WAVFH,* 118–122.

70. Deeth, H.C., and C.H. Fitz-Gerald. 1983. Lipolytic enzymes and hydrolytic rancidity in milk and milk products. *Dev. Dairy Chem.* 2:195–235.

71. Defigueiredo, M.P., and D.F. Splittstoesser (eds.). 1976. *Food Microbiology: Public Health and Spoilage Aspects.* Westport, Conn.: AVI Publishing.

72. De Giusti, M., and E. De Vito. 1992. Inactivation of *Yersinia enterocolitica* by nitrite and nitrate in food. *Food Addit. Contam.* 9:405–408.

73. Desrosier, N.W. 1977. *The Technology of Food Preservation.* 4th ed. Westport, Conn.: AVI Publishing.

74. Diaz-Cinco, M.E., O. Fraijo, P. Grajeda, et al. 1992. Microbial and chemical analysis of Chihuahua cheese and relationship to histamine and tyramine. *J. Food Sci.* 57:355–356.

75. Dickson, J.S., and M.E. Anderson. 1991. Control of *Salmonella* on beef tissue surfaces in a model system by pre- and post-evisceration washing and sanitizing, with and without spray chilling. *J. Food Prot.* 54:514–518.

76. Duncan, S.E., G.L. Christen, and M.P. Penfield. 1991. Rancid flavor of milk: Relationship of acid degree value, free fatty acids, and sensory perception. *J. Food Sci.* 56:394–397.

77. Eckles, C.H., W.B. Combs, and H. Macy. 1951. *Milk and Milk Products.* 4th ed. New York: McGraw-Hill.

78. Eckner, K.F., and E.A. Zottola. 1991. The behavior of selected microorganisms during the manufacture of high moisture Jack cheeses from ultrafiltered milk. *J. Dairy Sci.* 74:2820–2830.

79. Emmons, D.B., D. Tulloch, and C.A. Ernstrom. 1990. Product-yield pricing system. 1. Technological considerations in multiple-component pricing of milk. *J. Dairy Sci.* 73:1712–1723.

80. Escobar, G.J., and R.L. Bradley, Jr. 1990. Effect of mechanical treatment on the free fatty acid content of raw milk. *J. Dairy Sci.* 73:2054–2060.

81. Fagerstone, K.A., and R.D. Curnow (eds.). 1989. *Vertebrate Pest Control and Management Materials.* Philadelphia: American Society for Testing and Materials.

82. Faull, W.B., J.W. Hughes, and W.R. Ward. 1991. *A Mastitis Handbook for the Dairy Practitioner.* Liverpool, England: Liverpool University Press.

83. Fields, M.L. 1979. *Fundamentals of Food Microbiology.* Westport, Conn.: AVI Publishing.

84. Fischer, P.L., P.J. Jooste, and J.C. Novello. 1987. The seasonal distribution of psychrophilic bacteria in Bloemfontein raw milk supplies. *South Afr. J. Dairy Sci.* 19:73–76.

85. Flynn, P. 1984. *Quality Milk Production (and Milking Machine Maintenance).* Dublin: Council for Development in Agriculture.

86. Forsythe, K.W., Jr., and M.E. Bredahl. 1991. Effects of animal health regulations on market access for exports of livestock products. *J. Agribusiness* 9:41–51.

87. Foster, E.M. (chrmn). 1975. *Prevention of Microbial and Parasitic Hazards Associated with Processed Foods. A Guide for the Food Processor.* Washington, D.C.: National Academy of Sciences.

88. Foster, E.M., F.E. Nelson, M.L. Speck, et al. 1983. *Dairy Microbiology.* Atascadero, Calif.: Ridgewood Publishing.

89. Fraser, F.M. 1981. *Historical Overview and Potential of Food Irradiation.* Ottawa: Atomic Energy Commission of Canada.

90. Gersovitz, M., and C.H. Paxson. 1992. Institutional and intertemporal influences on the trade of developing countries. *Am. Econ. Rev.* 82:180–185.

91. Gibson, J.P. 1989. The effect of pricing systems, economic weights, and population parameters on economic response to selection on milk components. *J. Dairy Sci.* 72:3314–3326.

92. Gilliland, S.E. 1989. Acidophilus milk products: A review of potential benefits to consumers. *J. Dairy Sci.* 72:2483–2494.

93. Gillis, A. 1987. Codex committee: Labeling for veg protein in meat. *J. Am. Oil Chem. Soc.* 64:166–171.

94. Gillis, K.G., C.D. White, S.M. Ulmer, et al. 1985. The prospects for export of primal beef cuts to California. *Can. J. Agric. Econ.* 33:171–194.

95. Glaeser, H. 1992. The single European market and the quality of dairy products. *Br. Food J.* 94:3–6.

96. Godkin, G.F. 1986. The reindeer industry in Canada. *Can. Vet. J.* 27:488–490.

97. Goldberg, J.J., J.W. Pankey, P.A. Drechsler, et al. 1991. An update survey of bulk tank milk quality in Vermont. *J. Food Prot.* 54:549–553.

98. Goldblith, S.A., M.A. Joslyn, and J.T.R. Nickerson. 1961. *Introduction to Thermal Processing of Foods.* Westport, Conn.: AVI Publishing.

99. Gonzalez, R.N., D.E. Jasper, T.B. Farver, et al. 1988. Prevalence of udder infections and mastitis in 50 California dairy herds. *J. Am. Vet. Med. Assoc.* 193:323–328.

100. Gorman, W.D., and H. Mori. 1988. Relative prices of imported grain fed beef in Japan: The case of air-freighted chilled carcasses. *Agribusiness* 4:535–548.

101. Gorman, W.D., H. Mori, and B.-H. Lin. 1990. Beef in Japan: The challenge for United States exports. *J. Food Distrib. Res.* 21:17–29.

102. Gormley, T.R. (ed.). 1990. *Chilled Foods, The State of the Art.* London: Elsevier Applied Science.

103. Gottemoller, C.A. 1983. Dairy farm practices and their effect on preliminary incubation counts. *Dairy Food Sanit.* 3:376–385.

104. Gould, B.W., and S.N. Kulshreshtha. 1985. An input-output analysis of the impacts of increased export demand for Saskatchewan products. *Can. J. Agric. Econ.* 33:127–149.

105. Graham-Rack, B., and R. Binstead. 1973. *Hygiene in Food Manufacturing and Handling.* 2d ed. London: Food Trade Press.

106. Gregor, H.F. 1963. *Environment and Economic Life.* Toronto: D. Van Nostrand.

107. Guthrie, R.K. (ed.). 1988. *Food Sanitation.* 3d ed. Westport, Conn.: AVI Publishing.

108. Guyton, A.C. 1990. *Textbook of Medical Physiology.* 8th ed. Philadelphia: W.B. Saunders.

109. Hackney, C.R., and M.D. Pierson (eds.). 1994. *Environmental Indicators and Shellfish Safety.* New York: Chapman and Hall.

110. Hady, P.J., and J.W. Lloyd. 1992. Economic issues for dairy practitioners. *Compend. Contin. Educ. Pract.* 14:1641–1645.

111. Haley, S.L. 1990. Measuring the effectiveness of the export enhancement program for poultry. *Agribusiness* 6:97–108.

112. Hall, C.W., and T.I. Hedrick. 1977. *Drying Milk and Milk Products.* 2d ed. Westport, Conn.: AVI Publishing.

113. Hall, C.W., and G.M. Trout. 1968. *Milk Pasteurization.* Westport, Conn.: AVI Publishing.

114. Hammond, J.W. 1991. Agricultural price and income policy in Norway. *Food Policy* 16:342–344.

115. Harding, F. 1992. Bases and experiences of expressing the protein content of milk—England and Wales. *J. Dairy Sci.* 75:3218–3220.

116. Harling, K.F., and R.L. Thompson. 1985. Government intervention in poultry industries: A cross-country comparison. *Am. J. Agric. Econ.* 67:243–250.

117. Harris, D., and A. Dickson. 1989. Korea's beef market and demand for imported beef. *Agric. Resour. Q.* 1:294–304.

118. Hartshorn, T.A., and J.W. Alexander. 1988. *Economic Geography.* 3d ed. Englewood Cliffs, N.J.: Prentice-Hall.

119. Hastings, M.Y. 1991. *Ostrich Farming.* Winchelsea, Victoria, Australia: M.Y. Hastings.

120. Hawrysh, Z.J., S.R. Gifford, and M.A. Price. 1985. Cooking and eating-quality characteristics of dark-cutting beef from young bulls. *J. Anim. Sci.* 60:682–690.

121. Hayes, D.J., T.I. Wahl, and G.W. Williams. 1990. Testing restrictions on a model of Japanese meat demand. *Am. J. Agric. Econ.* 72:556–566.

122. Hayes, D.J., J.R. Green, H.H. Jensen, et al. 1991. Measuring international competitiveness in the beef sector. *Agribusiness* 7:357–374.

123. Heen, K., R.L. Monahan, and F. Utter. 1993. *Salmon Aquaculture.* Oxford, England: Fishing News Books.

124. Heid, J.L., and M.A. Joslyn (eds.). 1967. *Fundamentals of Food Processing Operations: Ingredients, Methods, and Packaging.* Westport, Conn.: AVI Publishing.

125. Henderson, J.L. 1971. *The Fluid Milk Industry.* 3d ed. Westport, Conn.: AVI Publishing.

126. Hinchy, M. 1978. The relationship between beef prices in export markets and Australian saleyard prices. *Q. Rev. Agric. Econ.* 31:83–105.

127. Hofmann, K., and R. Hamm. 1978. Sulhydryl and disulfide groups in meats. *Adv. Food Res.* 24:1–111.

128. Holley, R.A., and M. Proulx. 1986. Use of egg washwater pH to prevent survival of *Salmonella* at moderate temperatures. *Poultry Sci.* 65:922–928.

129. Hollowell, E.R. 1980. *Cold and Freezer Storage Manual.* 2d ed. Westport, Conn.: AVI Publishing.

130. Holmes, C.W., and G.F. Wilson. 1984. *Milk Production from Pasture.* Wellington, New Zealand: Butterworths Agricultural Books.

131. Holst, P.J., and R.A. Whitelaw. 1980. Export of goat meat from Australia. *Int.*

Goat Sheep Res. 1:48–54.

132. Hughes, K.L. 1990. Animal production and human health. In *Microbiology of Animals and Animal Products.* Vol. 6, *World Animal Science.* Elsevier, Amsterdam, 195–232.

133. International Dairy Federation. 1985. *Payment for Milk on the Basis of Quality.* Bull. 192. Brussels: International Dairy Federation.

134. ———. 1990. *Milk Collection in Warm Developing Countries.* Brussels: International Dairy Federation.

135. Jensen, J.M., J.H. Johnson, and S.T. Weiner. 1992. *Husbandry and Medical Management of Ostriches, Emus and Rheas.* College Station, Tex.: Wildlife and Exotic Animal Teleconsultants.

136. Johnson, E.A., J.H. Nelson, and M. Johnson. 1990. Microbiological safety of cheese made from heat-treated milk, Part III. Technology, discussion, recommendations, bibliography. *J. Food Prot.* 53:610–623.

137. Jones, T.C., and R.D. Hunt. 1983. *Veterinary Pathology.* 5th ed. Philadelphia: Lea and Febiger.

138. Jordan, D.C. 1985. The effect of quality and component premiums on mastitis awareness. *Annu. Meet. Natl. Mastitis Counc.* 24:86–91.

139. Josephson, E.S., and M.S. Peterson (eds.). 1983. *Preservation of Food by Ionizing Radiation.* Vols. 1 and 2. Boca Raton, Fla.: CRC Press.

140. Kaiser, H.M., P.A. Oltenacu, and T.R. Smith. 1988. The effects of alternative seasonal price differentials on milk production in New York. *Northeast J. Agric. Resour. Econ.* 17:46–55.

141. Kalogridou-Vassiliadou, D., K. Manolkidis, and A. Tsigoida. 1992. Somatic cell counts in relation to infection status of the goat udder. *J. Dairy Res.* 59:2128.

142. Kamau, D.N., S. Doores, and K.M. Pruitt. 1990. Enhanced thermal destruction of *Listeria monocytogenes* and *Staphylococcus aureus* by the lactoperoxidase system. *Appl. Environ. Microbiol.* 56:2711–2716.

143. Kargbo, J.M. 1992. Meat imports in Sierra Leone: Analysis, constraints and policy implications. *Food Policy* 17:361–370.

144. Katyega, P.M.J. 1982. Egyptian water buffalo in Tanzania. *World Anim. Rev.* 43:42–43.

145. Kelbert, D.P. 1989. Improving milk quality through cow management. *Annu. Meet. Natl. Mastitis Counc.* 1989:177–180.

146. Kelly, J.M., D.A. Whitaker, and E.J. Smith. 1988. A dairy herd health and productivity service. *Br. Vet. J.* 144:470–481.

147. Kelly, R.W., G.H. Moore, and K.R. Drew. 1980. Characteristics of red deer (*Cervus elaphus* L.) related to their performance as farmed animals. *Proc. Aust. Soc. Anim. Prod.* 13:198–202.

148. Kerr, W.A., and D.B. McGivern. 1991. A problem of multiple displacements— Canadian concerns over the importation of Irish beef. *Ir. J. Agric. Econ. Rur. Sociol.* 14:49–65.

149. Klausner, R.B., and C.W. Donnelly. 1991. Environmental sources of *Listeria* and *Yersinia* in Vermont dairy plants. *J. Food Prot.* 54:607–611.

150. Konieccko, E.S. 1985. *Handbook of Meat Analysis,* 2d ed. Wayne, N.J.: Avery Publishing.

151. Kosikowski, F. 1982. *Cheese and Fermented Milk Foods.* 3d ed. Ann Arbor, Mich.: Edwards Brothers.

152. Kotschwar, L., D. Simon, and E.W.F. Peterson. 1993. Laws governing the use of technical standards as barriers to trade: The case of trade in livestock products. *Agribusiness* 9:91–101.

153. Krissoff, B. 1989. The European ban on livestock hormones and implications

for international trade. *Natl. Food Rev.* 12:34–36.

154. Lawless, H.T., and M.R. Claassen. 1993. Validity of descriptive and defect-oriented terminology systems for sensory analysis of fluid milk. *J. Food Sci.* 58:108–112, 119.

155. Lawrie, R.A. (ed.). 1981. *Developments in Meat Science.* Development Series. Barking, England: Applied Science Publishers.

156. ———. 1991. *Meat Science.* Oxford: Pergamon Press.

157. Leary, W.I. 1983. International meat marketing: Decreasing regulatory trade barriers. *J. Am. Vet. Med. Assoc.* 183:1434–1436.

158. Lee, D.O., and J.F. Wickens. 1992. *Crustacean Farming.* London: Blackwell Scientific Publications.

159. Lee, J.H., D. Henneberry, and D. Pyles. 1991. An analysis of value-added agricultural exports to middle-income developing countries: The case of wheat and beef products. *South. J. Agric. Econ.* 23:141–154.

160. Ley, F.V. 1983. New interest in the use of irradiation in the food industry. *Soc. Appl. Bact. Symp. Ser.* 11:113–129.

161. Libby, J.A. (ed.). 1975. *Meat Hygiene.* 4th ed. Philadelphia: Lea and Febiger.

162. Lilius, E.M., and U. Pesonen. 1990. Use of inflammatory cell activities in bovine milk to diagnose mastitis. *Am. J. Vet. Res.* 51:1527–1533.

163. Lin, B.H., and H. Mori. 1991. Implicit values of beef carcass characteristics in Japan: Implications for the U.S. beef export industry. *Agribusiness* 7:101–114.

164. Liston, J. 1990. Microbial hazards of seafood consumption. *Food Technol.,* Dec., 56–62.

165. Livingston, G.E., and C.M. Chang. 1979. *Food Service Systems: Analysis, Design, and Implementation.* New York: Academic Press.

166. Lo, L.L., D.G. McLaren, F.K. McKeith, et al. 1992. Genetic analyses of growth, real-time ultrasound, carcass, and pork quality traits in Duroc and Landrace pigs: II. Heritabilities and correlations. *J. Anim. Sci.* 70:2387–2396.

167. Longree, K., and G. Armbruster. 1987. *Quantity Food Sanitation.* 4th ed. New York: John Wiley and Sons.

168. Lovell, A.T. 1991. Foods from Aquaculture. *Food Technol.* 45:87–96.

169. Lowe, F.R. 1981. *Milking Machines: A Comprehensive Guide for Farmers, Herdsmen, and Students.* New York: Pergamon Press.

170. Lu, C.D., M.J. Potchoiba, and E.R. Loetz. 1991. Influence of vacuum level, pulsation ratio and rate on milking performance and udder health in dairy goats. *Small Rumin. Res.* 5:1–8.

171. Maijala, K., and M. Wilhelmson. 1983. Meat production from moose and reindeer in Fennoscandia. *Proc. 5th World. Conf. Anim. Prod.* 1:51–58.

172. Maisi, P. 1990. Milk NAGase, CMT and antitrypsin as indicators of caprine subclinical mastitis infections. *Small Rumin. Res.* 3:493–501.

173. Malloy, D.L. 1991. The *Codex Alimentarius* provides international standards for food production and safety. *J. Agric. Taxation Law* 12:334–341.

174. Mann, I. 1960. *Meat Handling in Underdeveloped Countries.* Rome: Food and Agriculture Organization.

175. ———. 1962. *Processing and Utilization of Animal By-products.* Rome: Food and Agriculture Organization.

176. Meyer, A. 1973. *Processed Cheese Manufacture.* London: Food Trade Press.

177. Miller, G.Y., and P.C. Bartlett. 1991. Economic effects of mastitis prevention strategies for dairy producers. *J. Am. Vet. Med. Assoc.* 198:227–231.

178. Mitchell, J.R. 1980. *Guide to Meat Inspection in the Tropics.* 2d ed. Farnham Royal, Bucks, England: Commonwealth Agricultural Bureaux.

179. Mjelde, J.W., J.R. Conner, J.W. Stuth, et al. 1992. The emerging exotic ungu-

late livestock industry: A survey of current producers. *Agribusiness* 8:473–484.

180. Morin, D.E., G.C. Petersen, H.L. Whitmore, et al. 1993. Economic analysis of a mastitis monitoring and control program in four dairy herds. *J. Am. Vet. Med. Assoc.* 202:540–548.

181. Morris, J.B. 1982. Selling pork overseas. *Agrologist* 11:13–14.

182. Morrison, R.M., T. Roberts, and L. Wifucki. 1992. Irradiation of U.S. poultry— benfits, costs, and export potential. *Food Rev.* 15(Oct.–Dec.):16–21.

183. Morton, I.D., J.I. Gray, and P.T. Tybor. 1988. Edible tallow, lard and partially defatted tissues. *Adv. Meat Res.* 5:275–302.

184. Moschini, G., and K.D. Meilke. 1991. Tariffication with supply management: The case of the U.S.-Canada chicken trade. *Can. J. Agric. Econ.* 39:55–68.

185. ——. 1992. Production subsidy and countervailing duties in vertically related markets: The hog-pork case between Canada and the United States. *Am. J. Agric. Econ.* 74:951–961.

186. Mossel, D.A.A. 1982. *Microbiology of Foods: The Ecological Essentials of Assurance and Assessment of Safety and Quality.* Utrecht: Faculty of Veterinary Medicine, University of Utrecht.

187. Mountney, G.J. (ed.). 1987. *Practical Food Microbiology and Technology.* 3d ed. New York: Van Nostrand Reinhold.

188. ——. 1989. *Poultry Products Technology.* 2d ed. Binghamton, N.Y.: Food Products Press.

189. Murphy, J.P., G.C.W. Ames, J.E. Epperson, et al. 1992. International poultry trade and the GATT: The effects of increased protection on market potential oversees. *J. Int. Food Agribusiness Mark.* 4:57–76.

190. Murray, A.C. 1989. Factors affecting beef color at time of grading. *Can. J. Anim. Sci.* 69:347–355.

191. Nash, C.E. 1988. A global view of aquaculture production. *J. World Aquaculture Soc.* 19:51–58.

192. Nelson, G.C., and Y.K. Lee. 1991. Is a beef deficiency payment Pareto-superior in South Korea? *West. J. Agric. Econ.* 16:106–118.

193. Niinivaara, F.P., and E. Petaja. 1985. Problems in the production and processing of reindeer meat. *Trends Mod. Meat Technol.* 1985:115–120.

194. Norman, H.D., J.R. Wright, C.B. Covington, et al. 1991. Potential for segregrating milk: Herd differences in milk value for fluid and five manufactured products. *J. Dairy Sci.* 74:2353–2361.

195. North, C.E., and W.H. Park. 1927. Standards for milk pasteurization. *Am. J. Hyg.* 7:147–173.

196. North, M.O. 1990. *Commercial Chicken Production Manual.* 4th ed. New York: Van Nostrand Reinhold.

197. Nunez, M., M. Medina, and P. Gaya. 1989. Ewes' milk cheese: Technology, microbiology, and chemistry. *J. Dairy Res.* 56:303–321.

198. Packard, V.S. 1984. The components of milk: Some factors to consider in component pricing plans. *Dairy Food Sanit.* 4:336–347.

199. Pain, B., S. Jarvis, and B. Clements. 1991. Impact of agricultural practices on soil pollution. *Outlook Agric.* 20:153–160.

200. Palm, C.E. (chrmn.). 1969. *Principles of Plant and Animal Pest Control.* Vol. 3, *Insect-Pest Management and Control.* Publ. 1965. Washington, D.C.: National Academy of Sciences.

201. Parkhurst, C., and G.J. Mountney. 1987. *Poultry Egg and Meat Production.* New York: Van Nostrand Reinhold.

202. Pearson, A.M., and F.W. Tauber. 1984. *Processed Meats.* 2d ed. Westport, Conn.: AVI Publishing.

203. Peris, C., P. Molina, N. Fernandez, et al. 1991. Variation in somatic cell count, California mastitis test, and electrical conductivity among various fractions of ewe's milk. *J. Dairy Sci.* 74:1553–1560.

204. Peterson, E.B., P.V. Preckel, T.W. Hertel, et al. 1992. Impacts of growth stimulants in the domestic livestock sector. *Agribusiness* 8:287–307.

205. Petrey, L.A. 1980. Aspects of deer slaughter for human consumption. *Proc. Aust. Soc. Anim. Prod.* 13:202–204.

206. Pillay, T.V.R. 1992. *Aquaculture and the Environment.* Oxford, England: Fishing News Books.

207. ———. 1993. *Aquaculture: Principles and Practices.* Oxford, England: Fishing News Books.

208. Pope, C.W. 1991. Poultry production's environmental impact on water quality. *Poult. Sci.* 70:1123–1125.

209. Poste, L.M. 1990. A sensory perspective of effect of feeds on flavor in meats: Poultry meats. *J. Anim. Sci.* 68:4414–4420.

210. Potter, N.N. 1986. *Food Science.* 4th ed. Westport, Conn.: AVI Publishing.

211. Pringle, J.D., and D.L. Burke. 1993. The Canadian lobster fishery and its management, with emphasis on the Scotia Shelf and the Gulf of Maine. *Can. Bull. Fish Aquat. Sci.* 226:91–121.

212. Pringle, W.L., and J.R. Hunt. 1984. Meat from non-ruminants. *Agrologist* 13:10–11.

213. Qvist, F. 1984. Economic importance of swine production on an international basis. *Proc. Int. Pig Vet. Soc. Congr.* 1984:2–3.

214. Rafferty, G.C. 1980. The export of venison from Scotland. *State Vet. J.* 35:130–134.

215. Ragan, J.R. (chrmn.). 1992. *Proper Handling Techniques for Non-ambulatory Animals.* Madison, Wis.: Livestock Conservation Institute.

216. Rathore, G.S. 1986. *Camels and Their Care.* New Delhi: Indian Council of Agricultural Research.

217. Regester, G.O., G.W. Smithers, M.E. Mangino, et al. 1992. Seasonal changes in the physical and functional properties of whey protein concentrates. *J. Dairy Sci.* 75:2928–2936.

218. Reid, W.M. 1990. History of avian medicine in the United States. X. Control of coccidiosis. *Avian Dis.* 34:509–525.

219. Reineccius, G. 1991. Off-flavors in foods. *Crit. Rev. Food Sci. Nutr.* 29:381–402.

220. Reneau, J.K., and V.S. Packard. 1991. Monitoring mastitis, milk quality, and economic losses in dairy fields. *Dairy Food Environ. Sanit.* 11:4–11.

221. Rhodes, R.J. 1993. World aquaculture production continues climbing. *Feedstuffs,* May 31, 11–13.

222. Richardson, G.H. (ed.). 1985. *Standard Methods for the Examination of Dairy Products.* 15th ed. Washington, D.C.: American Public Health Association.

223. Richter, R. 1981. Hydrolytic rancidity: Its prevalence, measurement, and significance. *Am. Dairy Rev.* 43:18DD, 18HH.

224. Roberts, R.K., and W.J. Martin. 1985. The effects of alternative beef import quota regimes on the beef industries of the aggregate United States and Hawaii. *West. J. Agric. Econ.* 10:230–244.

225. Roberts, T., and D. Smallwood. 1991. Data needs to address economic issues in food safety. *Am. J. Agric. Econ.* 73:934–942.

226. Rogers, S.A., and G.E. Mitchell. 1989. The relationship between somatic cell count, composition, and manufacturing properties of bulk milk. 5. Pasteurized milk and skim milk powder. *Aust. J. Dairy Technol.* 44:57–60.

227. Rogers, S.A., G.E. Mitchell, and J.P. Bartley. 1989. The relationship between somatic cell count, composition and manufacturing properties of bulk milk. 4. Non-protein constituents. *Aust. J. Dairy Technol.* 44:53–56.

228. Rosson, C.P., III, M.D. Hammig, and J.W. Jones. 1986. Foreign market promotion programs: An analysis of promotion response for apples, poultry, and tobacco. *Agribusiness* 2:33–42.

229. Rosson, C.P., II, E.E. Davis, A. Angel, et al. 1993. Free trade impacts on U.S.–Mexican meat trade. *Agribusiness* 9:159–173.

230. Roswell, H.C. 1991. Transportation of animals—a global animal welfare issue. *Proc. Anim. Welfare Sess., XXIV World Vet. Congr.*, Animal Welfare Commission, World Veterinary Association, London, 69–83.

231. Sakata, R., and Y. Nagata. 1992. Heme pigment content in meat as affected by the addition of curing agents. *Meat Sci.* 32:343–350.

232. Sanderson, J.E., J.R. Consaul, and K. Lee. 1991. Nitrate analysis in meats: Comparison of two methods. *J. Food Sci.* 56:1123–1124.

233. Sandrey, R.A., and A.C. Zwart. 1986. Dynamics of herd build-up in a new industry: Commercial red deer production in New Zealand. *West. J. Agric. Econ.* 11:92–99.

234. Sapp, S.G., and G.W. Williams. 1988. The socio-economic issues of Japanese beef imports. *Agribusiness* 4:63–77.

235. Schalm, O.W., E.J. Carroll, and N.C. Jain. 1971. *Bovine Mastitis.* Philadelphia: Lea and Febiger.

236. Schrader, L.F., and D.J. Brown. 1992. U.S. poultry demand maturing, export demand bright spot. *Feedstuffs* 64(Oct.12):17, 20.

237. Schukken, Y.H., F.J. Grommers, D. van de Geer, et al. 1990. Risk factors for clinical mastitis in herds with a lowbulk milk somatic cell count. 1. Data and risk factors for all cases. *J. Dairy Sci.* 73:3463–3471.

238. Schukken, Y.H., F.J. Grommers, D. van de Geer, et al. 1991. Risk factors for clinical mastitis in herds with a lowbulk milk somatic cell count. 2. Risk factors for *Escherichia coli* and *Staphylococcus aureus*. *J. Dairy Sci.* 74:826–832.

239. Schukken, Y.H., A. Weersink, K.E. Leslie, et al. 1993. Dynamics and regulation of bulk milk somatic cell counts. *Can. J. Vet. Res.* 57:131–135.

240. Schultz, H.W., E.D. Day, and L.M. Libbey (eds.). 1967. *Symposium on Foods: The Chemistry and Physiology of Flavors.* Westport, Conn.: AVI Publishing.

241. Schultz, W.D., J.W. Smith, and P.D. Thompson (eds.). 1978. *Proceedings of the International Symposium on Machine Milking.* Washington, D.C.: National Mastitis Council.

242. Schwartz, H.J., and M. Dioli (eds.). 1992. *The One-Humped Camel (Camelus dromedarius) in Eastern Africa.* Weikersheim, Federal Republic of Germany: Verlag Josef Margraf.

243. Seamer, D.J. 1986. The welfare of deer at slaughter in New Zealand and Great Britain. *Vet. Rec.* 118:257–258.

244. Sharf, J.M. (ed.). 1966. *Recommended Methods for the Microbiological Examination of Foods.* 2d ed. Washington, D.C.: American Public Health Association.

245. Shelef, L.A., and Q. Yang. 1991. Growth suppression of *Listeria monocytogenes* by lactates in broth, chicken, and beef. *J. Food Prot.* 54:283–287.

246. Sikorski, Z.E. (ed.). 1990. *Seafood: Resources, Nutritional Composition and Preservation.* Boca Raton, Fla.: CRC Press.

247. Skala, J.H., E.L. McGown, and P.P. Waring. 1987. Wholesomeness of irradiated food. *J. Food Prot.* 50:150–160.

248. Skold, K.D., G.W. Williams, and M.L. Hayenga. 1987. Meat export marketing: Lessons from successful exporters. *Agribusiness* 3:83–97.

249. Skovgaard, N. 1992. Microbiological aspects and technological need: Techno-

logical needs for nitrates and nitrites. *Food Addit. Contam.* 9:391–397.

250. Smith, R. 1991. A game of dominoes: Meat exporters must play by set of tough rules. *Feedstuffs* 63(Oct. 28):1, 28–29.

251. Smith, R.F. (chrmn). 1976. *Pest Control: An Assessment of Present and Alternative Strategies.* Vol. 5, *Pest Control and Public Health.* Washington, D.C.: National Academy of Sciences.

252. Spill, M., and D. Harris. 1989. Recent developments in the North American beef market. *Agric. Resour. Q.* 1:305–311.

253. Spooncer, W.F. 1988. Organs and glands as human food. *Adv. Meat Res.* 5:197–217.

254. Stadelman, W.J., V.M. Olson, G.A. Shemwell, et al. 1988. *Egg and Poultry-Meat Processing.* Chichester, England: Ellis Horwood.

255. Stadelman, W.J., and O.J. Cotterill (eds.). 1990. *Egg Science and Technology.* 3d ed. Binghamton, N.Y.: Food Products Press.

256. Steele, J.H., and R.E. Engel. 1992. Radiation processing of food. *J. Am. Vet. Med. Assoc.* 201:1522–1529.

257. Stenholm, C.W., and D.B. Waggoner. 1991. Developing future-minded strategies for sustainable poultry production. *Poultry Sci.* 70:203–210.

258. Stewart, G.F., and J.C. Abbot. 1972. *Marketing Eggs and Poultry.* FAO Mark. Guide 4. Rome: Food and Agriculture Organization.

259. Stolle, F.A. 1986. Rodding in the West Berlin slaughterhouse: A possible method of improving hygiene in slaughterhouses or additional labour expenditure in the modern cattle slaughtering procedure. *Br. Vet. J.* 142:30–35.

260. Sutherland, A.K. 1991. Semi-automated slaughtering and dressing of cattle. *Aust. Vet. J.* 68:57.

261. Swift, D.R. 1993. *Aquaculture Training Manual.* 2d. ed. Oxford, England: Fishing News Books.

262. Tanson, R. 1986. Technical environment for caged laying hens. Dissertation. Rep. 154. Swedish University of Agricultural Sciences, Uppsala.

263. Tao, S. 1991. China's animal husbandry: The past ten years. *Outlook Agric.* 20:191–194.

264. Tarrant, P.V. 1989. Animal behaviour and environment in the dark-cutting condition in beef—a review. *Ir. J. Food Sci. Technol.* 13:1–21.

265. Thiel, C.C., and F.H. Dodd. 1977. *Machine Milking.* Reading, England: National Institute for Research in Dairying.

266. Thorne, S. (ed.). 1991. *Food Irradiation.* London: Elsevier Applied Science.

267. Thorogood, S.A., P.D.P. Wood, and G.A. Prentice. 1983. An evaluation of the Charm Test—a rapid method for the detection of penicillin in milk. *J. Dairy Res.* 50:185–191.

268. Timms, L.L. 1988. Influencing producers to use DHI-SCC. *Annu. Meet. Natl. Mastitis Counc.* 27:43–50.

269. Townsend, W.E., and L.C. Blankenship. 1989. Methods for detecting processing temperatures of previously cooked meat and poultry products—a review. *J. Food Prot.* 52:128–135.

270. Tsai, L.S., J.E. Schade, and B.T. Molyneux. 1992. Chlorination of poultry chiller water: Chlorine demand and disinfection efficiency. *Poultry Sci.* 71:188–196.

271. Tucker, C.S., and E.H. Robinson. 1990. *Channel Catfish Farming Handbook.* New York: Van Nostrand Reinhold.

272. Tulloh, N.M., and J.H.G. Holmes. 1992. *Buffalo Production.* World Anim. Sci. Amsterdam: Elsevier.

273. United Nations. Food and Agriculture Organization. 1970. *Joint FAO/WHO Expert Committee on Milk Hygiene, Third Report.* FAO Agric. Stud. 83. Rome: Food and Agriculture Organization.

274. ———. 1984. *Codex General Standards for Irradiated Food and Recommended International Code for Practice.* Codex Alimentarius 15. Rome: Food and Agriculture Organization.

275. ———. 1991. *Guidelines for Slaughtering, Meat Cutting, and Further Processing.* Rome: Food and Agriculture Organization.

276. ———. 1991. *FAO Animal Health Yearbook.* Rome: Food and Agriculture Organization.

277. United Nations. World Health Organization. 1972. *Vector Control in International Health.* Geneva: World Health Organization.

278. ———. 1974. *Equipment for Vector Control.* 2d ed. Geneva: World Health Organization.

279. ———. 1985. *Specifications for Pesticides Used in Public Health: Insecticides, Molluscicides, Repellents, Methods.* 6th ed. Geneva: World Health Organization.

280. ———. 1988. *Food Irradiation: A Technique for Preserving and Improving the Safety of Food.* Geneva: World Health Organization.

281. United Nations Conference on Trade and Development (UNCTAD). 1989. *Studies in the Processing, Marketing and Distribution of Commodities. The Marketing of Bovine Meat and Products: Areas for International Cooperation* (UNCTAD/COM/1). New York: United Nations.

282. U.S. Department of Agriculture. 1990. *Agricultural Handbook 685.* Washington, D.C.: U.S. Goverment Printing Office.

283. ———. 1991. *U.S. Inspected Meat and Poultry Packing Plants: A Guide to Construction and Layout.* Agric. Handb. 570. Washington, D.C.: U.S. Government Printing Office.

284. ———. 1992. *Agricultural Statistics.* Washington, D.C.: U.S. Government Printing Office.

285. ———. 1992. *List of Proprietary Substances and Nonfood Compounds Authorized for Use under USDA Inspection and Grading Programs.* Misc. Publ. 1419. Washington, D.C.

286. U.S. Department of Health, Education, and Welfare. 1970. *Screening and Confirmatory Tests for the Detection of Abnormal Milk.* Rev. Washington, D.C.: U.S. Government Printing Office.

287. U.S. Department of Health, Education, and Welfare. Food and Drug Administration. 1975. *Current Concepts in Food Protection.* Cincinnati, Ohio: Cincinnati Training Facility, Food and Drug Administration.

288. U.S. Department of Health and Human Services. Food and Drug Administration. 1991. *Grade "A" Pasteurized Milk Ordinance, 1989 Rev.: Recommendations of the United States Public Health Service, Food and Drug Administration.* PHS Publ. 229. Washington, D.C.: U.S. Government Printing Office.

289. University of California, Davis, Agricultural Extension Service. 1967. *Milking Management and Its Relationship to Milk Quality.* Agric. Ext. Serv. Publ. AXT-94. Davis: University of California.

290. Unnevehr, L.J., and J.S. Eales. 1991. U.S. beef product prices and trade liberalization. *J. Agribusiness* 9:13–22.

291. Urbain, W.M. 1978. Food irradiation. *Adv. Food Res.* 24:155–227.

292. ———. 1986. *Food Irradiation.* New York: Academic Press.

293. Van Arsdel, W.B., M.J. Copley, and A.I. Morgan, Jr. (eds.). 1973. *Food Dehydration.* 2d ed. Vol. 1, *Drying Methods and Phenomena*; Vol. 2, *Practices and Applications.* Westport, Conn.: AVI Publishing.

294. Van der Sluis, E., and D.J. Hayes. 1991. An assessment of the 1988 Japanese beef market access agreement on beef and feed-grain markets. *Rev. Agric. Econ.* 13:45–58.

295. Varnam, A.H., and J.P. Sutherland. 1994. *Milk and Milk Products: Technology,*

Chemistry, and Microbiology. London: Chapman and Hall.

296. Vickery, J.R. 1979. Centenary of the export of frozen meat and butter. *Food Technol. Australia* 31:486, 489–490.

297. Vigh-Larsen, F. 1988. Deer farming in Denmark, with special emphasis on the management and handling of fallow deer. *Curr. Top. Vet. Med. Anim. Sci.* 48:61–70.

298. Wahl, T.I., D.J. Hayes, and G.W. Williams. 1991. Dynamic adjustment in the Japanese livestock industry under beef import liberalization. *Am. J. Agric. Econ.* 73:118–132.

299. Waldroup, A.L., J.T. Skinner, R.E. Hierholzer, et al. 1992. Effects of bird density on *Salmonella* contamination of prechill carcasses. *Poult. Sci.* 71:844–849.

300. Warner, R.D., and J.C. Gordon. 1986. Integrated pest management of insects in food products. *Dairy Food Sanit.* 6:4–9.

301. Warriss, P.D., S.C. Kestin, S.N. Brown, et al. 1984. The time required for recovery from mixing stress in young bulls and the prevention of dark cutting beef J. Sci. Food Agric. 35(8): 863–868.

302. Webb, B.H., and E.O. Whittier. 1970. *By-products from Milk.* 2d ed. Westport, Conn.: AVI Publishing.

303. Weeks, P. 1979. The effect of the proposed U.S. meat import law on U.S. imports of meat. *Q. Rev. Rural Econ.* 1:297–305.

304. Weeks, P., and B. Turner. 1981. Effects of the proposed Canadian meat import law. *Q. Rev. Rural Econ.* 3:232–239.

305. Weersink, A., and L.W. Tauer. 1991. Causality between dairy farm size and productivity. *Am. J. Agric. Econ.* 73:1138–1145.

306. Wheaton, F.W., and T.B. Lawson. 1985. *Fishing Gear Products.* New York: John Wiley and Sons.

307. Williams, C.M. 1986. Meat on the table. Ours, theirs or both? *Agrologist* 15:16–17.

308. Williams, G.W. 1989. The case of U.S. meat exports. *J. Food Distrib. Res.* 20:12–16.

309. Wilson, A. 1985. *Practical Meat Inspection.* 4th ed. London: Blackwell Scientific Publications.

310. Wilson, N.R.P., E.J. Dyett, R.B. Hughes, et al. 1981. *Meat and Meat Products. Factors Affecting Quality Control.* London: Applied Science Publishers.

311. Wilson, R.T. 1984. *The Camel.* London: Longman.

312. Wong, N.P., R. Jenness, M. Keeney, et al. 1988. *Fundamentals of Dairy Chemistry.* 3d ed. New York: Van Nostrand Reinhold.

313. Wood, B.J. 1985. *Microbiology of Fermented Foods.* Westport, Conn.: AVI Publishing.

314. World Health Organization. 1974. *Fish and Shellfish Hygiene.* Report of a WHO Expert Committee convened in cooperation with the FAO. Rome: Food and Agriculture Organization.

315. Yagil, R. 1982. *Camels and Camel Milk.* FAO Anim. Prod. Health Pap. 26. Rome: Food and Agriculture Organization.

316. Yousif, O.K., and S.A. Babiker. 1989. The desert camel as a meat animal. *Meat Sci.* 26:245–254.

317. Zaglul, J.A., and R.G. Cassens. 1987. Post-mortem processing of export beef in Costa Rica. *Meat Sci.* 20:85–95.

318. Zhang, W., D.L. Kuhlers, and W.E. Rempel. 1992. Halothane gene and swine performance. *J. Anim. Sci.* 70:1307–1313.

3

Foodborne Disease

Objectives

- Define and identify primary and secondary sources of food contamination and distinguish between them.

- Differentiate between foodborne infection and intoxication in microbiologically caused foodborne disease.

- Describe the epidemiologic characteristics of foodborne diseases caused by bacteria, viruses, fungi, and parasites.

- Describe the mechanism of food-associated allergic reactions.

- Describe how chemical intoxicants may enter or become part of the food chain, resulting in foodborne disease.

- Identify and describe sources of food poisoning associated with consumption of naturally occurring poisons from animals.

- Describe the meat and milk residue problems associated with antibiotic therapy, use of growth promoters, and farm-associated chemicals (including radionuclides) in food animals.

Sources of Food Contamination[72,144,145,150,176,185,199,204,221,233,250, 253,254,257,327,338,386,405,417,427,429]

Types of Sources of Food Contamination

The source of food contamination may be primary, coming directly from an infected food animal or its secretions or excretions, or secondary, resulting from contamination in handling of food.

Primary Contamination

Infected Animals.   A food animal may be slaughtered while it is either infected with a microbial pathogen or contaminated with chemical or other residues. In some instances, this presents an occupational hazard to stockyard or abattoir workers, but more often it poses a threat to the consumer. Antemortem inspection reveals only a small percentage of these cases. Apparently healthy carrier animals also may shed pathogens in milk.

Fecal Pollution from Infected Animals.   A second way in which primary contamination may enter the food chain is through the discharges (primarily fecal) of infected animals. A few pigs excreting *Salmonella* that are traveling with uninfected pigs may result in a high prevalence of new infections at the abattoir. During the slaughtering process, great care must be taken to prevent fecal contamination of the carcass and knives when the viscera are removed.

Secondary Contamination

Secondary contamination may come from infected humans, other animals, fomites, or feed additives.

Infected Humans.[353]   Infected humans may be the source of contamination at any point in the food chain but are most frequently implicated when preparing food for the table. For instance, loving parents excreting large numbers of staphylococcal organisms, or restaurant cooks dispensing salmonellae along with the egg salad, may act as sources of food contamination.

Other Animals

Vertebrates.   Animals, rodents in particular, are a source of secondary food contamination. As with humans, this may take place at any stage in the production chain, contaminating stored animal feed, raw materials, and finished products in the abattoir or food stored in retail outlets, restaurants, or private homes. In addition to contaminating food with their feces and urine, on a worldwide basis, they are responsible for a staggering amount of food loss as a result of the amount of food they consume.

Invertebrates.   Invertebrate animals compete with humans for food at every stage of the food chain, right up to the consumer's lips. Because of their small size, they are capable of contaminating food that is safe from larger vermin. For the most part, they pose a threat to the health of consumers by acting as mechanical vectors of disease-producing microorganisms. As with rodents, they exacerbate the problem of an insufficient world food supply. In addition to consuming stored food, they also make large quantities of it unpalatable as a result of their activities.

Fomites. Water, soil, plants, and air may serve as vehicles for foodborne disease agents.[34]

Water. Because water is used as a cleansing agent in many food-processing steps, it can pose a serious threat to human health if contaminated. One *Salmonella*-infected carcass going through a machine in an abattoir can contaminate the water sufficiently to transmit this disease agent to a large percentage of the carcasses that follow.

Soil. Soil may be a vehicle for foodborne disease agents such as *Clostridium botulinum* and *Listeria monocytogenes*. Depending on the moisture, alkalinity, temperature, and organic material present, soil can support microorganisms for long periods and can serve not only as a source of secondary contamination of food but also as the means for primary contamination by food animals. This contamination may take place not only by ingestion but by other routes such as inhalation of dust, wound contamination, or entry through relaxed or injured teat sphincters.

Plants. Plants can serve as a source of food contamination in several ways. Ingestion of wild onions by dairy cows causes the milk from these cows to have a decided off flavor. Certain toxic plants can remove animals from the food supply by causing death or a disease condition severe enough to result in condemnation of the animal as unfit for human consumption. Other plants, although not toxic in themselves, can serve as concentrators of environmental pollutants.

Air. Staphylococcal foodborne disease may occur as the result of contamination from the classic skin pustule or sore but more often occurs as the result of airborne contamination. Coughs and sneezes are very efficient ways of disseminating these pathogens, and under normal conditions a person with a good load of staphylococci in the anterior nares can contaminate large areas in a kitchen or food-processing facility.

Chemical Adulterants. A food additive can serve as the source of food contamination if addition of a toxic chemical rather than a harmless material results from carelessness or mislabeling. Utilization of antibiotics as additives in animal-feed supplements also can be a source of chemical adulteration when sufficient withdrawal time is not observed, resulting in unacceptable residue concentrations.

Reported Foodborne Disease[36]

Morbidity and mortality statistics affect the lives of people worldwide. Reports of casualties in natural disasters, such as hurricanes, or in man-made events, such as plane crashes or smoking-related deaths, affect how we perceive risk associated with where we live and what we do. Reports of disease may come from news media as well as government or private agencies involved with disease control or prevention. The mechanism for reporting foodborne disease, as well as the diseases reported, varies from country to country because governmental structures, as well as criteria for disease reporting, differ. An understanding of this structure and these criteria is critical to an understanding of the relationship of reported foodborne disease data to individual and societal decisions about food.

Foodborne Disease Outbreaks and Cases Defined[439,441]

The following outbreak definition, used by the U.S. Centers for Disease Control (CDC), is an example of a definition devised by a national health agency. A foodborne disease outbreak is defined as

> an incident in which 1) two or more persons experience a similar illness after ingestion of a common food, and 2) epidemiologic analysis implicates the food as the source of the illness. A few exceptions exist; for example, one case of botulism or chemical poisoning constitutes an outbreak.
>
> Outbreaks of known cause are those for which laboratory evidence is obtained and specified criteria are met. Outbreaks of unknown etiology are subdivided into four subgroups by incubation period of the illnesses: <1h (probable chemical poisoning), 1-7h (probable *Staphyloccus* food poisoning), 8-14h (probable *C. perfringens* food poisoning), and >14h (other infectious or toxic agents).

Whereas a "generic" definition may be applied to outbreaks, the criteria for what constitutes a case must be developed disease by disease. The two major categories usually considered in developing a **case definition** are (1) a **clinical description**, including incubation period, signs, symptoms, response to therapy, and length of illness, and (2) **laboratory and epidemiologic criteria for diagnosis**, such as isolation of the agent from feces of patients and/or foods eaten by them and not from feces of healthy controls or foods eaten by the controls. For reporting purposes, a case may be classified as either **confirmed** or **probable**. A **laboratory-confirmed** case is one that is confirmed by isolating and identifying the causative agent. Although other laboratory methods may be used in clinical diagnosis, only agent isolation is accepted for laboratory confirmation for reporting purposes. In developing a **strict case definition**, it is often useful to divide the criteria into **major** (must be present), **minor** (at least some must be present), and **exclusionary** (must be negative if obtained). In addition, a less strict definition may be useful for early recognition, or **screening**, of cases and initiation of therapy that could prevent the development of more severe clinical manifestations. A screening definition also may facilitate retrospective epidemiologic surveys of the records of hospitalized patients to identify cases for more-extensive evaluation and application of more-strict definitions.

In contrast to a laboratory-confirmed case, any of the following categories may be used to characterize probable cases.

Clinically compatible case: a clinical syndrome generally compatible with the disease, but no special clinical criteria need to be met unless they are noted in the case classification.

Supportive laboratory results: specified laboratory results consistent with the diagnosis but not meeting the criteria for laboratory confirmation.

Epidemiologically linked case: a case in which the patient has had contact with one or more persons who have or had the disease, and transmission of the agent by the usual, plausible modes of transmission (i.e., foodborne). A case may be considered epidemiologically linked to a laboratory-confirmed case if at least one case in the chain of transmission is laboratory confirmed.

Defined clinical case: a case precisely meeting the clinical case definition. In clinical practice, diagnosis may be established with the use of other criteria, but for the purposes of reporting, the stated criteria must be met.

These categories of cases are useful in assisting investigations of clusters of illness suspected of being foodborne in origin. In the absence of such investigations, the occasional sporadic case of foodborne disease is less likely to be detected and reported. In addition, syndromes in which enteric signs are not dominant are also likely to be overlooked as foodborne in origin. For example, an acute vasculitis is sometimes associated with foodborne agents such as group A ß-hemolytic streptococci or *Yersinia pseudotuberculosis*.[14,50] Arthritis is associated with rheumatic fever, but it is also associated with *Y. enterocolitica* infection, which may be overlooked.[237] Brucellosis, with its insidious onset and lack of enteric signs, is notorious as a diagnostic challenge.

Reporting in Australia

Australia does not have a formal national reporting system for foodborne disease. Since 1977 it has published a fortnightly bulletin, *Communicable Diseases Intelligence,* which reports new cases of notifiable bacterial, viral, and parasitic diseases, together with laboratory reports on viral diseases encountered, grouped according to their clinical manifestations. Differences in reporting standards between states make these data incomplete; also some diseases are undoubtedly underreported. A small number of the diseases classified as notifiable are foodborne or commonly foodborne. Thus, in 1992, the number of **reported** cases of campylobacteriosis was 7,607. Other 1992 figures were listeriosis, 37; salmonellosis other than typhoid, 4,152; yersiniosis, 504; and botulism, 0. The bulletin publishes synopses of disease epidemics and brief reports of cases with unusual epidemiology.

Reporting in Canada[415,418]

A national foodborne disease reporting system was established in 1973. For statistical reporting, Health Canada defines incidents, cases, and outbreaks and establishes criteria for confirmation of etiologic diagnosis. *Incident* is a general and inclusive term that refers to an outbreak or a single case. *Case* is defined as a person who has been ill following consumption of food considered to be contaminated on the basis of epidemiologic evidence or laboratory analysis. *Single case* is defined as one case that is not linkable to other cases with respect to the consumption of food. *Outbreak* is an incident in which two or more cases experience a similar illness after ingestion of the same food or water from the same source, or after contact with water from the same source, and where epidemiologic evidence implicates food or water as the common vehicle of the illness. Etiology may be confirmed by laboratory or epidemiologic evidence. Criteria for both laboratory and epidemiologic diagnosis are established for each etiologic agent. As in the United States, etiology can be classified as laboratory confirmed, probable, or unknown, depending upon the evidence available for diagnosis.

Information for the national reporting system is gathered from multiple sources. Most of the information comes directly from provincial sources, some comes from federal sources, and some directly from municipalities. Information at these levels is collected from consumers, physicians, laboratories, and institutions. Underreporting is common, for the usual reasons.

Reporting in the United Kingdom

The reporting of cases of food poisoning in Scotland is separate from that in England and Wales, which are administered together. This section summarizes the reporting procedures in England and Wales.

Cases of food poisoning occurring in England and Wales may be notified in one of two ways. The first is the statutory notification of clinical cases to the local authority, and the second is the nonstatutory reporting of the isolation of foodborne pathogens to the Communicable Disease Surveillance Centre (CDSC) of the Public Health Laboratory Service (PHLS).

The statutory notification of communicable disease was introduced during 1889 as a means of identifying and preventing the spread of infectious disease. At present, the two principal pieces of legislation concerned with communicable disease control in England and Wales are the Public Health (Control of Disease) Act of 1984 and the Public Health (Infectious Diseases) Regulations of 1988. The Public Health (Control of Disease) Act of 1984 requires any medical practitioner to report cases of food poisoning to the local authority, usually a medical consultant in communicable disease control. A certificate is filled out and a copy sent to the District Health Authority. Food poisoning is not defined by the act, but the Chief Medical Officer of the Department of Health issued the following definition during 1992: "Food poisoning is any disease of an infectious or toxic nature caused or thought to be caused by the consumption of food or water." The definition includes (1) all foodborne or waterborne illnesses regardless of presenting signs and symptoms; thus it includes not only acute illnesses characterized by diarrhea and/or vomiting but also illnesses presenting manifestations not related to the gastrointestinal tract, and (2) illnesses caused by toxic chemicals irrespective of origin, whether these are naturally occurring inorganic compounds or of biological or man-made origin. Illnesses resulting from known allergies and food intolerances are excluded from the definition.

Under the provisions of the Public Health (Infectious Diseases) Regulations of 1988, the local authority may require, on receipt of a report of typhoid, paratyphoid and other salmonella infections, amoebic and bacillary dysentery, and staphylococcal infections likely to cause food poisoning, (1) the person concerned to refrain from engaging in any occupation connected with food until the risk causing the infection is removed, (2) measures to be taken for the protection of the public health, and (3) people to assist in securing compliance with these requirements.

All statutory notifications are sent by the local authority to the Office of Population Censuses and Surveys (OPCS) for collation. Since 1922, the summaries have been published each week in the *Register General's Weekly Returns*. The annual number of notifications in England and Wales for the years 1989-1991 were 52,557, 52,145, and 52,543. Of these, 185 (0.35%), 187 (0.35%), and 169 (0.32%) patients died.

The isolation of food-poisoning pathogens by laboratories within the PHLS (there are 52 in England and Wales) and outside are reported to the CDSC of the PHLS on a voluntary basis. The information is collated and published weekly in the *Communicable Disease Report* (Tables 3.1-3.4). The CDSC uses a number of definitions. An outbreak is defined as two or more related cases of foodborne infection or intoxication. Outbreaks are classified as family outbreaks when members of a single household are affected, or general outbreaks if more widespread. A sporadic case is a person with symptoms who has no known association with another case. Bacterial food poisoning includes illnesses associated with toxin produced in food or the gastrointestinal tract by *Bacillus* spp., *Clostridium* spp., and *Staphylococcus aureus*. It also includes illness caused by scombrotoxin as well as ciguatera and red welk toxins.

The two systems of reporting involving the OPCS and the PHLS lead to the publication of two separate, but related, sets of statistics. This is, in itself, rather confusing, and the methods of reporting are under review with the hope that a unified system will be developed ultimately. The increasing use of electronic reporting should allow development, in due course, of a single database for all reports of communicable disease in humans, including food poisoning.

Reporting in the United States[23,24]

In the United States, the criteria used in the definition of a **case** of a new disease are defined by epidemiologists from the 50 state health agencies together with CDC personnel. After that, however, each state must adopt criteria for its reporting system. For instance, the criteria for a case of *Escherichia coli* O157:H7 infection are still being developed, although the epidemiologists voted in 1993 to make the disease reportable. In contrast, all of the states use the CDC definition of a foodborne disease outbreak.

During 1983-1987 (the most recent 5-yr period for which data are available in 1994), bacterial agents were responsible for 66% of the 909 foodborne disease **outbreaks** reported to the CDC (see Table 3.5). In addition, outbreaks of bacte-

Table 3.1.   **Isolation of campylobacters and salmonellae from fecal samples examined in England and Wales and reported to the Communicable Disease Surveillance Centre of the Public Health Laboratory Service**

	1989	1990	1991	Total
Campylobacters	32,526	34,552	32,636	99,714
Salmonellae	24,653	25,300	22,627	72,580[a]
	(171 serotypes)	(186 serotypes)	(160 serotypes)	

Source: Adapted from 389.
   [a]Of these 43,017 (59%), 15,194 (21%), and 2,237 (3%) were *S. enteritidis, S. typhimurium* and *S. virchow* respectively.

Table 3.2.   **Outbreaks of bacterial food poisoning in England and Wales reported to the Communicable Disease Surveillance Centre of the Public Health Laboratory Service**

	1989		1990		1991		Total	
Pathogen	General	Family	General	Family	General	Family	General	Family
Bacillus spp.	23	4	23	10	12	2	58	16
Clostridium perfringens	52	3	51	2	43	1	146	6
Clostridium botulinum	1	—	—	—	—	—	1	—
Salmonella spp.	123	812	118	777	151	785	392	2,374
Staphylococcus aureus	7	1	5	4	4	4	16	9

Source: Adapted from 389.

Table 3.3.   **Places where implicated food was consumed in general outbreaks of food poisoning in England and Wales and reported to the Communicable Disease Surveillance Centre of the Public Health Laboratory Service, 1989–1991**

	Pathogen				
Location	*Bacillus* spp.	*Clostridium perfringens*	*Salmonella* spp.	*Staphylococcus aureus*	Total
Restaurant, hotel, or reception	40	58	202	5	305
Hospital	2	19	21	—	42
Institution	1	35	52	1	89
School	2	7	9	—	18
Canteen	3	9	9	1	22
Farm	—	—	3	—	3
Shop	1	1	32	1	35
Infected abroad	—	—	12	—	12
Others	3	5	44	5	57
Unspecified	6	12	8	3	29
Total	58	146	392	16	612

Source: Adapted from 389.

Table 3.4. Reported food vehicles in general outbreaks of food poisoning in England and Wales reported to the Communicable Disease Surveillance Centre of the Public Health Laboratory Service, 1989–1991

	Pathogen				
Type of food	*Bacillus* spp.	*Clostridium perfringens*	*Salmonella* spp.	*Staphylococcus aureus*	Total
Chicken	2	13	23	3	41
Turkey	2	5	21	—	28
Other poultry	—	—	—	1	1
Beef	—	22	7	—	29
Pork or ham	—	12	5	3	20
Lamb or mutton	—	9	—	—	9
Cold meats	—	—	7	—	7
Gravy or sauces	—	1	1	—	2
Shellfish	—	1	—	—	1
Milk	—	—	3	—	3
Dairy products (excluding milk)	1	—	1	—	2
Eggs	1	—	50	1	52
Rice	19	—	—	1	20
Vegetables or spices	—	2	1	—	3
Bakery products	—	—	7	—	7
Sweets or puddings	4	—	21	1	26
Mixed foods	27	19	47	2	95
Other/not stated	—	38	174	1	213
Total	58	146	392	16	612

Source: Adapted from 389.

rial origin accounted for 92% of the 54,540 total **cases** involved in the reported foodborne outbreaks, with *Salmonella* spp. being responsible for 57%. On the other hand, parasitic agents accounted for the fewest outbreaks (4.0%) and cases (0.4%). Outbreaks of viral origin tend to be relatively few (4.5%) but involve many cases (5.1%) when they occur.

Although these data provide an indication of the relative importance of various agents as causes of foodborne disease in the United States, they do not reflect the actual national risk of foodborne disease during the interval. The numbers cannot be used to calculate meaningful attack rates per 100,000 population. First, all outbreaks were not reported, and second, no sporadic cases (i.e., cases not defined as part of an outbreak) were included.

Summary of Foodborne Disease Outbreaks and Cases

Bacteria are the most frequently reported cause of foodborne disease. When the individual cases involved in foodborne outbreaks are tabulated, a preponderance of bacterial causes is demonstrated. Among all outbreaks of foodborne disease, bacteria and viruses are responsible for more than 90% of the cases. Occasionally, more than one agent may be involved in an outbreak, and this can complicate both diagnosis (e.g., bimodal distribution of incubation periods) and control (e.g., food contaminated from multiple sources).

During the period 1961-1982, factors that contributed to 1,918 reported foodborne disease outbreaks in the United States were identified (see Table 3.6).

Table 3.5.   Confirmed foodborne disease outbreaks and cases by etiologic agent, United States, 1983-1987

Etiologic agent	1983	1984	1985	1986	1987
Bacteria					
Bacillus cereus	0/0	3/23	7/42	4/187	2/9
Brucella	1/29	0/0	1/9	0/0	0/0
Campylobacter	8/162	4/126	9/174	4/227	3/39
Clostridium botulinum	13/46	11/16	17/33	22/27	11/18
Clostridium perfringens	5/353	8/882	6/1,016	3/202	2/290
Escherichia coli	3/157	2/76	1/370	1/37	0/0
Salmonella	72/2,427	78/4,479	79/19,660	61/2,833	52/1,846
Shigella	7/1,993	9/470	6/241	13/773	9/6,464
Staphylococcus aureus	14/1,257	11/1,153	14/421	7/250	1/100
Streptococcus, group A	1/535	2/83	1/12	2/248	1/123
Streptococcus, other	1/16	0/0	0/0	1/69	0/0
Vibrio cholerae	0/0	0/0	1/2	0/0	0/0
Vibrio parahaemolyticus	0/0	0/0	0/0	1/2	2/9
Other bacteria	2/107	0/0	1/152	0/0	0/0
Total	127/7,082	128/7,307	143/22,132	119/4,855	83/8,928
Chemicals					
Ciguatoxin	13/43	18/78	27/106	18/70	11/35
Heavy metals	4/97	3/44	3/13	1/3	2/19
Monosodium glutamate	0/0	2/7	0/0	0/0	0/0
Mushrooms	5/23	2/4	1/4	4/16	2/2
Scombrotoxins	13/27	13/67	15/57	20/60	22/95
Shellfish	0/0	0/0	2/3	0/0	0/0
Other chemicals	10/74	4/16	10/209	5/66	2/6
Total	45/264	42/216	58/392	48/215	39/157
Parasites					
Giardia	0/0	0/0	1/13	2/28	0/0
Trichinella spiralis	4/8	11/60	8/39	6/40	4/15
Total	4/8	11/60	9/52	8/68	4/15
Viruses					
Hepatitis A	10/530	2/29	5/118	3/203	9/187
Norwalk virus	1/20	1/137	4/179	3/463	1/365
Other viruses	0/0	1/444	1/114	0/0	0/0
Total	11/550	4/610	10/411	6/666	10/552
Confirmed total	187/7,904	185/8,193	220/22,987	181/5,804	136/9,652

Source: 454.
Note: Figures represent number of outbreaks/number of cases.

Distinct patterns of agent-factor relationship were observed; e.g., improper cooling, which involved either holding the food at room temperature or storing it in large or deep containers in the refrigerator, was frequently associated with *Clostridium perfringens* enteritis. Contaminated raw foods were most often shellfish, pork, or raw milk. An unsafe source included shellfish from sewage-polluted waters, raw milk, and wild mushrooms.

Foodborne Disease Mortality.   Overall, the case fatality rate associated with foodborne disease is low. Using the 1983-1987 CDC outbreak data, for example, there were only 135 deaths among 54,540 cases or a case fatality rate of 0.2%, although there were marked differences in case fatality rates among the diseases.

The data in Table 3.7 represent the deaths expected among affected persons in a generally healthy population. Foodborne diseases may be much more severe, however, in populations with less resistance. In famine-stricken populations, for example, infection with agents such as the common enteric pathogens and a measles virus can have devastating consequences.

Costs of Foodborne Disease.[267,388,416] The available foodborne disease morbidity and mortality data provide a conservative estimate of the problem because of underreporting. In addition, the data do not indicate the overall cost of foodborne disease both to individuals and society. Many estimates have been published based on such measurable costs as diagnosis, treatment, loss of work, and public control efforts, as well as costs less easy to measure such as pain, suffering, and legal actions. The impact on industry may be limited, as when a single restaurant is closed, or it may be more widespread, as when an outbreak oc-

Table 3.6   Factors contributing to reported foodborne disease outbreaks, United States, 1961–1982

Contributing factor	Percentage
Improper cooling	43.7
Lapse of 12 h or more between preparing and eating	22.6
Colonized person handling implicated food	18.1
Incorporating contaminated raw food/ingredient into foods that received no further cooking	15.8
Inadequate cooking/canning/heat processing	15.5
Improper hot holding	13.3
Inadequate reheating	10.6
Obtaining food from unsafe source	10.0
Cross-contamination	5.4
Improper cleaning of equipment/utensils	5.4
Use of leftovers (also lapse of 12 h or more)	3.3
Toxic containers/pipelines	3.2
Intentional additives	2.4

Source: 454.

Table 3.7.   Case fatality rates from foodborne diseases by etiologic agent, United States, 1983–1987

	Cases	Deaths	Case-fatality rate (%)
Bacterial agents			
Brucella	38	1	2.6
Campylobacter	727	1	0.14
Clostridium botulinum	140	10	7.1
Clostridium perfringens	2,743	2	0.073
Escherichia coli	640	4	0.6
Salmonella	31,245	39	0.12
Shigella	9,971	2	0.02
Streptococcus, other	85	3	3.5
Other bacteria	268	70	26.1
Chemical agents			
Other	371	1	0.27
Parasitic agents			
Trichinella spiralis	162	1	0.62
Viral agents			
Hepatitis A	1,067	1	0.094

Source: 454.

curs among passengers on an international airline flight or cruise ship. For example, estimates of the cost of acute foodborne illness in the United States during a single year in the 1980s ranged from $4.8 billion to $23 billion.

Microbiologic and Parasitic Causes of Foodborne Disease[27,66,94,132,136,170,291,305,374]

Foodborne Infection versus Foodborne Intoxication

Microorganisms may cause foodborne disease in either of two ways. **Intoxications** involve pathologic changes in the host caused by the ingestion of preformed toxins. Foodborne **infections**, on the other hand, result from replication of the microorganisms in the host after ingestion.

Bacteria Associated with Foodborne Intoxication

Two foodborne diseases, staphylococcal food poisoning and botulism (except for infant botulism), result from the action of preformed toxins. Intoxications are especially important because the preformed toxin may remain in the food if resistant to inactivation (e.g., by heat). Incubation period is important in differentiating foodborne chemical and staphylococcal intoxications (7 h or less) from foodborne disease caused by *Clostridium perfringens* and other infectious or toxic agents (8 h or more). Preformed toxins may be absorbed immediately by the intestinal tract whereas organisms must first multiply when disease is the result of infection.

Staphylococcal Toxin[10,43,133,171,180,272,421,440]

Source and Incubation Period. Staphylococcal food poisoning is associated most frequently with coagulase-positive *Staphylococcus aureus* phage types I or III, which produce any of seven serologically distinct protein enterotoxins: A, B, C_1, C_2, C_3, D, and E, with types A and D being more often involved because they can be produced over a wider range of microbial growth conditions. Staphylococcal organisms are ubiquitous. Most commonly, clinical isolates are from the respiratory tract and the skin (pimples, carbuncles, furuncles, suppurating wounds, etc.) of humans and animals, with food handlers being the most important source. The incubation period is very short (1-6 h, typically 2-4 h). Clinical signs include nausea, vomiting, diarrhea, and intestinal cramps.

Foods Often Involved in Staphylococcal Poisoning. Staphylococcal food poisoning is usually associated with meat, including beef, chicken, ham and other pork products, and turkey. Cheese and other dairy products, fish and shellfish, salads, and pastries with cream filling may also be implicated. *S. aureus* is capable of growing in brine (6-10% salt) and elaborating enterotoxin; hence salt-preserved foods are sometimes a source of intoxication.

Prevention. Two factors are important in prevention of staphylococcal intoxication. First, sanitation is necessary to prevent contamination, especially by food handlers. Persons with lesions containing purulent exudate should not be permitted to handle food until proper medical advice is sought. Second, outbreaks can be avoided by storing foods at the proper temperature to inhibit growth (below 4.4°C [40°F]) or to destroy the organisms by heating at 77°C (170°F) for 20 min. Contaminated food must be maintained at 121°C (250°F) for at least 60 min, however, to destroy the toxin. This is impractical for many foods, particularly milk and dairy products. Frequently, a problem occurs when contaminated milk is held at a temperature permitting toxin production before pasteurization or cheese manufacture from raw milk. The heat-stable toxin remains potent under these conditions. Although most staphylococci would have entered the milk from infected personnel, some may have originated in the udder of the cow or goat or may have entered the milk from the environment.

Botulinum Toxins[127,155,164,289,296,309,371,383,408,413,437,447]

Types. Botulism, caused by *Clostridium botulinum,* is another foodborne intoxication. Seven serotypes of polypeptide exotoxin (A-F and M) are produced by various strains of the organism. Only serotypes A, B, E, and F produce botulism in people. Type A is encountered most frequently in North America, whereas in Europe type B is most common. Type E is found in North America, Asia, and Europe. Type F toxin has been associated with meat canning in the home. Botulinum toxins are remarkably potent; people have died from merely tasting food to determine if it was "spoiled." Evidence of spoilage is not a valid indicator of the presence of botulinum toxin. Foodborne, infant, and wound botulism are the three clinical forms.

Source. *C. botulinum* is a ubiquitous saprophyte resident in soil. It is isolated also from the intestinal tract and contaminated wounds. Although type E is frequently associated with seafoods, it is terrestrial in origin, being carried to the sea, where the spores survive in the marine environment including in the gut of coastal fish.

Incubation Period. These toxins are highly neurotoxic, acting by interference with the synthesis and/or release of acetylcholine at nerve endings. Clinical signs are related, therefore, to action of the toxin at peripheral sites of cholinergic nerves. Although it is an intoxication, the incubation period ranges from 2 h to 8 d (average 1-2 d).

Foods Commonly Associated with Botulism. Botulism is commonly associated with certain foods that are canned in the home, such as vegetables and meats, whereas fruits canned in sugar syrup are not. Seafood also has been reported as a source of toxin. In an outbreak in Cairo in 1991 involving uneviscerated salted fish, 18 of 91 patients died. In Canada, 59% of recent outbreaks were associated with ingestion of either raw, parboiled, or "fermented" meats from marine mammals, with fermented salmon eggs or fish accounting for most of the remainder. In 1989, contaminated yogurt was implicated in an outbreak in the United Kingdom, with one death among 27 persons affected.

Control. *C. botulinum* is an anaerobic, spore-forming bacillus. Therefore, killing the vegetative organism by cooking is not sufficient. Canning time-temperature relationships are designed to ensure that *C. botulinum* spores are destroyed (121°C [250°F] for 20 min destroys spores, 80°C (176°F) for 5 min destroys the toxin). The absence of oxygen in canned products provides a favorable environment for sporulation and growth, and it is important that these are destroyed before the product is canned. Strict sanitation is helpful in preventing the introduction of the organism.

Homemade preserved fruit and vegetables were commonly consumed in rural communities until the postwar era, when a wider range of commercial food products became available. Rural recessions and consumer preference for home-grown "unadulterated" food have led to a resurgence in the making of home preserves by people less knowledgeable than their forebears about the need for strict adherence to hygiene and cooking procedures.

Botulism in Babies.[12]   Infant botulism is a form in which botulinal toxin is produced in vivo in the infant gut rather than being ingested preformed in food. The clinical spectrum ranges from mild transient constipation through a more prolonged "failure to thrive" condition to "sudden infant death syndrome." Almost all cases occur in infants 1-6 mo of age. Honey used in infant foods has been a high-risk factor and proven source of *C. botulinum* spores. The wide occurrence of cases suggests that there may be sources in addition to honey.

Bacteria Associated with Foodborne Infection

Many foodborne diseases result from infections. Organisms most commonly involved are *Clostridium perfringens, Campylobacter jejuni,* and the *Salmonella* group.

Clostridium perfringens[149,258,341,372]

Characteristics. *C. perfringens* (type A, C, or D) is an anaerobe, the spores of which may survive cooking (the heat may actually stimulate sporulation), whereas slow cooling allows multiplication. Meat and poultry, gravy, soups, and stews are commonly associated with foodborne outbreaks of *C. perfringens* infection. When the cooked meat or poultry product is taken from the refrigerator and allowed to stand at room temperature, further multiplication of the organisms may occur in the absence of air. The incubation period is 2-22 h (typically 6-8 h). Gastroenteritis is the typical clinical syndrome (usually caused by type A, occasionally type D) and is generally not life threatening in healthy persons, whereas necrotic enteritis (caused by type C) is unusual and is associated with a high case fatality rate. Clinical signs of gastroenteritis, which persist for 8-12 h, usually involve abdominal cramps and watery diarrhea.

Transmission. *C. perfringens* is ubiquitous, and its transmission to food cannot be entirely prevented. It is a common inhabitant of the intestinal tract of humans and animals. Food, therefore, can be contaminated from either source in the abattoir. It is also possible for the organism to be introduced into food by unhygienic food handlers. Foodborne outbreaks caused by *C. perfringens* occur primarily in institutions or commercial food operations where many people handle a food before it is served and where larger quantities (e.g., big roasts) are prepared.

Control. Although *C. perfringens* spores are usually destroyed when the medium is kept at a temperature of 100°C (212°F) for 30 min, this time-temperature relationship often may alter the aesthetic quality of the food. Whereas the pH of raw meat and poultry normally stabilizes at 5.4-5.7 (meat from stressed animals may be 6.0-6.2), *C. perfringens* will multiply at pH 5.0 and above, so this is a suitable medium. It also multiplies at warmer temperatures sufficient to retard growth of other organisms, thereby allowing it to thrive under some conditions associated with poor food handling. Reheating of the food must be adequate to kill the vegetative organisms, particularly if initial heating was inadequate. The basis for controlling foodborne outbreaks is to keep hot things hot and cold things cold; i.e., to prevent replication, cooked foods should not be held at temperatures between 4.4°C (40°F) and 60°C (140°F).

Salmonellae[74,78,79,86,105,130,137,141,142,152,159,163,169,172,189,198,212,252,256,261,282,290,295,300,323,377,388,404,406,409]

The Organisms. There are more than 2,000 *Salmonella* serovars based on somatic (O), capsular (Vi), and flagellar (H) antigens. A few serotypes are host-adapted, i.e. the reservoir hosts for them are limited, as follows:

Host(s)	Serovar(s)
Cattle	*S. dublin*
Humans	*S. paratyphi, S. typhi*
Turkeys	*S. arizona*
Pigs	*S. choleraesuis*
Poultry	*S. gallinarum, S. pullorum*
Sheep	*S. abortus ovis*

The epidemiology of the host-adapted serovars is distinct. Therefore, they are presented later under "Typhoid and Paratyphoid," "Infectious Diseases from Milk or Milk Products," "Eggborne Disease," and " 'Unusual' Foodborne Diseases." *S. typhimurium,* a non-host-adapted serovar, is associated most commonly with animal and human salmonellosis.

Although there are many *Salmonella* serovars, the number prevalent in a country at any one time is relatively small. Non-host-adapted *Salmonella* circulate in a community (human and animal population) by ingestion of foods contaminated with feces and by contact spread (animal-to-animal, animal-to-person, person-to-animal, person-to-person). During the 1980s, the following were some of the predominant non-host-adapted serovars isolated worldwide from people and animals: *S. typhimurium, S. enteritidis, S. heidelberg, S. agona, S. derby, S. infantis, S. saint paul, S. montevideo, S. virchow,* and *S. hadar.*

The Disease. Although most infections are subclinical, if clinical signs develop, the incubation period is usually less than 72 h (typically 12-14 h). Signs include abdominal cramps, diarrhea, vomiting, chills, and fever. The case fatality rate is usually no more than 1-2%, with greatest risk among the very young, old, and debilitated. Whereas no dose-severity relationship was found for *S. typhi* infection, it was evident among infections with several non-host-adapted serovars. Persons who consume more of the vehicle have shorter incubation periods and are affected more severely.

Source. Many foods are involved in outbreaks of salmonellosis including particularly poultry, meats, gravies, eggs, fish, shellfish, or milk. A few outbreaks of salmonellosis have been associated with contaminated cheese. Salmonellosis in infants tends to be sporadic and not associated with foodborne outbreaks, suggesting a source other than food. Nonhuman reservoirs are a source of non-host-adapted salmonellae for human foodborne disease. Salmonellae in rendered animal feeds are important in maintaining the cycle of infection in domestic animals. Beef, pork, and poultry carcasses are frequently contaminated at slaughter with feces. If the intestine of meat birds is ruptured when the "pluck" is removed, the rib cage can become contaminated with feces. "Stuffing" of large birds (turkeys) is thus easily prone to contamination with salmonellae. While the musculature may reach appropriate temperature during cooking, temperatures of the stuffing may only rise to 43°C (109°F), enabling the bacteria to multiply extensively. Infection may persist for years in dairy herds with shedding of the agent in feces and milk, especially where liver fluke is endemic. Antibiotic resistance is often attributed to the feeding of antibiotics to animals. It is of interest that, in New Zealand, multiple antibiotic-resistant strains of *Salmonella* from humans were three times more frequent than from other (animal) sources.

Control. Controlling and preventing outbreaks of salmonellosis involve three general approaches: breaking the cycle, educating the public, and preventing contamination.

Breaking the cycle involves (1) controlling rendering so that animal feeds containing recycled animal by-products are free of viable salmonellae, (2) shortening the interval between the time an animal leaves the farm and its slaughter because this reduces environmental buildup of organisms in stockyards and abat-

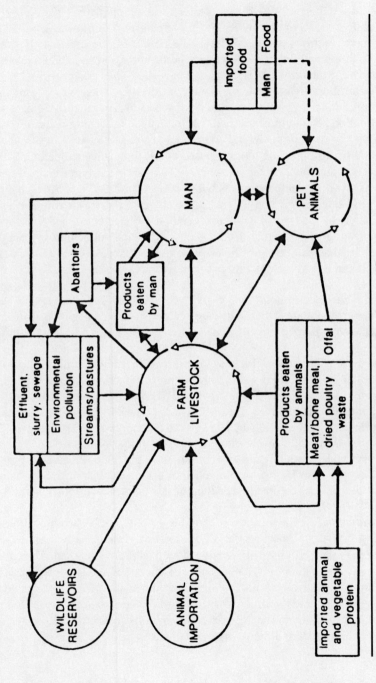

Fig. 3.1. Cycle of *Salmonella* transmission. Source: 17.

toirs (see Chapter 2, "Transport of Livestock and Poultry"), (3) using proper cooking temperatures for food, and (4) pasteurizing milk.

Public education of how *Salmonella* organisms are transmitted is vital to an effective control program. Education should also be concerned with the proper handling and preparation of foods in the home and should be included in the school curriculum.

Contamination may be prevented by making consumers aware that a cooked product can be recontaminated when handled improperly (e.g., a knife and cutting board should be cleaned before they are used for another food after cutting up raw meat or poultry). Foods should be covered and thawed under refrigeration because holding foods exposed to the air at room temperature allows for microbial buildup if contaminated as well as possible airborne contamination. Recently, growth at temperatures between 4° and 10°C (39° and 50° F) has been observed among several non-host-adapted serovars. This raises concern regarding the safety of food stored under refrigeration and then consumed without subsequent heating adequate to kill salmonellae. Although not a popular practice or practical in all areas, it is nevertheless wise to have employees who handle food examined periodically to determine if they are carriers of *Salmonella*.

Typhoid and Paratyphoid.[120,157,302,351,411,412] Typhoid and paratyphoid fevers are nonzoonotic foodborne diseases caused by the serovars *Salmonella typhi* and *S. paratyphi* A, B, and C. Each serovar comprises multiple phage types.

Typhoid. The case of Typhoid Mary provides a dramatic illustration of the potential danger associated with transmission of infectious agents through the food chain. (An account of this case is in *Eleven Blue Men,* by B. Roueche.) The incubation period is relatively long (2-3 wk), and healthy carriers (infection occurs particularly in the gallbladder) may shed the organism intermittently. Clinical signs of typhoid fever are often vague. The disease is suspected in patients with anorexia, headache, fever, and rose spots on the torso, especially if the onset of illness is slow. Diarrhea is not a helpful sign because constipation is more frequent. Because of the insidious nature of this disease, control of typhoid outbreaks is difficult.

Paratyphoid. The incubation period is usually 1-2 d. The clinical onset is abrupt, with fever, malaise, headache, splenomegaly, and usually diarrhea. Rose spots on the torso, and mesenteric lymphadenopathy may be seen. The case fatality rate is lower than for typhoid, and sporadic cases and small outbreaks are more common than in typhoid. As in typhoid, human carriers are a problem in control. Cattle infected with *S. paratyphi* B have been reported associated with abortion and shedding of the agent in feces and milk. Frozen chicken from a retail market in the United States was contaminated with *S. paratyphi* of an unspecified type.

Campylobacter jejuni.[38,52,123,134,151,184,211,230,301,340,373,380,407,446] Two species of *Campylobacter* cause disease in people: *C. fetus* subsp. *fetus* (subsp. *in-*

testinalis), associated with occasional cases of a severe systemic form; and *C. jejuni*, a widely recognized enteric pathogen. Since the 1970s, the latter organism, previously identified as *Vibrio fetus* subsp. *jejuni*, has become recognized as a major cause of foodborne disease. Failure to realize its significance has been the result, in large part, of its requirements for optimum growth, which caused it to be overlooked in procedures used routinely to isolate human pathogens.

The Organism. *Campylobacter* is a gram-negative, microaerophilic, thermophilic organism that grows best at a temperature of 43°C (109.4°F) and requires special media for cultivation. Although it does not replicate at temperatures less than 30°C (86°F), it survives refrigeration and freezing. Because of its pathogenicity (only 500 cells may produce clinical disease) and wide distribution in nature, it is a leading cause of foodborne disease.

The Disease. The incubation period is 2-10 d (typically 3-5 d). Clinical signs include fever, profuse diarrhea (sometimes bloody), abdominal pain, and nausea. The disease usually lasts for 3 d, but illness may reoccur for periods of up to 2 wk. Although it affects all ages, prevalence is highest during the first year of life followed by a second peak during young adulthood (Fig. 3.2). Bacteremia occurs in less than 1% of cases.

Reservoir. *C. jejuni* is ubiquitous in foods of animal origin. Contamination rates of raw poultry products may reach nearly 90%. Most clinically healthy poultry, swine, and cattle excrete *C. jejuni* in their feces and more than 10^6 cells/g may be present.

Transmission. In temperate regions, most cases occur during the warmer months. Although poultry is currently the predominant source, outbreaks have been traced to raw meat or milk. In outbreaks among children who drank raw milk, the attack rates were nearly 50%. The organism survives only briefly in fresh cheese. It is sensitive to drying, so viable cells are not found in dried milk or eggs. *C. jejuni* has been isolated from poultry, meat, water, shellfish, and sewage. In poultry abattoirs, cross-contamination of carcasses during defeathering is believed to be important in spread of the organism. Immature companion animals represent a source of infection, particularly for children. In daycare centers or mental institutions, person-to-person transmission may be common.

Control. Pasteurization of milk and milk products and proper treatment of drinking water would prevent a proportion of human campylobacteriosis cases but would do little to reduce the risk from other sources. Proper production, processing, and preparation of foods of animal origin will reduce the level of contamination, but because the organism is ubiquitous, proper cooking is the only practical control procedure. Handling of raw poultry to avoid cross-contamination of foods is very important. Microwave cooking may not completely remove contamination from meat. Interest has focused on low-level irradiation as an ef-

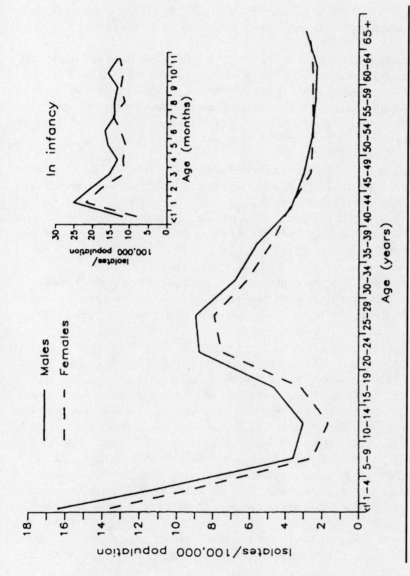

Fig. 3.2. Annual isolation rates of *Campylobacter*, by age and sex, in the United States, 1982-1986.

fective technique for killing *C. jejuni,* particularly in cut-up poultry. This procedure, when coupled with modern impervious packaging techniques, provides a commercially acceptable method for supplying consumers with a *C. jejuni*-free product.

Enterococcus (Streptococcus) faecalis.[128]   *E. faecalis,* a ß-hemolytic member of Lancefield group D, inhabits the intestinal tract of humans and animals. It may contaminate food or milk through fecal-contaminated water or unsanitary food handling and occurs commonly in milk and milk products. There is circumstantial evidence from trials in human volunteers that this organism may cause enteric disease. However, the number of cases reported has decreased drastically in recent years. It has been proposed as an indicator of fecal contamination comparable to the use of *E. coli.* Strict sanitation is the only practical means of controlling this organism because it is fairly resistant to heat treatment, being able to withstand 60°C (140°F) for 30 min.

Hemolytic Streptococci.[46,64,65,97,113,116,255,451]   Hemolytic *Streptococcus* spp. are typed serologically into Lancefield groups based on specific carbohydrate antigens. Interspecies spread is most common with Lancefield groups A (*S. pyogenes*), B (*S. agalactiae*), and D (*S. bovis*), although some consider human and animal strains of *S. agalactiae* to be different. Hemolytic streptococci inhabit the oropharyngeal and nasal cavities of some individuals. Milk and other foods become contaminated when infected individuals create an aerosol by coughing and sneezing. Mastitic cows also may transmit hemolytic streptococci through milk and cause the typical strep throat syndrome and scarlet fever (tonsillitis and pharyngitis). Rheumatic fever or glomerulonephritis may be sequelae, particularly with group A infections. Raw milk was a major vehicle of foodborne streptococcal infection and cause of scarlet fever before the widespread use of pasteurization. More recently, eggs used in a salad or sandwiches and contaminated by food handlers have been implicated in two large outbreaks of streptococcal pharyngitis. The eggs had been boiled and peeled at least 24 h before they were eaten, and the possibility of defective refrigeration existed in both incidents. A *Streptococcus* of the same type-specific T protein antigen was isolated from the patients and a food handler. One food handler had a "cold" and the other pharyngitis on the day the eggs were peeled. The incubation period ranged from <12 h to 3 d. Meat has been incriminated as a source of skin infection with group A and group D (*S. suis*) streptococci among abattoir workers.

Bacillus cereus.[200,258,376,382]   This aerobic spore-forming bacterium, a saprophytic gram-positive rod, has been associated with outbreaks of foodborne disease since the 1950s. It can be isolated from soil, water, and most types of vegetation. Foodborne disease outbreaks are encountered when excessively high numbers are present in food because of poor sanitation or improper handling. Different strains produce either of two toxins associated with their ability to hydrolyze starch (enterotoxin) or not (emetic toxin). The signs produced by the enterotoxin (diarrheal syndrome) are similar to *Clostridium perfringens* infections but are much milder, including nausea and abdominal cramps that persist for 6-12 h. The incubation period for this form is 8-16 h. The signs produced by the emetic toxin (emetic syndrome) are similar to *Staphylococcal aureus* intoxication, including sudden onset of nausea and vomiting in 1-5 h. Recovery from either form occurs within 12-24 h. Foods involved in these outbreaks include boiled and fried rice, pasta, cooked meat and poultry, milk, ice cream, vanilla pudding, and cream sauces. Control measures are essentially the same as for *C. perfringens* infections.

***Shigella* spp.**   Foodborne shigellosis results from unsanitary practices. In many countries, most cases are caused by *S. sonnei.* (Other species are *S. boydii, S. dysenteriae,* and *S. flexneri.*) Outbreaks of this disease are a problem in nurseries and daycare centers where fecal-oral transmission is common among young children. The highest incidence of disease caused by shigellas is frequently in young children not yet toilet trained. Shigellosis has been diagnosed in monkeys and their handlers. A 1-3 d incubation period usually follows direct transmission. The outstanding signs associated with *Shigella* infections include bloody diarrhea and a persistent fever, especially in untreated patients. A carefully taken history or inspection of premises often reveals the possibility of feces-contaminated water, vegetables, or dairy products. Effective control methods involve strict sanitary practices.

***Brucella* spp.**[284,344,392,394]
    The Organisms.   Three species of *Brucella,* each comprising multiple biotypes, infect food-producing animals. The principal reservoirs are cattle (*B. abortus*), goats (*B. melitensis*), and swine (*B. suis*).

    The Disease.   The incubation period is usually between 3 and 21 d, but several months may elapse before clinical signs occur. The recurrent nature of the fever over several months gave rise to the term *undulant* fever. The onset of illness is usually gradual. The typical signs and symptoms are fever, myalgia, sweating, chills, headache, joint pain, depression, and weight loss. Complications may involve arthritis, orchitis, myocarditis, and vertebral osteomyelitis. Subclinical infections occur commonly. Although all of the *Brucella* spp. may produce severe illnesses, they are most frequent among those infected with *B. melitensis* and least frequent among those infected with *B. abortus.* Because *B. suis* produces a more invasive disease than *B. abortus,* it results more often in severe human illness.

    Occurrence.   Worldwide, brucellosis results from ingesting raw milk or milk products from infected cows or goats. In some regions, camel, reindeer, or sheep milk may be a source. Brucellosis is primarily an occupational disease among persons exposed to the reproductive organs of infected animals in areas where milk and milk products are usually pasteurized. High-risk groups include farmers, veterinarians, and abattoir workers. Infectious aerosols are especially important in abattoirs. In the United States, for example, the number of human cases reported annually decreased from more than 5,000 in the late 1940s to less than 100 in the early 1990s as the bovine brucellosis eradication program progressed. The risk of infection from swine is principally from occupational exposure. However, infected pork eaten as undercooked meat or sausage is an important source of *B. suis* infection. Infection as a result of ingesting fresh raw caribou placenta has been reported. In the Sudan, infection with both *B. abortus* and *B. melitensis* has been associated with the practice of eating raw liver and other offal (tissues other than muscle) from cattle and goats. Where the live attenuated *B. abortus* strain 19 vaccine is used in cattle, the organism (which can cause disease in people especially if inoculated) may be shed in milk. Since the eradication of bovine brucellosis from Australia, *B. suis* accounts for all locally acquired cases of brucellosis, predominantly among those hunting feral swine and processing feral pork for export markets.

Control. *Brucella* spp. are killed by pasteurization. All milk, whether to be consumed as liquid milk or to be used in the manufacture of a milk product, should be pasteurized. Wherever brucellosis in animals is prevalent, soft cheeses made from raw milk are a particular hazard because *Brucella* can survive in them for as long as 6 mo. *Brucella* do not survive well in sour milk, sour cream, or cheese ripened for more than 3 mo (hard cheeses). Although the milk and dairy product-consuming public can be protected by pasteurization, eradication of brucellosis in domestic animals is essential as this eliminates occupational exposure.

Mycobacterium spp.[109,158,231,293,297,306,345]
The Organisms. The principal agents of human infection are *M. tuberculosis* (human reservoir) and *M. bovis* (bovine reservoir). *M. africanum* (related to *M. bovis*) affects wild primates and humans in tropical Africa. The *M. avium* complex affects swine and poultry as well as humans, with infection acquired from the environment.

The Disease. A period of 4-12 wk is required for the development of an observable lesion (e.g., lung nodule) or significant tuberculin reaction. Although the greatest risk is within 1-2 yr of infection, it may be decades before progressive disease is evident. Many subclinical infections occur. The two clinical syndromes, pulmonary and extrapulmonary, are related to the two principal routes of transmission—inhalation and ingestion. Prominent signs associated with pulmonary disease include productive cough, fever, weight loss, fatigue, night sweats, chest pain, and hemoptysis. Signs of extrapulmonary disease include cervical lymphadenitis (scrofula), meningitis, osteomyelitis, pericarditis, as well as signs associated with infection of most other organs. Severe outcome is more likely when primary infection occurs in the young or immunodeficient (e.g., AIDS).

Occurrence. Cattle and buffalo are the principal domestic animal reservoir of *M. bovis* infection. Sheep may become infected when grazed on the same pasture with infected cattle, but infection does not persist in sheep flocks. Goats also may become infected when exposed to infected cattle, and on occasion, goat milk may present a hazard if there are udder lesions. Intraspecies transmission of tubercle bacilli among cattle and people is mainly by inhalation; thus pulmonary infection predominates in both hosts. Pulmonary infections with *M. bovis,* presumably from inhalation, have been reported recently among abattoir workers slaughtering cattle in Australia and a veterinarian in Canada who had necropsied an infected farm-reared elk. Early in the 1900s, before milk pasteurization became a common practice, extrapulmonary tuberculosis caused by *M. bovis* was common, particularly in young children. As late as the 1930s, *M. bovis* was the agent involved in 5-6% of all human deaths caused by tuberculosis in the United Kingdom. In a recent survey in Nigeria, *M. bovis* was found in milk from 5% of seminomadic Fulani cattle. This would not be an unusual finding in areas where bovine tuberculosis is still prevalent.

Control. Similar to brucellosis, pasteurization of milk and eradication of infection in domestic cattle are the primary means of eliminating human exposure to *M. bovis.* In fact, milk pasteurization was introduced as a way to protect against milkborne tuberculosis before the hazard of *Brucella* from cows' milk was recognized and the time-temperature standards adopted were based on those required to kill *M. bovis.* In areas where both brucellosis and tuberculosis are endemic in animals, and liquid milk is boiled, but raw milk is used to make soft cheese, human *M. bovis* infection is not as common as brucellosis. Eradication of *M. bovis* in cattle may be complicated by infection in wildlife, e.g., badgers and opossums in the United Kingdom and New Zealand, respectively. As a result of a persistent testing regimen accompanied by a rigid policy of locating and slaughtering TB reactors, Australia is now provisionally free of *M. bovis* infection.

Francisella tularensis. Tularemia may be contracted by the oral, respiratory, and cutaneous routes, especially as a result of the bite of an infected tick or other arthropod vector. Risk of exposure to infected ticks may be associated with occupation (e.g., sheep shearers) or recreation. Exposure may also occur through handling and dressing wild game (rabbits, squirrels, raccoons), ingestion of infected meat, or drinking contaminated stream water. Of the two types (biovars) of *F. tularensis,* type A is more virulent and associated with terrestrial species (e.g., rabbits) whereas type B is less virulent and associated with aquatic species (e.g., muskrats). The incubation period is 1-10 d (typically 3-5 d). The typhoidal form is associated with ingestion of the agent and presents gastroenteritis, headache, chills, fever, vomiting, toxemia, lymphadenitis, and often bronchopneumonia. Strict hygienic habits while handling small game animals aid in preventing infection. Cooking the meat for 10 min at 57°C (135°F) prevents foodborne transmission.

Vibrio cholerae[117,190,240,250,288,357,399,431]

The Agent. *V. cholerae* consists of three serogroups (O:1, atypical O:1, and non-O:1) and two biotypes (*cholerae* and *El Tor*). *V. cholerae* O:1 is the only serotype that consistently produces cholera toxin and is considered the agent of classic cholera. Although not required for growth, the organism will grow in the presence of up to 6% NaCl.

The Disease. The incubation period is from a few hours to 5 d (typically 2-3 d). Clinical signs include profuse diarrhea with abdominal pain, vomiting, headache, and fever. Death is the result of profound dehydration that accompanies clinical illness. Cholera is the only foodborne disease required by International Health Regulations to be reported to the World Health Organization (WHO).

Source. The agent is shed from the body primarily in the feces. In endemic areas, contamination of food as a result of poor hygiene perpetuates the cycle of transmission. Foods associated with cholera include vegetables, fish, and pork products. Sewage (night soil) is used to fertilize vegetables in some countries whereas sewage contamination of drinking water is a major cause of cholera outbreaks. *V. cholerae* O:1 may be found in water of coastal estuaries, and shellfish harvested from these areas are significant sources of infection. Recent evidence indicates that *V. cholerae,* as well as other pathogenic vibrios associated with seafood, lives in estuarine environments adhering to zooplankton.

India exports more than 3,000 tons of frozen frog legs annually, primarily to the United States and France, a high percentage of which have been found to be contaminated with salmonellae. The frog legs are produced in endemic cholera areas; however, there is no surveillance for *V. cholerae.* The organism will survive for more than 28 d at −20°C (−4°F), but no cases of cholera associated with consumption of imported frog legs have been reported.

Control. The provision of safe drinking water has been synonymous worldwide with cholera control following John Snow's classic investigation of cholera associated with London's Broad Street pump. Where combined with adequate sewage treatment, epidemic cholera has disappeared. Careful food handling is needed to prevent contamination by fecal carriers or contaminated water. Water from coastal seafood-harvesting sites should be tested for contamination.

Vibrio parahaemolyticus[13,35,37,59,95,139,206,235,241,286,298,360,431,445]

   The Agent.   *V. parahaemolyticus* is an obligate halophile (i.e., requires NaCl for growth), and several serotypes, involving O and K antigens, are recognized. Clinical isolates produce a thermostable hemolysin. It occurs worldwide in unpolluted coastal water especially where the temperature exceeds 15°C (59°F). It does not flourish in estuary waters.

   The Disease.   Enteric disease develops after an incubation period of 4 h to 14 d (typically 12-14 h). The course of clinical illness is usually about 3 d.

   Source.   Outbreaks of food poisoning result from consumption of large doses of organisms (100,000) in raw, improperly cooked, and/or improperly handled seafood. This organism is commonly associated with seafood and has been isolated from virtually every type of shellfish consumed by humans. Although infections have caused death of certain shellfish, *V. parahaemolyticus* is considered a member of their normal flora. High numbers of hemolysin-producing organisms have been found in sediments of coastal estuaries during the summer months.

   Control.   Control of *V. parahaemolyticus* is fairly simple. Currently, no spread of infection has been recognized from human to human. In areas where raw seafood is consumed, special emphasis must be placed on proper handling and storage. However, numbers are not greatly reduced by depuration methods if oysters come from saltwater areas. The organism is sensitive to refrigeration, freezing, and heat. If products are cooked, thorough penetration of heat >55°C (>131°F) (perhaps up to 65°C [149°F]) is necessary. Also, seafood should not be held without refrigeration for extended periods after it is cooked. Outbreaks of *V. parahaemolyticus* infection have been associated with shrimp and crab "boils" when the batches of shrimp or other shellfish were too large to be boiled at once. In these instances, the shellfish in the center of the cooker were not cooked adequately, thereby allowing organisms to survive and produce foodborne disease.

Vibrio vulnificus.[63,76,205,206,227,268,318,402,443]   Another halophilic vibrio, *V. vulnificus,* is perhaps the most virulent of the vibrios found in estuarine environments. This organism may produce wound infections among healthy persons handling raw shellfish. When persons with pre-existing illness (e.g., liver disease) eat raw seafood containing *V. vulnificus,* septicemia with a high case fatality rate (>50%) may result. The organism evidently has a commensal or symbiotic relationship with oysters.

Vibrio alginolyticus.[2,60,139,175,219,241,242,286,343,367]   *V. alginolyticus,* a halophilic vibrio found in estuarine environments, is frequently isolated from seafood, particularly shellfish, in markets around the world. It is considered a fish spoilage organism and is usually associated with wound infections in people and marine mammals. Only a few human cases of enteric disease have been reported to date.

Aeromonas hydrophila.[29,32,39,53,70,85,182,183,262,263,264,275,313,314,420] Although found in aquatic environments, *A. hydrophila* has been isolated from several other sources. Some strains of the organism produce a hemolysin and an enterotoxin and may cause disease in humans ranging from mild diarrhea to life-threatening febrile gastroenteritis. It was reported that a patient on steroid therapy developed septicemia after eating cockles whereas two healthy persons who ate the cockles did not become ill. Severe illness has occurred in other compromised persons after eating crabs and oysters. The organism grows at 5°C (41°F) on vegetable produce and has been isolated from lamb carcasses and feces (but not wash water) in an abattoir, as well as from catfish and their pond water. The ability of the agent to grow and produce toxin at refrigeration temperatures and its recovery from raw milk, beef, lamb, pork, chicken, fish, and shellfish indicate its potential as a foodborne pathogen. Infections also are associated with the therapeutic use of medicinal leeches.

Yersinia enterocolitica.[95,183,191,207,216,228,244,245,401] Diarrhea, fever, and abdominal pain, usually persisting for several days, are typical of *Y. enterocolitica* infections. The condition can be life threatening in neonates and immunocompromised patients. Asymptomatic infections are not uncommon. In some areas, the incidence of yersiniosis is second only to salmonellosis. Although many *Y. enterocolitica* serotypes exist, only a few, such as O:3 and O:8, tend to be associated with human illness in a given region. Most human infections are associated with ingestion of raw or undercooked pork. Outbreaks, particularly in children, have been associated with exposure to raw swine intestine during preparation of chitterlings in the home. Although outbreaks of enteric disease caused by pasteurized milk are unusual, the ability of *Y. enterocolitica* to grow at refrigeration temperatures makes pasteurized milk a potential vehicle if contaminated. One large outbreak involved chocolate milk. Recently, nosocomial infections have been reported among patients who received transfused blood from donors who had a history of gastrointestinal symptoms within 3 wk of donation. The incubation period ranges from 1 to several days.

Listeria monocytogenes.[41,111,112,131,140,173,183,210,215,270,279,303,315,319,322,364,365,368,369,370,375,397,422,423,438] Listeriosis, and its association with ensilage, has long been familiar to veterinarians as a cause of "circling disease" and abortion in cattle and sheep. More recently, it has attracted attention as a serious human foodborne disease associated with meningitis, septicemia, and abortion. Immunocompromised persons are at high risk of acquiring listeriosis from foodborne exposure. In perinates, the case fatality rate exceeds 25%. The prognosis is poor in untreated advanced cases of meningitis or in the aged. Illness in neonates is characterized as either early onset (<7 d > birth) or late onset (approximately 14 d > birth), and prepartum or postpartum exposure may be inferred. Most of the former are recognized as septicemia and the latter as meningitis. More than half of all intrauterine infections result in early onset neonatal illness with the remainder resulting in fetal death.

Source. *L. monocytogenes* is a non-spore-forming, motile, gram-positive bacillus that is widespread in the environment and has been isolated from most foods of animal origin. There are at least 7 serotypes and 14 subtypes among which a few (1/2a, 1/2b, and 4b) cause a high proportion of human clinical infections. The organism is carried asymptomatically in the intestinal tract with carriage rates of 10-29% found in human population samples.

Epidemiology. Epidemiologic evidence suggests that both outbreaks and sporadic cases of listeriosis are associated primarily with food. The incubation period is 1 d to 2 mo (typically a few days to 3 wk). The organism has been isolated from meat, poultry, and seafood as well as milk and milk products (e.g., soft cheeses). The latter are increasingly associated with outbreaks of human illness. During cheesemaking the organism is concentrated in the curd with relatively few cells remaining in the whey. Although the pH eventually drops from 6.8 to 4.8 or less, the warm milk is a suitable medium for growth of *Listeria* during the first several hours of curd formation. In some soft cheeses, the organism may survive or even multiply during the various stages of ripening whereas most cells are killed in cottage cheese making because, after the curd is cut, it is heated to 55°C (131°F) over 90 min before the whey is drained. Differences between serotypes isolated from meat and those isolated from cattle suggest much of the meat contamination comes from the environment after slaughter. In 1990 in Australia, 16% of cooked chicken and chicken pate sampled was contaminated, whereas none was found in small goods (e.g., salami, ham, bologna). Although raw beef and poultry from refrigerators of sporadic cases had the highest rate of contamination with *Listeria,* a higher correlation occurred between the serotypes from patients and ready-to-eat foods, particularly soft cheeses and foods from store delicatessen counters. During 1987-1989, pate from a single manufacturer in the United Kingdom was associated with an increase in listeriosis involving two phage types of serotype 4b.

Control. Available evidence indicates that *L. monocytogenes* can multiply at pH above 5.5 and survive for long periods in refrigerated foods, as well as multiply in them. Vegetables consumed raw should be thoroughly washed before storing in a refrigerator. An outbreak occurred after consumption of refrigerated shredded cabbage (coleslaw) that had been fertilized with sheep manure. Although numbers may increase in ground beef or sausage during refrigerated storage, *L. monocytogenes* grows well on refrigerated ham and poultry products. Some organisms have been reported to survive temperatures used in high-temperature short-time (HTST) pasteurization. Avoid undercooking of any food of animal origin. There was no effect on viable numbers of *L. monocytogenes* when sausage was heated to a core temperature of 60°C (140°F), but none was found when it was heated to 68.3°C (155°F). Ensure adequate pasteurization of milk and dairy products as well as prevent postpasteurization contamination. Do not consume after manufacturer's expiration date.

Escherichia coli[26,31,92,93,95,96,148,259,308,311,321,335,352,400]

The Organism.   Until the late 1950s, *E. coli* organisms were considered nonpathogenic, gram-negative, usually motile members of the normal flora of the lower intestine of warm-blooded animals. Since then, gastroenteritis in humans has been associated with *E. coli* producing four types of pathogenesis: enteropathogenic *E. coli* (EPEC), enteroinvasive *E. coli* (EIEC), enterotoxigenic *E. coli* (ETEC), and enterohemorrhagic *E. coli* (EHEC). Each type has certain associated virulence factors, as follows: EPEC adheres to intestinal epithelium and may produce cytotoxins; EIEC invades intestinal epithelium and produces Shigalike toxins; ETEC produces heat-labile (LT) and heat-stable (ST) enterotoxins and specific adhesion fimbria; EHEC colonizes the intestine and produces Shigalike toxin. Shigalike toxin (SLT) is also called *verotoxin* (*VT*) because of its toxicity for Vero cells. Therefore, *E. coli* strains producing this toxin are referred to as *VTEC*. Serologic typing of *E. coli* strains is based on somatic (O), capsular (K), flagellar (H), and fimbrial (F) antigens.

The Diseases.   The incubation period for most serotypes is 0.5-5.0 d (typically 12-72 h), whereas for the EHEC serotype *E. coli* O157:H7 the period is longer, between 4 and 8 d. All strains produce diarrhea, which may be complicated by other syndromes often associated with certain specific serotypes. EIEC strains cause fever and dysentery; ETEC strains, dehydration and shock; and EHEC strains, hemolytic uremic syndrome (HUS), hemorrhagic colitis (HC), or thrombotic thrombocytopenic purpura (TTP). Diarrhea and cramps are the usual presenting signs of HC followed later by bloody diarrhea and vomiting. Except for diarrhea in neonates, the case fatality rate is usually low unless the patient has some other underlying disease or a nonenteric syndrome such as HUS or TTP ensues. HUS presents as acute renal failure that may require dialysis and may develop without prior enteric signs.

Source.   Most infections occur as a result of person-to-person contact or ingestion of food contaminated by infected persons. Epidemic multiple antibiotic-resistant *E. coli* (EMREC) has been associated with nosocomial outbreaks that may have been introduced by contaminated food but were particularly effective at subsequent person-to-person spread. Among the four types, three are host-specific, spread solely by infected humans. Only EHEC affecting humans also has a reservoir in animals. Several outbreaks, in which *E. coli* O157:H7 was isolated from sick persons, have been associated with consumption of undercooked ground meat and raw milk. Contaminated cheese or water have been sources of ETEC and EIEC outbreaks. *E. coli* O157:H7 has been found in 1-2%

of retail samples of beef, lamb, pork, and poultry in Canada. For reasons as yet unknown, subclinical infection with *E. coli* O157:H7 occurs only in a small percentage of cattle on relatively few farms. Although undercooked meat has gained publicity as a source of *E. coli* O157:H7 for children with HUS, contact with another child with diarrhea between 2 and 10 d before onset presents a significantly greater overall risk of infection and the subsequent development of HUS.

Control. Sanitary food-handling practices are essential to avoid fecal contamination, particularly excluding anyone with diarrhea from food preparation. Outbreaks of *E. coli* O157:H7 have been associated with contaminated vegetables and fresh apple cider. Particular care is needed in handling and cooking ground meat. Meat trimmings from external portions of a carcass more likely to be contaminated are often used in preparing ground meat. It should be heated to an internal temperature of at least 63°C (145°F) or greater 68.3°C (155°F) because excessive contamination is not unlikely. Milk must be pasteurized because raw milk is not sterile and fecal contamination with *E. coli* O157:H7 from healthy carrier cows is possible.

Viruses and Rickettsiae Associated with Foodborne Disease[9,67]

Infectious Hepatitis Virus.[449]   Type A or infectious viral hepatitis (IVH) of humans is differentiated from type B (serum) hepatitis on the basis of incubation period and other features. The incubation period for type A is 15-50 d, for type B, 2-6 mo. Signs and symptoms for either type include fever, jaundice, anorexia, gastrointestinal irritation, and a low case fatality rate. Type A has been contracted through dairy products, shellfish, and homemade bakery items, but almost any food unhygienically prepared may be a source of infection. Bivalve molluscs (e.g., clams, mussels, and oysters) are particularly important because, being filter feeders, they concentrate the virus when growing in sewage-polluted waters. Water is also an important vehicle for transmitting the agent. IVH is almost always associated with poor personal hygiene and poor sanitation.

Small Round-structured Viruses.[143,161,320]   A caliciviruslike agent was first implicated as the cause of an outbreak of gastroenteritis with low-grade fever in Norwalk, Ohio. This agent, the Norwalk virus, was responsible for nearly half of all U.S. foodborne disease cases in 1982, even though associated with only two outbreaks—an indication of its infectivity given the proper conditions. Several other serotypes have been identified since, including Hawaii and Snow Mountain, resulting in a situation similar to human influenza in which cross-immunity is limited. The incubation period is 12-48 h. Both foodborne and waterborne transmission can occur via the fecal-oral route. Patients shed the viruses in feces for a few days after onset of illness. When bivalves from polluted waters are involved, more than one virus may be transmitted resulting in gastroenteritis followed by hepatitis a few weeks later.

Poliomyelitis Enterovirus. Poliomyelitis is potentially a foodborne disease. However, direct fecal-oral contact was the principal mode of transmission in outbreaks until the disease ceased to be a major problem after the introduction of immunization in the 1950s. Raw milk or unsanitary handling of pasteurized milk had been suspected in some polio outbreaks, but the most plausible explanation of foodborne poliomyelitis is contamination by infected persons (fecal shedders).

Coxiella burnetii.[28,118] This organism causes Q fever, a respiratory disease that is widely distributed in the world. Reservoirs include cattle, sheep, and goats. The organism, which is highly resistant to drying, may be shed by ruminants in uterine fluids at parturition. People become infected by inhalation. The incubation period is 2-3 wk. Infected animals may harbor *C. burnetii* in the mammary gland and excrete it in milk for long periods (200 d reported). Raw milk has been implicated as a source of human infection in some cases. *C. burnetii* is relatively heat-resistant and may survive pasteurization by the long time holding (LTH) method, but not by the HTST method. Unless the number of infective particles is very high, however, pasteurization by the LTH method reduces the number of infective particles below that of an infective dose. To ensure safety, some countries have increased the temperature of the LTH method from 61.7° to 62.8°C (143° to 145°F) (see Chapter 2).

Fungi Associated with Foodborne Disease[234,346,433]

Many fungal species have been demonstrated to produce toxins (mycotoxins) able to affect laboratory animals experimentally. Under natural conditions, however, only a few are known to cause illness in people and animals. Ergotism and mushroom poisoning have long been recognized, whereas aflatoxin and ochratoxin are mycotoxins more recently associated with illness.

Ergotism. *Claviceps purpurea*, growing on cereal grains, produces the mycotoxin ergotamine, the cause of ergotism. The toxin causes vasoconstriction, tissue necrosis, and death. In the past, human outbreaks were particularly associated with contaminated rye, but the condition has largely disappeared with modern cleaning and milling of grain used for flour. Cases are still seen occasionally in livestock fed ergot-containing hay or grain. The heat produced by baking bread or cooking pancakes denatures most, if not all, of the alkaloid toxin.

Mushroom Poisoning. Some mushrooms, members of the basidiomycetes, are edible and some are toxic (e.g., *Amanita* spp.). Edible mushrooms grown commercially are nontoxic. Most poisoning, which may result in convulsions and death, is the outcome of misidentification of mushrooms collected in the wild. Because of this mode of collection, only a few people are affected in a single incident. Cattle may be affected when they graze on toxic species. This form of foodborne disease can be prevented only by careful differentiation between edible and toxic types of mushrooms.

Aflatoxin.[342,444] Some species of *Aspergillus,* particularly *A. flavus* and *A. parasiticus,* produce aflatoxins (AFs), hepatotoxins that cause disease in poultry, ruminants, swine, and people. The most common sources have been contaminated peanuts, cottonseed meal, corn, and various grains. Icterus and other signs of hepatic disease have been associated with acute aflatoxicosis in people who have consumed moldy grain. AFs have also been incriminated as a factor in the etiology of kwashiorkor in young children as well as primary liver cancer. Although several AFs are usually found together on contaminated foods of plant origin, AFB_1 is generally the most abundant and is the most toxic. When AFB_1 is absorbed by lactating mammals, it can be converted to another toxic metabolite, AFM_1, and excreted in the milk within 48 h. AFM_1 is stable in dairy products and not inactivated by pasteurization. When the surface of cheese is contaminated with *Aspergillus,* particularly while it is stored under refrigeration, it generally does not support growth of the fungus and subsequent toxin production. Muscle meat does not maintain significant residue concentrations, but gross signs of hepatopathy are often visible in the livers of affected animals. Suspected foods can be analyzed for the presence of aflatoxin by laboratory procedures.

Ochratoxin. Certain species of *Aspergillus* and *Penicillium* produce ochratoxins, the most common naturally occurring form being ochratoxin A, which causes renal disease in swine, poultry, and people. Balkan endemic neuropathy is a chronic kidney disease usually affecting people 30 yr of age and older, and females more than males. The human and porcine nephropathies are similar. In ruminants, ochratoxin A is converted by gut microorganisms to nontoxic ochratoxin a. Although significant concentrations of ochratoxin A may be found in kidneys of affected pigs, muscle concentrations are considerably less. The toxin may be found in the muscle and eggs of affected poultry.

Parasites Associated with Foodborne Disease [25,56,273,361,379]

Taeniasis.[84,87,156,359] The adult stage of two cyclophyllidian tapeworms, *T. saginata* and *T. solium,* parasitizes the small intestines of humans. The larval stage (cysticerci) of these parasites occurs in meat from cattle (a condition called *beef measles*) and swine (*pork measles*), respectively. (See Figs. 3.3. and 3.4.)

Taenia saginata. *T. saginata,* the beef tapeworm of humans, is one of the largest cestodes, with an average length of 5-10 m. Cattle become infected when their feed or pastures are contaminated with feces of humans infected with *T. saginata.* Risk of infection is greatest when cattle are grazed on pastures irrigated with inadequately treated human sewage and when cattle are fed for extended periods in feedlots. Embryophores (ova) swallowed by cattle hatch in the duodenum and liberate onchospheres. These enter the lymphatics or blood vessels of

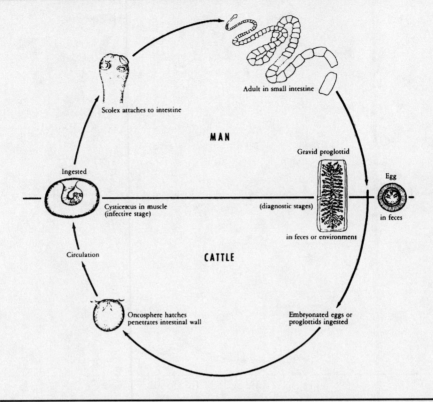

Fig. 3.3.   Life cycle of *Taenia saginata*. Source: 273.

the hepatic portal system and spread to the muscles via the general circulation. They develop to oval infective cysticerci in 60-75 d. After consumption by humans of inadequately cooked beef, cysticerci attach to the intestinal mucosa and grow to sexually mature tapeworms in about 3 mo. Gravid (ova-containing), actively motile segments detach from the strobilus in the intestines and are passed in the feces. For this reason, individual ova outside a segment are difficult to find in feces by flotation procedures. The number of ova available for cattle infection may be great, inasmuch as an infected person often expels 8 or 9 segments daily, each of which contains 80,000-100,000 ova. Ova do not survive desiccation, but retain infectivity in moist pastures for 60-70 d at 20°C (68°F) and 80 d at 10°C (50°F). (See Fig. 3.3.) The only known ways of preventing contamination of pasture irrigated with human sludge containing *T. saginata* ova are by slow filtration of effluent through sand and microstraining techniques. In practice, some ova survive treatment and infect cattle. Carcasses are judged fit for human consumption by rigid meat inspection procedures.

Taenia solium.   The life history of *T. solium* is similar to *T. saginata* except that it involves man and swine. In contrast to *T. saginata,* however, em-

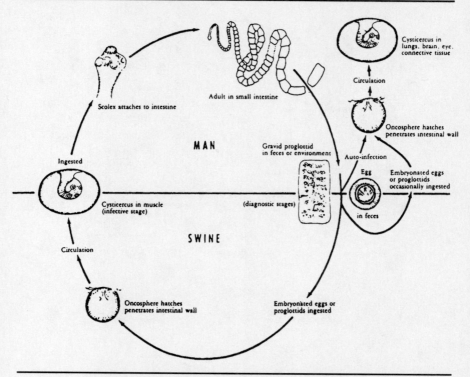

Fig. 3.4.   Life cycle of *Taenia solium*. Source: 273.

bryophores of *T. solium* in human feces ingested by people hatch and the larvae invade tissue, causing cysticercosis. Humans are commonly infected by the consumption of inadequately cooked pork containing cysticerci, resulting in the development of *T. solium* tapeworms in the human intestine. Alternatively, humans may be infected by the ingestion of *T. solium* ova from an infected (carrier) person or may result from autoinfection, resulting in a more severe disease than that produced by *T. saginata.* The parasite then develops the cystic manifestations of disease with cysticerci invading muscles throughout the body and may result in heart failure when the myocardium is involved. Brain involvement (neurocysticercosis) usually results in severe neurologic disturbance and may be fatal. Removal of the cysticerci is a frequent reason for brain surgery in endemic areas. Massive infections are usual in swine, and cysticerci show little predilection for specific muscle sites. In light infections, greater numbers are found in muscles of the upper portions of the limbs, abdomen, and diaphragm. *Cysticercus cellulosae* is used sometimes to refer to the larval stage in pork measles. (See Fig. 3.4.)

Echinococcosis.[71,310]   Hydatid disease, caused by the cysts of *Echinococcus granulosus,* may occur in domestic ungulates and swine where the parasite is endemic. When cysts develop in the liver, it is considered unfit for human food. Typically, human infection with this parasite occurs as a result of direct ingestion of ova in feces from infected dogs, the reservoir host of the adult tapeworm. Dogs become infected from eating uncooked tissues, particularly lungs and other offal, of infected intermediate hosts. A fatal case of hepatic hydatid disease in the United Kingdom caused by *E. multilocularis,* a nonendemic species of *Echinococcus,* was attributed to consumption of a Swiss cheese imported from an endemic area where the milk was not pasteurized and poor hygienic practices were associated with milking. Presumably, the milk was contaminated with feces from an infected carnivore. Ova of a related taeniid, *T. taeniaformis,* experimentally survived the cheesemaking process and subsequently infected mice.

Trichinosis.[6,16,220,223,307,328,329,330,348,391,450] *Trichinella spiralis* is a nematode parasite of carnivorous and omnivorous animals. Pigs, rats, bears, and humans are hosts. Adults occur in the small intestine, and larval stages occur in musculature. Trichinosis is more prevalent in the temperate northern hemisphere and arctic than in the tropics and southern hemisphere.

Life Cycle. Transmission of *T. spiralis* is primarily by consumption of muscle containing encysted larvae. They are released, develop to maturity in the duodenum and jejunum, copulate, and produce new larvae, which invade the lacteals and lymph spaces 5 d after infection for about 40 d. Larvae migrate via lymphatics and the general circulation to striated muscles. Larvae encyst and must then be ingested by a new host for the cycle to continue. (See Fig. 3.5.)

Five species of *Trichinella* have been proposed, based on geographic distribution, host species, and other characteristics. *T. spiralis* is cosmopolitan and associated with domestic swine, whereas the others are associated with bears or other sylvatic hosts and have more-limited geographic distribution (e.g., Africa). They all cause clinical disease in humans.

Human Infection. Most *T. spiralis* infections in humans are subclinical. Estimates from cadaver sampling indicated that the prevalence in the United States decreased considerably from 15-20% in the 1940s to 2-4% in the 1970s. Outbreaks continue to occur, however, particularly among Asian and European ethnic groups who consume raw or undercooked pork dishes. The case fatality rate ranges from 1 to 2%. Even a few larvae (1-50/g) in human muscle are considered a heavy infection, and 1,000/g are considered critical. Clinical infections in domestic animals occur but are not diagnosed except at necropsy. The most common sources of *Trichinella* are pork (including wild boar) and bear meat whereas in Europe trichinosis has been associated with horse meat. However, cysts have not been demonstrated in horse meat associated with outbreaks, and unless proven otherwise, trichinosis associated with beef or horse meat is most likely the result of cross-contamination with infected pork.

Toxoplasmosis.[1,44,98,99,100,101,115,324,339,378] *Toxoplasma gondii* is a protozoan parasite with a life cycle involving a tissue phase (tachyzoites, bradyzoites in

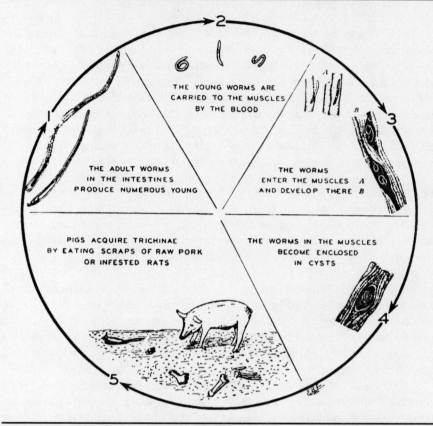

THE YOUNG WORMS ARE
CARRIED TO THE MUSCLES
BY THE BLOOD

THE ADULT WORMS
IN THE INTESTINES
PRODUCE NUMEROUS YOUNG

THE WORMS
ENTER THE MUSCLES   A
AND DEVELOP THERE   B

PIGS ACQUIRE TRICHINAE
BY EATING SCRAPS OF RAW PORK
OR INFESTED RATS

THE WORMS IN THE MUSCLES
BECOME ENCLOSED
IN CYSTS

Fig. 3.5. Life cycle of *Trichinella spiralis*. Source: 455.

cysts) found in many host species and an intestinal phase (oocysts shed in feces) recognized only in felids. The parasite is usually transmitted to humans through the ingestion of undercooked meat from infected food animals or unwashed vegetables contaminated by cat feces. Persons handling cat feces in domestic kitty litter trays may also be exposed to infection. Infection may also occur through ingestion of raw milk. Early developing fetuses of exposed mothers without *Toxoplasma* antibody are most vulnerable to disease. The results of serologic sampling indicate *T. gondii* infection is very common (up to 50% or more of adults have antibody), but disease is uncommon. Antibody prevalence is high in some ethnic groups who do not eat pork and who have virtually no exposure to cats; hence it is likely that unidentified sources of *T. gondii* infection exist. Sheep, swine, cattle, and goats have been infected experimentally with oocysts of *T. gondii,* and the cysts remained viable in the tissues for many months. Clinical illness (e.g., abortion) as a result of natural infection is often detected in sheep and swine, whereas toxoplasmosis is not known to be a problem in cattle. Antibody prevalence may exceed 30% in sheep and swine, whereas it seldom exceeds 1%

in cattle. (These differences cannot be explained simply on the basis of differing risk of exposure to cat feces.) Nevertheless, when digestion or other techniques have been used to isolate the parasite, similar infection rates have often been found in the three species. Therefore, we must assume a similar risk of human infection from the meat or milk of these species until adequate evidence to the contrary is presented.

Human disease can range from inapparent subclinical infection to debilitating systemic involvement producing central nervous system damage, chorioretinitis, and abortion. Surveys have revealed that certain individuals (farmers, veterinarians, butchers) have a higher frequency of antibodies to *T. gondii* than the general population, indicating that infection may be an occupational hazard.

Prevention consists of educating the public not to consume raw or undercooked meat or unpasteurized milk and to wash and peel fruit and vegetables before consuming them raw. Smoking and brine curing destroy bradyzoites in infected pork, the value of freezing is disputed. Inasmuch as *T. gondii* oocysts shed in the feces of cats require 36-48 h development before they become infective, indoor kitty litter trays should be emptied and sterilized daily. Pregnant women without *T. gondii* antibody should avoid cleaning litter pans and avoid contact with cats of unknown feeding history.

Hands should be washed after handling raw meat and especially before eating, when there has been potential for skin contamination with cat feces, such as gardening. Gloves should be worn by pregnant women while gardening or preparing meat.

Cryptosporidiosis.[165,179] Cryptosporidiosis, caused by the coccidian *Cryptosporidium parvum,* is usually an enteric disease (principally diarrhea) in immunocompetent persons but may result in a life-threatening systemic condition in immunocompromised patients. Although cryptosporidiosis outbreaks are most often associated with unfiltered water supplies (the organism is resistant to chlorination), ingestion of raw sausage, tripe, and milk as well as contact with farm animals have been implicated. Subclinical enteric infection is common in cattle and sheep.

Sarcocystosis.[350] Two species of the coccidian *Sarcocystis* spp. are known to infect people, one of the definitive hosts, following the consumption of undercooked beef (*S. hominis*) or pork (*S. suihominis*) containing cysts. Ultimately, sporocysts are shed in the feces, which can infect cattle or swine, the intermediate hosts, when ingested. Human infection may result in diarrhea, but most infections are asymptomatic. Cats and dogs are other definitive hosts.

Endogenous Parasites of Fish. Fish are a source of several trematode, cestode, and nematode parasites that may affect people.

Trematodes. Trematode infections occur as a result of eating raw or salted fish, bivalve shellfish (such as clams or mussels), or snails, all of which may be secondary intermediate hosts containing encysted metacercariae. Various fish-eating mammals and birds serve as definitive reservoir hosts. Light infections are often asymptomatic. Occurrence of trematode syndrome and its symptoms follow:

Amphimerus pseudofelineus (Western Hemisphere) fever, diarrhea, and icterus

Clinostomum complenatum (Japan and Middle East) cough and pharyngitis

Clonorchis sinensis (Southeast Asia) fever, diarrhea, anorexia, and biliary obstruction

Echinostoma spp. (Asia and the Philippines) colic and diarrhea

Haplorchis spp. (Australia and the Mediterranean) diarrhea

Heterophyes spp. (Asia and the Pacific Islands) abdominal pain and mucoid

Metagonimus yokogawai (Asia) mucoid and diarrhea

Opisthorchis spp. (Europe and Asia) fever and icterus (*O. felineus, O. viverrini*)

Paragonimus westermani (Southeast Asia, Philippines, Africa, and the Pacific Coast of South America) cough, hemoptysis, and sometimes CNS effects

***Diphyllobothrium* spp.** Although freshwater fish, e.g., walleyes and northern pike, are the most frequent intermediate hosts of broad tapeworms, *D. latum* or *D. pacificum*, some marine species, e.g., salmon, may also be infected. Eating undercooked fish is the primary reason why humans become infected with *Diphyllobothrium* spp. Infections may also result from the preparation of gefilte fish if the mixture is tasted before it is cooked or from the consumption of raw fish (sushi). (See Fig. 3.6.)

***Angiostrongylus* spp.** Fish and land crabs may be paratenic hosts of the nematodes *A. cantonensis* (Pacific and Southeast Asia) and *A. costaricensis* (Western Hemisphere). Rodents are the definitive hosts. Humans become infected from eating undercooked paratenic or intermediate hosts (such as snails or slugs). Unwashed salad is a source of exposure. Infection in humans may result in fever, abdominal pain, anorexia, vomiting, and diarrhea (abdominal form). Invasion of the CNS occurs commonly, resulting in headache, stiff neck and back, and parasthesia.

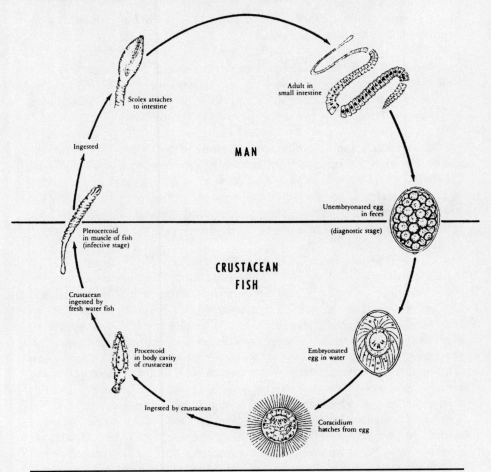

MAN

Scolex attaches
to intestine

Adult in
small intestine

Ingested

Unembryonated egg
in feces

(diagnostic stage)

Plerocercoid
in muscle of fish
(infective stage)

CRUSTACEAN
FISH

Crustacean
ingested by
fresh water fish

Embryonated
egg in water

Procercoid
in body cavity
of crustacean

Ingested by crustacean

Coracidium
hatches from egg

Fig. 3.6. Life cycle of *Diphyllobothrium latum*. Source: 273.

Anisakidae. Marine mammals (such as dolphins, porpoises, or seals) serve as the normal definitive host for the adult stage of anisakid nematodes. Marine fish are either intermediate or paratenic hosts of *Anisakis simplex* (*A. marina*), *Contraceacum* spp., and *Pseudoterranova dicipiens*. The parasites cause severe eosinophilic granulomas or ulceration in the human gastrointestinal tract 4 h to 10 d after ingestion of encapsulated larvae (2 cm long) in the viscera and flesh of raw, salted, or pickled herring, cod, mackerel, or other fish. Larvae are killed by deep-freezing (−20°C [4°F] for 24 h) or heating to 55°C (101°F) for at least 10 s. The popularity of sushi has increased the number of reported cases.

Capillaria philippinensis. Freshwater fish in Southeast Asia are intermediate hosts of *C. philippinensis* and humans become infected from eating raw or undercooked fish containing infective larvae. Land animal hosts have not been found. Human infection may be life threatening, with signs of abdominal pain, intermittent diarrhea, and weight loss.

Dioctophyma renale. Humans may become infected with *D. renale* by eating fish containing infective larvae, e.g., bullhead fish in the United States. The dog is the definitive host for this parasite, and humans are an accidental host. Transmission from human to human has not been demonstrated.

Gnathostoma spp. Humans may develop visceral larva migrans from eating undercooked fish infected with larvae of *G. spinigerum* or, less commonly, *G. hispidum.* Although diagnosed most often in Asia, cases have been reported in Europe, Australia, and the western hemisphere. Fish-eating mammals are the definitive hosts. Infection may also occur by third-stage larvae penetrating the skin where fish or frog flesh is used as a poultice.

Infectious Diseases from Milk or Milk Products[91,126,337,453]
Bacterial Diseases.[7,8,344,363,397,407,446]   The principal diseases associated with the consumption of milk or milk products are caused by members of several bacterial genera. The reservoir for these bacteria may be milk-producing mammals (either cows or goats), humans, or contamination from the environment. In diseases such as brucellosis, salmonellosis, listeriosis, campylobacteriosis, or tuberculosis, milk may be already contaminated with the causative bacteria when it leaves the udder. Because milk is an excellent medium for replication of some bacteria (propagative vehicle), human carriers of pathogenic bacteria (such as *Streptococcus* spp., which cause scarlet fever and rheumatic fever) are potential sources of contamination. Milkborne typhoid is now fortunately rare, but some epidemics have been catastrophic. One such episode occurred in Melbourne, Australia, in 1943 when 440 cases were attributed to a single typhoid carrier working in a dairy that was bottling raw milk. Milk contaminated with rat excreta has been a vehicle for *Streptobacillus moniliformis,* a cause of rat-bite fever. Milkborne transmission of the causative agents of anthrax (*Bacillus anthracis*), nocardiosis (*Nocardia asteroides*), and pasteurellosis (*Pasteurella multocida*) is considered a possibility, albeit remote under the usual course of infection. Melioidosis occurs in humans and animals as a result of infection with *Pseudomonas pseudomallei,* an organism that resides in the soil of certain tropical and subtropical environments. The occurrence of *Ps. pseudomallei* in goat milk in northern Australia has raised the possibility that raw goat milk could be a vehicle. (See Tables 3.8 and 3.9.)

Salmonella dublin infection.[108,260,384]   In cattle, the reservoir host *S. dublin* causes acute febrile disease with severe diarrhea (particularly in calves) and abortion when initial infection occurs during pregnancy. The agent is shed in feces of carrier cattle, as well as in milk from lactating cows. The disease in people typically involves fever, abdominal pain, diarrhea, and nausea, with an acute

Table 3.8. Bacterial diseases associated with milk or milk products

	Reservoir			
Diseases caused by gram-negative agents	Cow	Goat	Human	Environment
Brucellosis				
Brucella abortus	+			
Brucella melitensis		+		
Campylobacteriosis				
Campylobacter jejuni	+			
Enteric disease				
Enteropathogenic *Escherichia coli*	+	+	+	
Salmonella dublin	+			
Salmonella typhi			+	
Salmonella spp.	+	+	+	
Shigella spp.			+	
Melioidosis				
Pseudomonas pseudomallei		+		(soil reservoir)
Rat-bite fever				
Streptobacillus moniliformis				+ (rat)
Yersiniosis				
Yersinia enterocolitica	+			+ (feces)

Table 3.9. Bacterial diseases associated with milk or milk products

	Reservoir			
Diseases caused by gram-positive agents	Cow	Goat	Human	Environment
Clostridial foodborne disease				
Clostridium botulinum (intoxication)				+ (soil)
Clostridium perfringens (infection)	+ (feces)	+ (feces)	+ (feces)	+ (soil)
Diphtheria				
Corynebacterium diphtheriae			+	
Listeriosis				
Listeria monocytogenes	+	+	+	+
Q fever				
Coxiella burnetii	+	+		
Staphylococcal food poisoning				
Enterotoxigenic *Staphylococcus aureus*	+	+	+	
Streptococcal infection				
Streptococcus pyogenes			+	
Other Group A *Streptococcus* spp.			+	
Tuberculosis				
Mycobacterium bovis	+	+		
Mycobacterium tuberculosis			+	

episode of 7 d (range of 1-28 d). An underlying illness reducing resistance (e.g., cancer, hepatic disease, diabetes, and steroid therapy) is a major risk factor. Most patients have a history of consuming raw milk or milk products; however, ingestion of raw calf liver extract intended as a therapeutic nutritional supplement has been implicated in some cases.

Infectious Diseases Other than Bacterial.[396] Risk of milkborne human infection by agents other than bacteria is generally remote. *Toxoplasma gondii* has been discovered in raw cows' and goats' milk and has been implicated as a source of infection for humans. The foot-and-mouth disease (FMD) virus may be shed in milk. The tickborne encephalitis virus (a group B arbovirus) is shed in goat milk, and cases resulting from milkborne transmission have occurred in Asia. Although the rickettsial agent of Q fever, *Coxiella burnetii,* may be shed in milk (particularly colostrum), milkborne transmission is of minor importance.

Relative Importance of Foodborne versus Milkborne and Waterborne Transmission.   Although milk is a potential vehicle for transmission of infection to humans, its importance has been reduced drastically in many countries by careful sanitation and pasteurization. Today, disease outbreaks of microbial origin are associated most often with raw or inadequately pasteurized milk and manufactured dairy products. Soft cheeses present a greater hazard (particularly if made from raw milk) than hard cheeses because ripening of hard cheeses is more effective in killing pathogens. Similarly, the quality of municipal water supplies has been improved so that potable water is the norm in many areas. Waterborne disease outbreaks still occur, however, when there are breakdowns in the distribution system allowing contamination.[194] (See Fig. 3.7.)

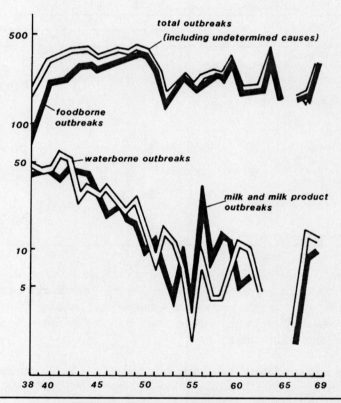

Fig. 3.7.   The relative importance of foodborne versus milk- and waterborne disease outbreaks in the United States, 1938-1968.

Eggborne Disease[209,232,442]

Use of Hatchery Eggs.[11,40,125,153,160,225,229,424] Throughout this century, chicks for poultry and egg production have come from commercial hatcheries, in each of which many hundreds or thousands of eggs are set daily. These eggs are initially incubated at 37.5°C (99.6°F) and 50-60% relative humidity (29°C [82°F] wet bulb). Usually after 4 d, the eggs are candled to identify those without a developing embryo. The "infertile" eggs are removed at this time. The numbers of these eggs are not inconsequential, and their disposal can present a problem. Over the years, infertile eggs have been recycled as livestock or poultry feed. Until the 1960s, it was also a common practice in many countries to sell infertile eggs to bakeries for use in recipes requiring eggs. Recycling infertile hatching eggs to poultry has caused outbreaks of pullorum disease. This is not surprising inasmuch as *Salmonella pullorum* is eggborne and incubator conditions are ideal for its replication. It is more remarkable, however, that millions of infertile hatching eggs were recycled for human consumption for many years without associated outbreaks of foodborne disease being reported despite the fact the egg is a good medium for microbial growth. Some of the bacterial foodborne pathogens may increase infertility in hatching eggs and a consequent increase in the numbers of infertile eggs discarded by the hatchery. In 1966, an estimated 15 million infertile hatching eggs were produced in the Federal Republic of Germany where dipping hatching eggs in an antibiotic solution was a common practice. There was concern because antibiotic-treated eggs delivered to bakeries could not be distinguished from untreated eggs. Although this practice reduced viable bacteria on the eggshell, it did not affect pathogens within the eggshell membranes. An episode of *S. hessarek* infection in humans arising from the use of inadequately sterilized egg powder derived from "incubator clears" has been reported in Australia.

Salmonellosis.[4,5,17,19,20,21,22,62,68,73,81,89,102,103,104,110,162,168,186,187,192,222,269,283,325,336,347,354,355,356,387,393,395,409,432,434] The clinical features of eggborne salmonellosis are similar to those described earlier associated with other foods. Its epidemiologic features (particularly those involving host-adapted *Salmonella* spp.) provide the clues to recognition and prevention of eggborne transmission.

Salmonella enteritidis. In the late 1970s, the number of reported foodborne disease outbreaks caused by *S. enteritidis* began to increase in Europe and North America. Foods containing raw or partially cooked eggs were often incriminated, and by the late 1980s, eggborne *S. enteritidis* had become a major cause of foodborne disease in several countries, with catastrophic results to industry. In 1988, for example, 46 outbreaks of *S. enteritidis* infection involving more than 1,000 people occurred in the United Kingdom. A 50-70% decrease in egg sales occurred. The market took many months to recover, but rigid enforcement of health and sanitary regulations for production and marketing have benefited consumers. Episodes have occurred even though quality eggs and accepted egg-handling and storage practices were used. Frequently, it has been possible to trace the incriminated eggs back to a specific source supplied by a limited number of laying flocks. Some outbreaks have been restricted geographically, as in the northeastern United States. The eggs may become contaminated internally via the transovarian route or as a result of an ascending infection from the cloaca, whereas external contamination of the eggshell may occur from enteric infection. The percentage of shell eggs from a flock infected with *S. enteritidis* with organisms in the internal contents is estimated to be from <0.1% to as high as 8.0%. Clinical manifestations are not evident in infected flocks. In 1990, the United States began testing flocks implicated in outbreaks and prohibited the sale of fresh eggs from *S. enteritidis*-positive flocks. Serologic tests of flocks provide presumptive evidence followed by culture to confirm the presence of the agent. Risk of infection is somewhat greater among children because eggs occur more frequently in their diet. Persons with diabetes mellitus are also at greater risk, possibly because of decreased gastric acidity and intestinal motility allowing greater opportunity for penetration of the gut and thereby systemic infection. The factors responsible for the recent international evolution of eggborne *S. enteritidis* infection in poultry have not been elucidated.

Salmonella pullorum and Salmonella gallinarum. *S. pullorum* and *S. gallinarum* are adapted to avian hosts, and the egg is the principal means of transmission as a result of ovarian infection. The two serotypes share antigens detected by agglutination, the diagnostic test used to eradicate pullorum disease and fowl typhoid from domestic poultry (chickens, ducks, and turkeys) in many countries since its introduction in the 1930s. Biochemical and epidemiologic differences exist between them although in Europe both are designated as *S. gallinarum*. Both serotypes are nonmotile and grow more slowly on culture medium than other salmonellae, thereby suggesting possible reasons for underreporting by laboratories involved primarily in examination of specimens from human patients. In spite of the paucity of reports, both organisms may cause serious human illness. For example, 29 of 45 persons who ate a wheat grits pudding made with raw eggs infected with *S. gallinarum* became very ill within 12 h. In an outbreak of *S. pullorum* gastroenteritis among military personnel, 172 of 423 affected were hospitalized with serious febrile illness. A rice pudding was incriminated in which hot steamed rice was added to a milk-raw egg mixture and allowed to "set." The incubation period for most cases was 4-12 h, with a maximum of 35 h. Experimentally, the number of *S. pullorum* organisms required to produce illness in human volunteers was considerably greater than for other *Salmonella* spp. tested. A single infected egg, however, can have numbers of *Salmonella* far in excess of those reported to be the minimum infectious dose.

Other Salmonella spp. During the 1940s-1950s, spray-dried or frozen egg products were principal vehicles in the international movement of the most common *Salmonella* spp. involved in foodborne outbreaks. Eggs are still an important source of infection in both outbreaks of salmonellosis, especially in restaurants and other public institutions, and sporadic infections. Enteric infections with non-host-adapted *Salmonella* spp. continue to occur among poultry flocks, with the list of *Salmonella* spp. involved similar to that found in mammalian host species. Feces containing salmonellae is the principal source of eggshell contamination. Shell contamination can be amplified when contaminated eggs are washed in a commercial egg washer with defective sanitation. Contamination is further enhanced if the eggshell is cracked, thus exposing the outer shell membrane. The contents of the egg may become contaminated whenever eggs with contaminated shells are broken out. Evidently, ovarian infections occur in flocks infected with non-host-adapted *Salmonella* spp., but their relation to eggborne disease has not been established as it has been for host-adapted species.

Prevention.   The time-temperature relationships typically involved in boiling, frying ("sunny-side up" is less effective than "over easy"), or scrambling eggs are insufficient to kill all eggborne salmonellae. Sale of soft-cooked eggs maintained in the shell at ambient temperature for extended periods should be prohibited. Boiling for at least 6-10 min is needed to kill salmonellae in the egg. Even more time will be needed if the eggs were kept under refrigeration rather than at room temperature immediately before cooking. However, eggs should be stored continuously under refrigeration before use (preferably no longer than 3 wk). Misuse of equipment, particularly microwave and other ovens, is another reason for insufficient cooking. Pasteurization helps to reduce contamination in liquid eggs, which may be particularly important when they are used in hospitals and other institutions where the consumers are likely to have lowered resistance to disease. To avoid eggborne salmonellosis, the eggs used in food preparation should be limited to those obtained from *Salmonella*-free flocks.

Other Zoonotic Pathogens.   Many zoonotic agents other than *Salmonella* spp. may infect birds; however, none is known to be eggborne. Although Newcastle disease virus can be isolated from eggs of infected hens and *Chlamydia psittaci* has been recovered from eggs of ducks, there is no evidence of eggborne transmission among birds or to people. Recovery of *Toxoplasma gondii* from an egg has been reported, but passage to the egg is unusual.[317] None of more than 100 eggs from hens with reproductive tracts infected with *T. gondii* was found to be infected.[201]

Nonmicrobial Food Poisoning[188]

The two major categories of foodborne disease not of microbial or parasitic origin are **physiologic adverse reactions to food** and **foodborne chemical intoxications**, which are presented in the following. The diseases in these categories are clearly described clinical entities with well-defined diagnostic criteria. Recently, illnesses referred to by various terms, including *food* and *chemical sensitivities* or *environmental illness,* have been reported.[410] These illnesses lack clinical definition, and consequently the therapy used has not always proven effective.

Physiologic Adverse Reactions to Food[90,248,249]

Some people have an intolerance to certain foods that may be immune-mediated or an idiosyncrasy.[448]

Many foods containing proteins are allergenic, and certain people develop a hypersensitivity reaction to them.[57,403] Repeated exposure often results in more serious reactions, as a result of histamine release, varying from a mild pruritic rash to severe abdominal cramps and even death. Milk and seafood are regularly associated with allergy, in contrast to meat and poultry.[16] When soya milk was substituted for cows' milk in the formula of newborn infants with a family history of allergic disease, no difference was noted in the frequency of allergy or wheezing later in childhood, whereas there was a lower incidence among children who had been breast-fed.[51] It is important to recognize, therefore, that cows' milk in the infant formula per se may not be responsible for the childhood allergy.

Food idiosyncrasy is the term used to describe adverse reactions to food that do not have an immunologic basis, including reactions that may be metabolic or pharmacologic in origin. An example is the syndrome of burning, numbness, and a tight sensation in the upper torso experienced by some persons who consume large quantities of the food additive monosodium glutamate, used to enhance flavor in Chinese food. Although widely publicized, "Chinese restaurant syndrome" is not easily reproducible experimentally. Lactase deficiency, affecting the ability to digest lactose in milk, is another well-known food idiosyncrasy.[129,146]

Foodborne Chemical Intoxications [236,247,277,278,280,287,299,312,333,425]

Hypervitaminosis A.[30,55,82,107,174,193,246,281,358,362,366,381,398,452] Excessive ingestion of vitamin A may result in acute or chronic toxicity. Signs of the former include abdominal pain, nausea, vomiting, severe headaches, dizziness, sluggishness, and irritability followed within a few days by generalized desquamation of the skin. Signs of chronic toxicity include bone and joint pain, hair loss, dryness and fissures of the lips, anorexia, benign intracranial hypertension, weight loss, and hepatomegaly. Acute toxicity may occur in adults who consume, in a brief period, several hundred times more than the recommended daily intake of 4,000-5,000 International Units (IU), whereas chronic toxicity results from consumption, over an extended period, from an excess of 10 times the recommended daily intake. Vitamin A toxicity has been known since ancient times among Eskimos who ate polar bear or seal liver containing several thousand IU/g of vitamin A. In the early 1900s, antarctic explorers who ate the livers of their sled dogs were similarly affected. In recent decades, cases of toxicity have been associated with consumption of excessive quantities of vitamin products and cod liver oil, as well as regular consumption of quantities of chicken or beef liver over extended periods.

Nitrate Salts. Nitrates have been used commonly in foods to increase shelf life. Miscalculations and mistaken identity are the usual reasons for excessive amounts being added during processing. The use of nitrate salts is controversial because they may, through a series of reactions, result in the production of nitrosamines that are carcinogenic precursors. The carcinogenic potential of any preservative must be considered before it is used in food.

Sodium Nicotinate and Sodium Sulfite.[332] Nicotinic acid, or its salts such as sodium nicotinate, has been used to maintain the bright red color in meat by its reaction with hemoglobin and myoglobin. If consumed in quantities of 50 mg or more, it can cause generalized vasodilatation with cutaneous symptoms of itching, burning, heat, and redness.

Sodium sulfite has been used to mask deterioration in meat. It can be very dangerous if consumed by persons who are allergic to it.

Plant Sprays (Pesticides).[42,47,106,122,124,166,197,304,428] Organic and inorganic insecticides, rodenticides, or herbicides are potential foodborne disease agents. Public education about the dangers of these compounds (often containing mercury or arsenic) is the best way to reduce risks of toxicity. All fruits and vegetables should be assumed to be treated and not safe unless properly handled (washed, peeled, etc.).

Zinc, Cadmium, and Copper.[48,69,426] Metals may contaminate a food by a leaching process between the food and its container. Old refrigerator shelves containing cadmium have been used as barbecue grills. Acids from meat cause the cadmium to leach and adhere to the meat, enabling it to be subsequently ingested. Similarly, galvanized garbage cans may be sources of zinc when used to store fruit drinks at large gatherings. Zinc and cadmium may cause severe abdominal pain and diarrhea. Acidic fruit drinks in copper tubing also may be a source of copper toxicity.

Mercury.[77,426] Heavy metal buildup in animals is an important example of disease agents being transmitted through the food chain. Mercury in fungicides may contaminate runoff water and eventually be concentrated in fish. Often, the mercury is taken up by microflora that serve as food for more-complex organisms in a chain leading to fish and humans (e.g., Minamata disease). Similar intoxications may occur when animals fed chemically treated feeds are eaten by humans (e.g., organic mercury poisoning resulting from eating pork from swine fed treated seed corn). Effluent from an industrial procedure has caused mercury poisoning in humans as a result of drinking water contamination. Similar signs occur in swine and humans as a result of demyelination of the central nervous system.

Drugs.[45,147,167,326,333] Veterinarians must be aware that the use of drugs, such as growth hormones, antibiotics, and insecticides, may leave residues in the tissues of food animals, and these may cause illness in humans. Cheese makers may have problems with milk from antibiotic-treated cows, whereas people who are sensitive to antibiotics, especially penicillin, may develop allergic responses after drinking milk. Fewer cases have been reported since the introduction of stringent controls on antibiotic usage in milking cows. On the other hand, confirmed antibiotic sensitivity reactions have not been reported in association with consumption of red meat or poultry. In fact, foodborne intoxication from drug residues in meat or poultry was unknown until the recent outbreaks in Europe in which people developed tremors, headache, tachycardia, and dizziness 1-3 h after eating beef liver containing the ß-agonist clenbuterol, which had been used illegally as a growth promoter. The concentrations in beef liver needed to produce human illness would probably have resulted from an overdose in cattle sufficient to cause clinical signs.

Plants.[203,208,226] Animals may concentrate toxic substances in their tissues or milk from ingested plants. When the tissues or milk are eaten by other animals, foodborne disease occurs. White snakeroot is an example of this phenomenon; cattle eat the plants and concentrate the toxic principle in the milk, and it is then consumed by humans. The clinical illness, referred to as *milk sickness* or *tremetol poisoning,* has a geographic distribution associated with the range of these plants.

Naturally Occurring Poisons from Animals[3,61,138,427,430]

Most of the naturally occurring poisons in foods of animal origin are toxins of marine finfish and shellfish. The major toxic foodborne diseases associated with fish are ciguatera, scombroid fish poisoning, paralytic shellfish poisoning (PSP), neurotoxic shellfish poisoning (NSP), amnesic shellfish poisoning (ASP), diarrheic shellfish poisoning (DSP), and puffer fish poisoning (PFP). All of these are heat-stable toxins; i.e., they are not destroyed by cooking. In all cases except for scombroid poisoning, the toxin is present at the time of harvest and is present in food as a result of accumulation and biomagnification through the food chain. Scombroid develops in certain fish species as a postmortem change. None of these toxins is detectable by organoleptic inspection. All of these except scombroid fish poisoning are limited geographically and occur in a limited number of fish species. Human illness can be widespread, however, because of tourism and the international seafood market.

Ciguatera.[213]   Ciguatera is a disease caused by eating certain fish species from tropical waters, particularly waters near reefs and islands. It causes acute gastrointestinal illness within 3-5 h after ingestion. Neurologic signs follow 12-18 h after ingestion and may persist for months or even years. Paralysis and death are rare, but the neurologic signs and symptoms may result in long-term disability.

The toxin is believed to be derived from a dinoflagellate algae, *Gambierdiscus toxicus,* which is ingested by the fish. Other algae and bacteria, including a marine vibrio, may also be involved in the production of ciguatoxins. Fish eating the algae become toxic, and the effect is magnified through the food chain. More than 400 species of fish have been implicated, including amberjack, snapper, grouper, barracuda, goatfish, and reef fish belonging to the Carrangidae. Many of these fish are no longer sold commercially because they have been frequently associated with ciguatera.

The worldwide importance of ciguatera as a cause of foodborne illness is hard to estimate, although it is not uncommon in the Caribbean and in the Pacific Islands. The risk of contracting ciguatera is low for most consumers outside of endemic areas. In the United States, ciguatera occurs primarily in Florida and Hawaii and is responsible for about half of the outbreaks of foodborne illness caused by seafood toxins. Its occurrence in Canada is more sporadic in nature and is due exclusively to the importation of affected fish.

Scombroid (Histamine) Fish Poisoning.[88]   Scombrotoxin is actually histamine that is produced by the bacterial decarboxylation of histidine in the muscle of fish after harvest. Histidine is naturally present at high levels in so-called scombroid fish, including tuna, mackerel, bonito, and saury. Mahimahi (dolphin), bluefish, and other fish with dark flesh are also associated with outbreaks of human disease. The toxicity of histamine is potentiated by other amines, such as cadaverine and putrescine, that are present in spoiled fish.

Clinical signs of scombroid poisoning in humans include labial edema, urticaria, an oral burning sensation (a peppery taste), nausea, vomiting, facial flushing, headache, and gastric pain. The disease is mild and self-limiting, with rapid onset and duration of only a few hours. Because of this, it may be significantly underreported.

The bacteria responsible for histamine production are primarily Enterobacteriaceae, *Clostridium, Lactobacillus,* and possibly *Vibrio* spp. In particular, *Morganella (Proteus) morganii* and *Klebsiella pneumoniae* are the most frequently implicated. These are mesophyllic bacteria that require temperatures greater than 15°C (29°F) for growth. Thus, scombroid poisoning in fish is a consequence of improper handling and temperature abuse postharvest.

Histamine is present in fish at toxic levels long before spoilage is obvious to taste or odor. Because of this, histamine content may be used as an index of spoilage in certain species of fish. The FDA has set 20 mg histamine per 100 g of flesh (200 ppm) as an indicator of spoilage in tuna. Levels of 50 mg per 100 g (500 ppm) are potentially hazardous to human health.

Shellfish Toxins.[54,349,419]   Dinoflagellates and other photosynthesizing plankton are the primary nutrients, either directly or indirectly, of all marine life. Under certain conditions, certain types of plankton multiply tremendously, resulting in an accumulation of millions and millions of organisms at the sea surface. These algal "blooms" vary in color from white to red (the "red tide") and, because some of these organisms produce toxins, are followed by increased morbidity in animals that feed directly or indirectly upon them. Fish and bird kills are often coincident with these events.

Bivalve molluscs are filter feeders, meaning that they gather food by trapping fine particles, including dinoflagellates and other small marine organisms from seawater. Some of these organisms, which produce toxins, are assimilated and temporarily stored in molluscan shellfish. The toxin-containing particles removed from the environment are concentrated up to 1,000 times or more and accumulate in different compartments of the mollusc. The mollusc, unlike many of its predators including humans, is not affected by the toxins it accumulates. Commercially harvested bivalves, including clams, oysters, mussels, and scallops, can contain lethal quantities of toxins even in the absence of a visible bloom.

The major known shellfish toxins are associated with four distinct syndromes: paralytic shellfish poisoning, neurotoxic shellfish poisoning, diarrheic shellfish poisoning, and amnesic shellfish poisoning.

Paralytic Shellfish Poisoning.   Paralytic shellfish poisoning (PSP), caused by a potent neurotoxin, is potentially life threatening to humans. Ingestion of a single clam, if contaminated heavily enough, can kill a person. Signs of PSP appear within an hour of eating shellfish and, in nonlethal cases, subside within a few days. Signs include numbness and tingling of the mouth and fingertips, ataxia, incoherence, aphasia, rash, fever, and respiratory paralysis. Death occurs in a small percentage of cases. There is no known antidote. The disease is caused by a group of heat-stable toxins known as *saxitoxins,* produced by species of *Gonyaulox* and *Alexandrium.* These species are prevalent on the east and west coasts of North America, respectively. Coastal shellfish-monitoring programs, which prohibit harvest during periods when toxin levels are high, have been effective in controlling this potent health hazard.

Neurotoxic Shellfish Poisoning.   Neurotoxic shellfish poisoning (NSP) is associated with the classic "red tide" organism *Gymnodinium breve,* which occurs in the Gulf of Mexico and can be carried north in the Gulf Stream. It produces classic fish kills, but human fatalities are rare. NSP is much less frequently reported in humans (at least in North America) than paralytic shellfish poisoning.

Diarrheic Shellfish Poisoning.   Diarrheic shellfish poisoning (DSP) is, as its name implies, a gastrointestinal illness caused by a family of at least five toxins of dinoflagellates and is associated with mussels, scallops, and clams. DSP is common in Japan and becoming a problem in Europe. It has not been reported in North America.

Amnesic Shellfish Poisoning. Amnesic shellfish poisoning (ASP), associated with eating blue mussels contaminated with domoic acid produced by the diatom *Nitzschia pungens,* was reported in Canada in 1988. In the initial outbreak, domoic acid was associated with gastroenteritis, confusion, and short-term memory loss. Three deaths (all among elderly patients) were recorded out of 107 cases in the original outbreak. Toxic mussels were traced to two estuaries with commercial grow out facilities on Prince Edward Island. The toxin has since been found in razor clams and crabs harvested from the Pacific coast of the United States, and testing programs for domoic acid have been instituted in both countries.

Puffer Fish Poisoning. Tetrodotoxin is found primarily in puffer fish (hence the common name for the disease, *puffer fish poisoning* [PFP]), but it may also be present in the octopus. The disease in humans is similar to paralytic shellfish poisoning. There are many species of puffer fish, and toxicity is quite variable between and within species. The origin of the toxin is widely believed to be the fish itself, although there is some recent evidence that it may be of food chain origin. Like other diseases related to seafood toxins, its distribution is sporadic. No cases of puffer fish poisoning have been reported in the continental United States since 1974. Puffer fish toxicity is a major cause of fatal food toxicity in Japan, where puffer fish, or fugu, is consumed as a delicacy. Deaths occur in Japan in spite of strict government regulations on the marketing and restaurant preparation of this fish. There is some controversy about the toxicity of Atlantic species found off the coast of North America. The Food and Drug Administration (FDA) has recently lifted a ban on importation of puffer fish from Japan, but the safety of puffer fish for human consumption is still a murky issue.

Adulterants in Milk
Antibiotics.[265,285] Whenever a lactating cow is treated by any route with an antibiotic, measurable levels of the antibiotic are usually detectable in the milk for a few days after the last treatment. This is of economic importance because antibiotics affect the microorganisms used to ripen cultured dairy products. Persons with hypersensitivity to antibiotics, particularly penicillin, may develop allergic reactions if they consume contaminated milk. Penicillin is not inactivated by pasteurization or drying, and levels as low as 0.03 IU/ml have caused skin rashes. Another concern is the development of antibiotic-resistant strains of disease-producing microorganisms as a result of the use of antibiotics in food animals. Legally, milk offered for sale must be obtained from healthy cows and must not contain unwholesome substances. The presence of antibiotics in milk constitutes adulteration, and milk must be withheld from sale for a specific period of time after therapy (usually 72-96 h) to ensure that no residues persist.

Even with the possibility of a recognized withholding period for the milk, few drugs are authorized for use in lactating cows.

Chemicals.   Milk may be contaminated with **toxic chemicals** either before or after it is removed from the cow.[239]

Vitamin D Toxicity.[178,202]   For more than 50 yr, vitamin D has been added to milk as an economical and effective means of preventing rickets, with 400 IU/quart the current standard. Outbreaks of hypervitaminosis D have occurred in persons of all ages, however, when milk containing excessive quantities of vitamin D was consumed, resulting in hypercalcemia. The condition is also associated with retarded growth in infants. Because milk fortified with vitamin D is important in the diet, careful surveillance of the fortification process is needed to ensure that the vitamin D concentration in milk is correct. In the United Kingdom, vitamin D fortification of milk is no longer permitted because of outbreaks of toxicity. This decision has led to a greater frequency of rickets and osteomalacia.

Insecticides.[266]   Insecticides, particularly chlorinated hydrocarbons, may be absorbed after direct exposure during application or from ingestion of foodstuffs treated in the field. As much as 20% of an ingested chlorinated hydrocarbon may be excreted in milk. Organic phosphates, however, are not usually excreted in milk. Indiscriminate use of insecticides on farms where milk is handled, or in processing plants, may be a source of contamination. From 25 to 62% of market milk supplies sampled during the 1950s contained significant levels of DDT. Because chlorinated hydrocarbons are concentrated in milk fat, butter may contain higher levels of these insecticides. The tolerance level for insecticides in milk products is **zero**. Therefore, these chemicals must not be applied to lactating cows and must be prevented from contaminating their feed.

Other Drugs Excreted in Milk.[316]   Many plants and therapeutic agents contain compounds that affect milk's color, odor, or taste, as well as its safety. Milk from cows fed phenothiazine will develop a pinkish color in a few hours after milking. Cattle that have consumed either white snakeroot (*Eupatorium rugosum*) or jimmyweed (*Haplopappus heterophyllus*) may develop a neurologic disease (trembles) from the plant toxin tremetol. Tremetol poisoning, or milk sickness, may be fatal in persons who consume milk from affected cows.

Radionuclides.[243]   Contamination of the milk supply with **radionuclides** has become a concern of health officials during the last 50 yr. Strontium-90 (radiostrontium), with a half-life of 28 yr, becomes incorporated in growing bone; iodine-131 (radioiodine), with a half-life of 8 d, is concentrated primarily in the thyroid. These two radionuclides are of major concern in milk, and others such as cesium-137 and barium-140 may also find their way into milk. The biologic effects of radionuclides depend on the tissue in which localization occurs and the type of radiation (α, β, or γ). Radionuclides may present a health hazard through the milk supply whenever they contaminate the environment in which milking animals are kept or their feed is produced. Significant environmental contamination occurred in several countries as a result of atmospheric testing of nuclear devices. In England, an area of 200-300 mile^2 (300-450 km^2) was contaminated with iodine-131 as a result of a reactor fire. Because of this, approximately 250,000 gal (1,000,000 l) milk from 600 exposed herds was condemned. In 1986, the Chernobyl disaster in the Ukraine focused worldwide attention on the hazards of a nuclear accident. Widespread contamination of reindeer milk in the Arctic region, as well as domestic milk supplies in several European countries, occurred as a result of the fallout. If a contaminating radionuclide has a short half-life (such as I-131), it is possible to store the contaminated product until the radioactivity decreases to an acceptable level. Decontamination by methods such as ion exchange is also possible.

Chemical Residues in Eggs[58,75,80,83,119,121,135,177,181,195,196,214,218,224,274,276,294,331,385,390,414,435,436]

Laying hens may be exposed to chemicals from feed or water, from sprays or dusts applied to them or their environment, as well as from chemically treated bedding if they are housed on the floor. Chemicals from all of these sources have been absorbed by hens and subsequent residues passed in eggs. Organochlorine pesticides, such as DDT, dieldrin, heptachlor, and lindane, in feed are absorbed and within a few weeks reach peak concentrations in the egg yolk. When fish meal containing polychlorinated biphenyls (PCBs) is used as a protein source, the PCBs appear in the egg yolk. If seed grain treated with methyl mercury fungicide is used in poultry feed, the compound appears in eggs. When antibiotics, such as tetracyclines, the fumigant ethylene dibromide, or feed grown where the herbicide atrazine is used are added to poultry rations, residues of these compounds will appear in eggs. Because of the potential for residues in eggs, many therapeutic drugs and pesticides should not be administered to laying hens. For some drugs used to treat pullets, an extended withdrawal period may be needed before they begin to lay. In some instances, it may be necessary to test bedding as well as feed and water to detect sources of exposure.

"Unusual" Foodborne Diseases

Foodborne disease is usually associated with endemic etiologic agents and commonly eaten foods. If either the pathogen and its food vehicle or the person consuming the food involved is "exotic" to that environment, the ensuing foodborne disease would be an "unusual" event. Even if the "food" consumed were found locally, but was not a common food item, the foodborne disease would still qualify as unusual. Common events occur commonly and are often easier to recognize.

One reason people travel to foreign lands is to partake of exotic foods prepared in native surroundings. Although this may be romantic, and international tourism is a huge industry, it is not without hazard.[36,163,353] The quality of food hygiene practice varies greatly around the world, largely in relation to the effectiveness of the enforcement of appropriate hygienic standards. (In many instances, the only serious effort is tourist industry driven.) The old advice to tourists, "Don't drink the water!," is well-known and sound. Less well-known, but equally sound, is, "avoid having ice in your drinks!" Contaminated water supplies (including water for ice making) are a common source of human illness in many areas and a significant source of food contamination in these same areas. Potable water is essential to safe food preparation. In these areas, the food animals are more likely to be infected with many of the potentially foodborne zoonotic agents. There is also a greater risk of contamination of food because of unsanitary practices at harvest (or at milking or slaughter) compounded by unsanitary postharvest handling, particularly inadequate refrigeration. To avoid a foodborne misadventure, eat only foods that have been well cooked and served **hot immediately** to ensure that the pathogens have been killed. Food items such as ice cream, soft cheeses, and meringue are notorious sources of foodborne illness wherever raw milk may have been an ingredient (even local pasteurization is sometimes faulty) or refrigeration may be questionable. The infectious and parasitic pathogens mentioned earlier in the chapter are also the agents most likely to be involved in foodborne episodes where unsanitary conditions are common. The vehicle may be unusual whereas the agent may not be, e.g., monkey meat as the source of *Salmonella enteritidis*.[238] When foreign travelers are involved, these episodes are more likely to be considered "unusual" and sometimes become newsworthy such as when cholera occurs among passengers who ate a meal on an international airline flight.

The intestinal parasite *Giardia intestinalis* is a common diarrheal agent worldwide, with contaminated water the usual vehicle for the transmission of its cysts.[194,217] Although infection in cattle and sheep is common, reports of spread via foods from these species are few. In one report, family members in Turkey ate a soup prepared from sheep tripe, which was prepared from the rumen or reticulum and was probably contaminated with gut contents. Although tripe from ruminants is a relatively common food ingredient, it is usually cooked sufficiently to preclude survival of intestinal pathogens.

Larvae of a nematode parasite of rats in the Pacific region and Southeast Asia, *Angiostrongylus cantonensis,* infect people when they consume undercooked land snails, one of the intermediate hosts of the parasite. The larvae invade the human central nervous system, resulting in eosinophilic lesions and CNS signs. Halzoun is an infrequent human pharyngeal infection with *Linguatula serrata;* it arises from ingestion of ova or nymphs of parasites present in raw or undercooked liver and lymph nodes of ruminants. Human infection with the liver fluke *Fasciola hepatica* occurs after consumption of encysted cercariae on plants such as watercress from pastures contaminated with ruminant feces in endemic areas. Several zoonotic parasites are transmitted to people by foods of plant origin or water that has been contaminated with feces of infected animals. These diseases may present a considerable diagnostic challenge when they involve persons from nonendemic areas or occur only infrequently.

Sometimes, animal tissues are consumed in the belief that they have medicinal or aphrodisiac properties. Although organisms harbored by these animals may be pathogenic when ingested, they are not always what we consider to be usual foodborne pathogens. *Salmonella arizona,* for example, is associated with disease in ethnic groups that consume snake meat as medicine.[33,154] Human "unusual" infections with several parasite species are associated with ingestion of various amphibia or reptiles as aphrodisiacs.

Significance of Food as a Vehicle for Disease

The significance of food as a vehicle for disease agents has been recognized since antiquity. Adam and Eve were forced to depart from Eden because of a food-associated problem, and Socrates drank hemlock. The results of offering poisoned food to one's enemies were widely chronicled in the Middle Ages. Parsley on the dinner plate is symbolic of the medieval sign left by the household food taster indicating that the food was safe to eat. Oftentimes, diseases were mistakenly blamed on ingestion of infectious agents or toxins. For example, it was only a few decades ago that inhalation (rather than ingestion) was finally recognized as the principal mode of spread of *Mycobacterium bovis* among cattle.[293] Today, we continue to identify "new" foodborne pathogens and characterize various dietary factors that present risks to health. Also recently, speculations on foodborne spread of oncogenic viruses[292] and *Borrelia burgdorferi* from animals to people[114] have been published. However stimulating they may be intellectually, speculations should be carefully assessed so they are not added unnecessarily to the growing list of unwarranted public concerns about food.

Bibliography

1. Abdel-Hameed, A.A. 1991. Sero-epidemiology of toxoplasmosis in Gezira, Sudan. *J. Trop. Med. Hyg.* 94:329–332.

2. Aggarwal, P., M. Singh, and S. Kumari. 1986. Isolation of *Vibrio alginolyticus* from two patients of acute gastroenteritis. *J. Diarrhoeal Dis. Res.* 4:30.

3. Ahmed, F.E. (ed.). 1991. *Seafood Safety.* Washington, D.C.: National Academy Press.

4. Ahmed, S., A.E. Jephcott, R.E. Stanwell-Smith, et al. 1992. Salmonellosis associated with "Combi-oven" cooked egg. *J. Public Health Med.* 14:68–71.

5. Altekruse, S., J. Koehler, F. Hickman-Brenner, et al. 1993. A comparison of *Salmonella enteritidis* phage types from egg-associated outbreaks and implicated laying flocks. *Epidemiol. Infect.* 110:17–22.

6. Ancelle, T., J. Dupouy-Camet, M.-E. Bougnoux, V., et al. 1988. Two outbreaks of trichinosis caused by horsemeat in France in 1985. *Am. J. Epidemiol.* 127:1302–1311.

7. Anderson, P.H.R., and D.M. Stone. 1955. Staphylococcal food poisoning associated with spray-dried milk. *J. Hyg. (Cambridge)* 53:387–397.

8. Anonymous. 1977. Milk-borne outbreaks of *Salmonella typhimurium. Br. Med. J.* 1(6076):1606.

9. Appleton, H. 1990. Foodborne viruses. *Lancet* 336:1362–1364.

10. Arbuthnott, J.P., D.C. Coleman, and J.S. de Azavedo. 1990. Staphylococcal toxins in human disease. *J. Appl. Bacteriol. Symp. Suppl.* 19:101S–107S.

11. Arhienbuwa, F.E., H.E. Adler, and A.D. Wiggins. 1980. A method of surveillance for bacteria on the shell of turkey eggs. *Poult. Sci.* 59:28–33.

12. Arnon, S.S., K. Damus, B. Thompson, et al. 1982. Protective role of human milk against sudden death from infant botulism. *J. Pediatr.* 100(4):568–573.

13. Aryanta, R.W., G.H. Fleet, and K.A. Buckle. 1991. The occurrence and growth of microorganisms during the fermentation of fish sausage. *Int. J. Food Microbiol.* 13:143–155.

14. Baba, K., N. Takeda, and M. Tanaka. 1991. Cases of *Yersinia pseudotuberculosis* infection having diagnostic criteria of Kawasaki disease. *Contrib. Microbiol. Immunol.* 12:292–296.

15. Bahna, S.L. 1978. Control of milk allergy: A challenge for physicians, mothers and industry. *Ann. Allergy* 41:1–12.

16. Bailey, T.M., and P.M. Schantz. 1990. Trends in the incidence and transmission patterns of trichinosis in humans in the United States: Comparisons of the periods 1975–1981 and 1982–1986. *Rev. Infect. Dis.* 12:5–11.

17. Baird-Parker, A.C. 1990. Foodborne salmonellosis. *Lancet* 336:1231-1235.

18. Baker, R.C., J.P. Goff, and J.F. Timoney. 1980. Prevalence of salmonellae on eggs from poultry farms in New York State. *Poult. Sci.* 59:289–292.

19. Baker, R.C., R.A. Qureshi, T.S. Sandhu, et al. 1985. The frequency of salmonellae on duck eggs. *Poult. Sci.* 64:646–652.

20. Barnes, G.H., and A.T. Edwards. 1992. An investigation into an outbreak of *Salmonella enteritidis* phage-type 4 infection and the consumption of custard slices and trifles. *Epidemiol. Infect.* 109:397–403.

21. Barnhart, H.M., D.W. Dreesen, R. Bastien, et al. 1991. Prevalence of *Salmonella enteritidis* and other serovars in ovaries of layer hens at time of slaughter. *J. Food Prot.* 54:488–491, 495.

22. Barrow, P.A., A. Berchieri, Jr., and O. Al-Haddad. 1992. Serological response of chickens to infection with *Salmonella gallinarum—S. pullorum* detected by enzyme-linked immunosorbent assay. *Avian Dis.* 36:227–236.

23. Bean, N.H., and P.M. Martin. 1990. Foodborne disease outbreaks in the United States, 1973–1987: Pathogens, vehicles, and trends. *J. Food Prot.* 53:804–817.

24. Bean, N.H., P.M. Griffin, J.S. Goulding, et al. 1990. Foodborne disease outbreaks, 5-year summary, 1983–1987. *MMWR* 39(SS-1):15–57.

25. Beaver, P.C., and R.C. Jung. 1985. *Animal Agents and Vectors of Human Disease.* 5th ed. Philadelphia: Lea and Febiger.

26. Belongia, E.A., K.L. MacDonald, G.L. Parham, et al. 1991. An outbreak of *Escherichia coli* O157:H7 colitis associated with consumption of precooked meat patties. *J. Infect. Dis.* 164:338–343.

27. Benenson, A.S. (ed.). 1990. *Control of Communicable Diseases in Man.* 15th ed. Washington, D.C.: American Public Health Association.

28. Benson, W.W., D.W. Brock, and J. Mather. 1963. Serologic analysis of a penitentiary group using raw milk from a Q fever infected herd. *Public Health Rep.* 78:707–710.

29. Bernardeschi, P., I. Boechi, and G. Cavallini. 1988. *Aeromonas hydrophila* in-

fection after cockles ingestion. *Haematologica Pavia* 73:548–549.

30. Bernstein, R.E. 1984. Liver and hypervitaminosis. *N. Engl. J. Med.* 311:604.

31. Besser, R.E., S.M. Lett, J.T. Weber, et al. 1993. An outbreak of diarrhea and hemolytic uremic syndrome from *Escherichia coli* O157:H7 in fresh-pressed apple cider. *J. Am. Med. Assoc.* 269:2217–2220.

32. Beuchat, L.R. 1991. Behavior of *Aeromonas* species at refrigeration temperatures. *Int. J. Food Microbiol.* 13:217–224.

33. Bhatt, B.D., M.J. Zuckerman, J.A. Foland, et al. 1989. Disseminated *Salmonella arizona* infection associated with rattlesnake meat ingestion. *Am. J. Gastroenterol.* 84:433–435.

34. Biggar, J.W., and J.N. Seiber (eds.). 1987. *Fate of Pesticides in the Environment.* Publ. 3320. Oakland: Division of Agriculture and Natural Resources, University of California.

35. Binta, G.M., T.B. Tjaberg, P.N. Nyaga, et al. 1982. Market fish hygiene in Kenya. *J. Hyg. Camb.* 89:47–52.

36. Black, R.E. 1990. Epidemiology of travelers' diarrhea and relative importance of various pathogens. *Rev. Infect. Dis.* 12(Suppl. 1):S73–S79.

37. Blake, P.A., R.E. Weaver, and D.G. Hollis. 1980. Diseases of humans (other than cholera) caused by vibrios. *Ann. Rev. Microbiol.* 34:341–367.

38. Blaser, M.J., D.N. Taylor, and R.A. Feldman. 1983. Epidemiology of *Campylobacter jejuni* infections. *Epidemiol. Rev.* 5:157–176.

39. Bloom, H.G., and E.J. Bottone. 1990. *Aeromonas hydrophila* diarrhea in a long-term care setting. *J. Am. Geriatr. Soc.* 38:804–806.

40. Board, R.G. 1965. Bacterial growth on and penetration of the shell membranes of the hen's egg. *J. Appl. Bacteriol.* 28:197–205.

41. Boerlin, P., and J.-C. Piffaretti. 1991. Typing of human, animal, food, and environmental isolates of *Listeria monocytogenes* by multilocus enzyme electrophoresis. *Appl. Environ. Microbiol.* 57:1624–1629.

42. Bond, E.J. (ed.). 1985. *Manual of Fumigation for Insect Control.* 3d ed. New York: Unipub.

43. Bone, F.J., D. Bogie, and S.C. Morgan-Jones. 1989. Staphylococcal food poisoning from sheep milk cheese. *Epidemiol. Infect.* 103:449–458.

44. Bowerman, R.J. 1991. Seroprevalence of *Toxoplasma gondii* in rural India: A preliminary study. *Trans. R. Soc. Trop. Med. Hyg.* 85:622.

45. Boyer, A.C., P.W. Lee, and J.C. Potter. 1992. Characterization of fenvalerate residues in dairy cattle and poultry. *J. Agric. Food Chem.* 40:914–918.

46. Breton, J., W.R. Mitchell, and S. Rosendal. 1986. *Streptococcus suis* in slaughter pigs and abattoir workers. *Can. J. Vet. Res.* 50:338–341.

47. Brown, A.W.A. 1978. *Ecology of Pesticides.* New York: John Wiley and Sons.

48. Brown, M.A., J.V. Thom, G.L. Orth, et al. 1964. Food poisoning involving zinc contamination. *Arch. Environ. Health* 8: 657–660.

49. Bryan, F.L. 1988. Risks of practices, procedures and processes that lead to outbreaks of foodborne diseases. *J. Food Prot.* 51:663–673.

50. Burns, J.C., W.H. Mason, M.P. Glode, et al. 1991. Clinical and epidemiologic characteristics of patients referred for evaluation of possible Kawasaki disease. *J. Pediatr.* 118:680–686.

51. Burr, M.L., E.S. Limb, M.J. Maguire, et al. 1993. Infant feeding, wheezing, and allergy: A prospective study. *Arch. Dis. Child.* 68:724–728.

52. Butzler, J.-P., and J. Oosterom. 1991. *Campylobacter*: Pathogenicity and significance in foods. *Int. J. Food Microbiol.* 12:1–8.

53. Callister, S.M., and W.A. Agger. 1987. Enumeration and characterization of *Aeromonas hydrophila* and *Aeromonas caviae* isolated from grocery store produce. *Appl. Environ. Microbiol.* 53:249–253.

54. Canzonier, W.J. 1988. Public health component of bivalve shellfish production and marketing. *J. Shellfish Res.* 7:261–266.

55. Carpenter, T.O., J.M. Pettifor, R.M. Russell, et al. Severe hypervitaminosis A in siblings: Evidence of variable tolerance to retinol intake. *J. Pediatr.* 111:507–512.

56. Casemore, D.P. 1990. Foodborne protozoal infection. *Lancet* 336:1427–1432.

57. Catsimpoolas, N. (ed.). 1977. *Immunological Aspects of Foods.* Westport, Conn.: AVI Publishing.

58. Cecil, H.C., R.W. Miller, and C. Corley. 1981. Feeding three insect growth regulators to White Leghorn hens: Residues in eggs and tissues and effects on production and reproduction. *Poult. Sci.* 60:2017–2027.

59. Chan, K.-Y., M.L. Woo, L.Y. Lam, et al. 1989. *Vibrio parahaemolyticus* and other halophilic vibrios associated with seafood in Hong Kong. *J. Appl. Bacteriol.* 66:57–64.

60. Chapman, P.A. 1987. *Vibrio alginolyticus* and diarrhoeal disease. *J. Diarrhoeal Dis. Res.* 5:40.

61. Chichester, C.O., and H.D. Graham (eds.). 1975. *Microbial Safety of Fishery Products.* New York: Academic Press.

62. Ching-Lee, M.R., A.R. Katz, D.M. Sasaki, et al. 1991. *Salmonella* egg surveillance in Hawaii: Evidence for routine bacterial surveillance. *Am. J. Public Health* 81:764–766.

63. Chuang, Y.-C., C.-Y. Yuan, C.-Y. Liu, et al. 1992. *Vibrio vulnificus* infection in Taiwan: Report of 28 cases and review of clinical manifestations and treatment. *Clin. Infect. Dis.* 15:271–276.

64. Claesson, B.E.B., N.G. Svensson, L. Gotthardsson, et al. 1992. A foodborne outbreak of group A streptococcal disease at a birthday party. *Scand. J. Infect. Dis.* 24:577–586.

65. Clifton-Hadley, F.A. 1983. *Streptococcus suis* type 2 infections. *Br. Vet. J.* 139:1–5.

66. Cliver, D.O. (ed.). 1990. *Foodborne Diseases.* New York: Academic Press.

67. ———. 1994. Viral foodborne disease agents of concern. *J. Food Prot.* 56:176–178.

68. Cloud, O.E. 1943. Perforation with peritonitis from *Shigella gallinarum* (variety of Duisberg). *Med. Bull. Veterans Adm.* 19:335–336.

69. Coleman, M.E., R.S. Elder, and P. Basu. 1992. Trace metals in edible tissues of livestock and poultry. *J. Assoc. Off. Anal. Chem. Int.* 75:615–625.

70. Condon, S., M.L. Garcia, A. Otero, et al. 1992. Effect of culture age, pre-incubation at low temperature and pH on the thermal resistance of *Aeromonas hydrophila. J. Appl. Bacteriol.* 72:322–326.

71. Cook, B.R. 1991. *Echinococcus multilocularis* infestation acquired in UK. *Lancet* 337:560–561.

72. Cooke, E.M. 1990. Epidemiology of foodborne illness: UK. *Lancet* 336:790–793.

73. Cowden, J.M., and N.D. Noah. 1989. Salmonellas and eggs. *Arch. Dis. Child.* 64:1419–1420.

74. Cowden, J.M., M. O'Mahony, C.L.R. Bartlett, et al. 1989. A national outbreak of *Salmonella typhimurium* DT 124 caused by contaminated salami sticks. *Epidemiol. Infect.* 103:219–225.

75. Cummings, J.G., K.T. Zee, V. Turner, et al. 1966. Residues in eggs from low level feeding of five chlorinated hydrocarbon insecticides to hens. *J. Assoc. Offic. Anal Chem.* 49:354–364.

76. Cunningham, L.W., R.A. Promisloff, and A.V. Cichelli. 1991. Pulmonary infiltrates associated with *Vibrio vulnificus* septicemia. *J. Am. Osteopath. Assoc.* 91:84–86.

77. Curley, A., V.A. Sedlak, E.F. Girling, et al. 1971. Organic mercury identified as the cause of poisoning in humans and hogs. *Science* 172:65–67.

78. D'Aoust, J.-Y. 1991. Pathogenicity of foodborne *Salmonella. Int. J. Food Micro-*

biol. 12:17–40.

79. D'Aoust, J.-Y. 1991. Psychrotrophy and foodborne *Salmonella. Int. J. Food Microbiol.* 13:207–215.

80. Daghir, N.J., and N.N. Hariri. 1977. Determination of total arsenic residues in chicken eggs. *J. Agric. Food Chem.* 25:1009–1010.

81. daSilva, E.N. 1985. The *Salmonella gallinarum* problem in Central and South America. *Proc. Int. Symp. Salmonella,* ed. G.H. Snoeyenbos, New Orleans, July 19–20, 1984, 150–156.

82. Davidson, R.A. 1984. Complications of megavitamin therapy. *South. Med. J.* 77:200–203.

83. Deema, P., E.C. Naber, and G.W. Ware. 1965. Residues in hen eggs from vaporizing insecticide tablets. *J. Econ. Entomol.* 58:904–906.

84. deKaminsky, R.G. 1991. Taeniasis-cysticercosis in Honduras. *Trans. R. Soc. Trop. Med. Hyg.* 85:531–534.

85. DePaola, A., P.A. Flynn, R.M. McPhearson, et al. 1988. Phenotypic and genotypic characterization of tetracycline- and oxytetracycline-resistant *Aeromonas hydrophila* from cultured channel catfish (*Ictalurus punctatus*) and their environments. *Appl. Environ. Microbiol.* 54:1861–1863.

86. Devi, S.J.N., and C.J. Murray. 1991. *Salmonella* carriage rate amongst school children—a three year study. *Southeast Asian J. Trop. Med. Public Health* 22:357–361.

87. Diaz, F., H.H. Garcia, R.H. Gilmam, et al. 1992. Epidemiology of taeniasis and cysticercosis in a Peruvian village. *Am. J. Epidemiol.* 135:875–882.

88. Dickinson, G. 1982. Scombroid fish poisoning syndrome. *Ann. Emerg. Med.* 11:487–489.

89. Ditchfield, J., and R.J. Julian. 1958. A case report: Egg borne pullorum infection in adult hens. *Can. J. Comp. Med.* 22:181–183.

90. Dobbing, J. (ed.). 1987. *Food Intolerance.* London: Bailliere Tindall.

91. Donnelly, C.W. 1990. Concerns of microbial pathogens in association with dairy foods. *J. Dairy Sci.* 73:1656–1661.

92. Dorn, C.R. 1993. Review of foodborne outbreak of *Escherichia coli* O157:H7 infection in the western United States. *J. Am. Vet. Med. Assoc.* 203:1583–1587.

93. Dorn, C.R., and E.J. Angrick. 1991. Serotype O157:H7 *Escherichia coli* from bovine and meat sources. *J. Clin. Microbiol.* 29:1225–1231.

94. Doyle, M.P. (ed.). 1989. *Foodborne Bacterial Pathogens.* New York: Marcel Dekker.

95. ——. 1990. Pathogenic *Escherichia coli, Yersinia enterocolitica,* and *Vibrio parahaemolyticus. Lancet* 336:1111–1115.

96. ——. 1991. *Escherichia coli* O157:H7 and its significance in foods. *Int. J. Food Microbiol.* 12:289–301.

97. Drusin, L.M., J.C. Ribble, and B. Topf. Group C streptococcal colonization in a newborn nursery. *Am. J. Dis. Child.* 125:820–821.

98. Dubey, J.P. 1982. Repeat transplacental transfer of *Toxoplasma gondii* in dairy goats. *J. Am. Vet. Med. Assoc.* 180:1220–1221.

99. ——. 1983. Distribution of cysts and tachyzoites in calves and pregnant cows inoculated with *Toxoplasma gondii* oocysts. *Vet. Parasitol.* 13:199–211.

100. ——. 1984. Experimental toxoplasmosis in sheep fed *Toxoplasma gondii* oocysts. *Int. Goat Sheep Res.* 2:93–104.

101. Dubey, J.P., K.D. Murrell, and R. Fayer. 1984. Persistence of encysted *Toxoplasma gondii* in tissues of pigs fed oocysts. *Am. J. Vet. Res.* 45:1941–1943.

102. Ebel, E.D., M.J. David, and J. Mason. 1992. Occurrence of *Salmonella enteritidis* in the U.S. commercial egg industry: Report on a national spent hen survey. *Avian Dis.* 36:646–654.

103. Edwards, P.R., and D.W. Bruner. 1943. The occurrence and distribution of *Sal-*

monella types in the United States. *J. Infect. Dis.* 72:58–67.

104. Edwards, P.R., D.W. Bruner, and A.B. Moran. 1948. Further studies on the occurrence and distribution of *Salmonella* types in the United States. *J. Infect. Dis.* 83:220–231.

105. Eld, K., A. Gunnarsson, T. Holmberg, et al. *Salmonella* isolated from animals and feedstuffs in Sweden during 1983–1987. *Acta Vet. Scand.* 32:261–277.

106. Fairchild, E.J. (ed.). 1978. *Agricultural Chemicals and Pesticides.* Fort Lee, N.J.: Jack K. Burgess, Inc.

107. Fallon, M.B., and J.L. Boyer. 1990. Hepatic toxicity of vitamin A and synthetic retinoids. *J. Gastroenterol. Hepatol.* 5:334–342.

108. Fang, F.C., and J. Fierer. 1991. Human infection with *Salmonella dublin. Medicine (Balt.)* 70:198–207.

109. Fanning, A., and S. Edwards. 1991. *Mycobacterium bovis* infection in human beings in contact with elk (*Cervus elaphus*) in Alberta, Canada. *Lancet* 338:1253–1255.

110. Fantasia, M., E. Filetici, M.P. Anastasio, et al. 1991. Italian experience in *Salmonella enteritidis,* 1978–1988: Characterization of isolates from food and man. *Int. J. Food Microbiol.* 12:353–362.

111. Farber, J.M. 1991. *Listeria monocytogenes. J. Assoc. Off. Anal. Chem.* 74:701–704.

112. Farber, J.M., and P.I. Peterkin. 1991. *Listeria monocytogenes,* a food-borne pathogen. *Microbiol. Rev.* 55:476–511.

113. Farley, T.A., S.A. Wilson, F. Mahoney, et al. 1993. Direct inoculation of food as the cause of an outbreak of group A streptococcal pharyngitis. *J. Infect. Dis.* 167:1232–1235.

114. Farrell, G.M., and E.H. Marth. 1991. *Borrelia burgdorferi:* Another cause of foodborne illness? *Int. J. Food Microbiol.* 14:247–260.

115. Fayer, R. 1981. Toxoplasmosis update and public health implications. *Can. Vet. J.* 22:344–352.

116. Fehrs, L.J., K. Flanagan, S. Kline, et al. 1987. Group A ß-hemolytic streptococcal skin infections in a U.S. meat-packing plant. *J. Am. Med. Assoc.* 258:3131–3134.

117. Fields, P.I., T. Popovic, K. Wachsmuth, et al. 1992. Use of polymerase chain reaction for detection of toxigenic *Vibrio cholerae* O1 strains from the Latin American cholera epidemic. *J. Clin. Microbiol.* 30:2118–2121.

118. Fishbein, D.B., and D. Raoult. 1992. A cluster of *Coxiella burnetii* infections associated with exposure to vaccinated goats and their unpasteurized dairy products. *Am. J. Trop. Med. Hyg.* 47:35–40.

119. Fishwick, F.B., E.G. Hill, I. Rutter, et al. 1980. Gamma-HCH in eggs and poultry arising from exposure to thermal vaporisers. *Pestic. Sci.* 11:633–642.

120. Francis, S., J. Rowland, K. Rattenbury, et al. 1989. An outbreak of paratyphoid fever in the UK associated with a fish-and-chip shop. *Epidemiol. Infect.* 103:445–448.

121. Frank, R., J. Raspar, H.E. Braun, et al. 1985. Disappearance of organochlorine residues from abdominal and egg fats of chickens, Ontario, Canada, 1969–1982. *J. Assoc. Off. Anal. Chem.* 68:124–129.

122. Frank, R., H.E. Braun, K.I. Stonefield, et al. 1990. Organochlorine and organophosphorus residues in the fat of domestic farm animal species, Ontario, Canada, 1986–1988. *Food Addit. Contam.* 7:629–636.

123. Fricker, C.R., and R.W.A. Park. 1989. A two-year study of the distribution of "thermophilic" campylobacters in human, environmental and food samples from the Reading area with particular reference to toxin production and heat-stable serotype. *J. Appl. Bacteriol.* 66:477–490.

124. Fries, G.F. 1978. Distribution and kinetics of polybrominated biphenyls and selected chlorinated hydrocarbons in farm animals. *J. Am. Vet. Med. Assoc.* 173:1479–1484.

125. Funk, E.M., and M.R. Irwin. 1955. *Hatchery Operation and Management.* New York: John Wiley and Sons.

126. Galbraith, N.S., P. Forbes, and C. Clifford. 1982. Communicable disease associated with milk and dairy products in England and Wales, 1951–80. *Br. Med. J.* 284:1761–1765.

127. Gao, Q.Y., Huang Y.F., Wu J.G., et al. 1990. A review of botulism in China. *Biomed. Environ. Sci.* 3:326–336.

128. Garg, S.K., and B.K. Mital. 1991. Enterococci in milk and milk products. *Crit. Rev. Microbiol.* 18:15–45.

129. Garza, C., and N.S. Scrimshaw. 1976. Relationship of lactose intolerance to milk intolerance in young children. *Am. J. Clin. Nutr.* 29:1902–1906.

130. Gay, J.M., and M.E. Hunsaker. 1993. Isolation of multiple *Salmonella* serovars from a dairy two years after a clinical salmonellosis outbreak. *J. Am. Vet. Med. Assoc.* 203:1314–1320.

131. Gellin, B.G., C.V. Broome, W.F. Bibb, et al. 1991. The epidemiology of listeriosis in the United States—1986. *Am. J. Epidemiol.* 133:392–401.

132. Genigeorgis, C.A. 1981. Factors affecting the probability of growth of pathogenic microorganisms in foods. *J. Am. Vet. Med. Assoc.* 179(12):1410–1417.

133. ———. 1989. Present state of knowledge on staphylococcal intoxication. *Int. J. Food Microbiol.* 9:327–360.

134. Genigeorgis, C., M. Hassuneh, and P. Collins. 1986. *Campylobacter jejuni* infection on poultry farms and its effect on poultry meat contamination during slaughtering. *J. Food Prot.* 49:895–903.

135. Getzendaner, M.E. 1965. Bromide residues in chicken tissues and eggs from ingestion of methyl bromide-fumigated feed. *J. Agric. Food Chem.* 13:349–352.

136. Gilchrist, A. 1981. *Foodborne Disease & Food Safety.* Monroe, Wis.: American Medical Association.

137. Giles, N., S.A. Hopper, and C. Wray. 1989. Persistence of *S. typhimurium* in a large dairy herd. *Epidemiol. Infect.* 103:235–241.

138. Gill, O.N., W.D. Cubitt, D.A. McSwiggan, et al. 1983. Illness associated with fish and shellfish in England and Wales. *Br. Med. J.* 287 (6401):1284–1285.

139. Gjerde, J., and B. Boe 1981. Isolation and characterization of *Vibrio alginolyticus* and *Vibrio parahaemolyticus* from the Norwegian coastal environment. *Acta Vet. Scand.* 22:331–343.

140. Glass, K.A., and M.P. Doyle. 1989. Fate of *Listeria monocytogenes* in processed meat products during refrigerated storage. *Appl. Environ. Microbiol.* 55:1565–1569.

141. Glynn, J.R., and D.J. Bradley. 1992. The relationship between infecting dose and severity of disease in reported outbreaks of salmonella infections. *Epidemiol. Infect.* 109:371–388.

142. Glynn, J.R., and S.R. Palmer. 1992. Incubation period, severity of disease, and infecting dose: Evidence from a *Salmonella* outbreak. *Am. J. Epidemiol.* 136:1369–1377.

143. Gordon, S.M., L.S. Oshiro, W.R. Jarvis, et al. 1990. Foodborne Snow Mountain agent gastroenteritis with secondary person-to-person spread in a retirement community. *Am. J. Epidemiol.* 131:702–710.

144. Graham, H.D. (ed.). 1980. *The Safety of Foods.* 2d ed. Westport, Conn.: AVI Publishing.

145. Graham-Rack, B., and R. Binsted. 1973. *Hygiene in Food Manufacturing and Handling.* 2d ed. London: Food Trades Press.

146. Graivier, L., N.E. Harper, and G. Currarino. 1977. Milk-curd bowel obstruction in the newborn infant. *J. Am. Med. Assoc.* 238:1050–1052.

147. Greenberg, R.A. 1985. *Antibiotics in Animal Feed: A Threat to Human Health?* 2d ed. Summit, N.J.: American Council on Science and Health.

148. Griffin, P.M., and R.V. Tauxe. 1991. The epidemiology of infections caused by *Escherichia coli* O157:H7, other enterohemorrhagic *E. coli,* and the associated hemolytic uremic syndrome. *Epidemiol. Rev.* 13:60–98.

149. Gross, T.P., L.B. Kamara, C.L. Hatheway, et al. 1989. *Clostridium perfringens* food poisoning: Use of serotyping in an outbreak setting. *J. Clin. Microbiol.* 27:660–663.

150. Guthrie, R.R. (ed.). 1988. *Food Sanitation.* 3d ed. New York: Van Nostrand Reinhold.

151. Haba, J.H. 1993. Incidence and control of *Campylobacter* in foods. *Microbiologia Sem.* 9:57–65.

152. Haddock, R.L. 1993. The origins of infant salmonellosis. *Am. J. Public Health* 83:772.

153. Haigh, T., and W.B. Betts. 1991. Microbial barrier properties of hen egg shells. *Microbios* 68:137–146.

154. Hall, M.L.M., and B. Rowe. 1992. *Salmonella arizonae* in the United Kingdom from 1966 to 1990. *Epidemiol. Infect.* 108:59–65.

155. Hambleton, P. 1992. *Clostridium botulinum* toxins: A general review of involvement in disease, structure, mode of action and preparation for clinical use. *J. Neurol.* 239:16–20.

156. Hancock, D.D., S.E. Wikse, A.B. Lichtenwalner, et al. 1989. Distribution of bovine cysticercosis in Washington. *Am. J. Vet. Res.* 50:564–570.

157. Harbourne, J.F., C.J. Randall, K.W. Luery, et al. *Salmonella paratyphi* B infection in dairy cows: Part I. *Vet. Rec.* 91:112–114.

158. Hardie, R.M., and J.M. Watson. 1992. *Mycobacterium bovis* in England and Wales: Past, present, and future. *Epidemiol. Infect.* 109:23–33.

159. Harris, A.A., C. Cherubin, R. Biek, et al. 1990. Frequency of *Salmonella typhimurium* the year after a massive outbreak. *Diagn. Microbiol. Infect. Dis.* 13:25–30.

160. Hartman, R.C., and G.S. Vickers. 1953. *Hatchery Management.* New York: Orange Judd Publishing.

161. Haruki, K., Y. Seto, T. Murakami, et al. 1991. Pattern of shedding of small, round-structured virus particles in stools of patients of outbreaks of food-poisoning from raw oysters. *Microbiol. Immunol.* 35:83–86.

162. Hasenson, L.B., L. Kaftyreva, V.G. Laszlo, et al. 1992. Epidemiological and microbiological data on *Salmonella enteritidis. Acta Microbiol. Hung.* 39:31–39.

163. Hatakka, M. 1992. Salmonella outbreak among railway and airline passengers. *Acta Vet. Scand.* 33:253–260.

164. Hauschild, A.H.W. 1985. Food-borne botulism in Canada, 1971–84. *Can. Med. Assoc. J.* 133:1141–1146.

165. Hayes, E.B., T.D. Matte, T.R. O'Brien, et al. 1989. Large community outbreak of cryptosporidiosis due to contamination of a filtered public water supply. *N. Engl. J. Med.* 320:1372–1376.

166. Hayes, W.J., Jr. 1975. *Toxicology of Pesticides.* Baltimore: Williams and Wilkins.

167. Hays, W.H. (chrmn.). 1981. *Antibiotics in Animal Feeds.* Rep. 88. Ames, Iowa: Council for Agricultural Science and Technology.

168. Hedberg, C.W., M.J. David, K.E. White, et al. 1993. Role of egg consumption in sporadic *Salmonella enteritidis* and *Salmonella typhimurium* infections in Minnesota. *J. Infect. Dis.* 167:107–111.

169. Hedberg, C.W., J.A. Korlath, J.-Y. D'Aoust, et al. 1992. A multistate outbreak of *Salmonella javiana* and *Salmonella oranienburg* infections due to consumption of contaminated cheese. *J. Am. Med. Assoc.* 268:3203–3207.

170. Hedberg, C.W., K.L. MacDonald, and M.T. Osterholm. 1994. Changing epidemiology of food-borne disease: A Minnesota perspective. *Clin. Infect. Dis.* 18:671–680.

171. Hedman, P., O. Ringertz, M. Lindstrom, et al. 1993. The origin of *Staphylococ-*

cus saprophyticus from cattle and pigs. *Scand. J. Infect. Dis.* 25:57–60.

172. Heffernan, H.M. 1991. Antibiotic resistance among salmonella from human and other sources in New Zealand. *Epidemiol. Infect.* 106:17–23.

173. Hefnawy, Y.A., S.I. Moustafa, and E.H. Marth. 1993. Behavior of *Listeria monocytogenes* in refrigerated and frozen ground beef and in sausage and broth with and without additives. *Lebensm. Wiss. u. Techol.* 26:167–170.

174. Henderson, M.J., and R.G. Jones. 1987. Cod liver oil or bust. *Lancet* 2(8553):274–275.

175. Hiratsuka, M., Y. Saitoh, and N. Yamane. 1980. The isolation of *Vibrio alginolyticus* from a patient with acute entero-colitis. *Tohoku J. Exp. Med.* 132:469–472.

176. Hobbs, B.C., and J.H.B. Christian (eds.). 1973. *Microbiological Safety of Foods.* London: Academic Press.

177. Hobson-Frohock, A. 1982. Residues of ethoxyquin in poultry tissues and eggs. *J. Sci. Food Agric.* 33:1269–1274.

178. Holick, M.F., Q. Shao, W.W. Liu, et al. 1992. The vitamin D content of fortified milk and infant formula. *N. Engl. J. Med.* 326:1178–1181.

179. Hoskin, J.C., and R.E. Wright. 1991. *Cryptosporidium:* An emerging concern for the food industry. *J. Food Prot.* 54:53–57.

180. Holmberg, S.D., and P.A. Blake. 1984. Staphylococcal food poisoning in the United States: New facts and old misconceptions. *J. Am. Med. Assoc.* 251(4):487–489.

181. Howell, J. 1969. Mercury residues in chicken eggs and tissues from a flock exposed to methylmercury dicyandiamide. *Can. Vet. J.* 10:212–213.

182. Hudson, J.A. 1992. Variation in growth kinetics and phenotype of *Aeromonas* spp. from clinical meat processing and fleshfood sources. *Int. J. Food Microbiol.* 16:131–139.

183. Hudson, J.A., and S.M. Mott. 1993. Growth of *Listeria monocytogenes, Aeromonas hydrophila* and *Yersinia enterocolitica* in pate and a comparison with predictive models. *Int. J. Food Microbiol.* 20:1–11.

184. Hudson, S.J., N.F. Lightfoot, J.C. Coulson, et al. 1991. Jackdaws and magpies as vectors of milkborne human campylobacter infection. *Epidemiol. Infect.* 107:363–372.

185. Hughes, J.M., and C.O.Tacket. 1983. Sausage poisoning revisited. *Arch. Intern. Med.* 143:425.

186. Humphrey, T.J. 1990. Public health implications of the infection of egg-laying hens with *Salmonella enteritidis* phage type 4. *World's Poult. Sci. J.* 46:5–13.

187. Humphrey, T.J., M. Greenwood, R.J. Gilbert, et al. 1989. The survival of salmonellas in shell eggs cooked under simulated domestic conditions. *Epidemiol. Infect.* 103:35–45.

188. Humphreys, D.J. 1988. *Veterinary Toxicology.* 3d ed. Philadelphia: Saunders.

189. Hunter, P.R., and J. Izsak. 1990. Diversity studies of salmonella incidents in some domestic livestock and their potential relevance as indicators of niche width. *Epidemiol. Infect.* 105:501–510.

190. Huq, A., E.B. Small, P.A. West, et al. 1983. Ecological relationships between *Vibrio cholerae* and planktonic crustacean copepods. *Appl. Environ. Microbiol.* 45:275–283.

191. Ichinohe, H., M. Yoshioka, H. Fukushima, et al. 1991. First isolation of *Yersinia enterocolitica* serotype O:8 in Japan. *J. Clin. Microbiol.* 29:846–847.

192. Imai, C. 1981. Preservability of soft-cooked eggs. *Poult. Sci.* 60:1436–1442.

193. Inkeles, S.B., W.E. Connor, and D.R. Illingworth. 1986. Hepatic and dermatologic manifestations of chronic hypervitaminosis A in adults. *Am. J. Med.* 80:491–496.

194. Isaac-Renton, J.L., and J.J. Philion. 1992. Factors associated with acquiring giardiasis in British Columbia residents. *Can. J. Public Health* 83:155–158.

195. Ivey, M.C., J.A. Devaney, G.W. Ivie, et al. 1982. Residues of stirofos (Rabon®) in eggs of laying hens treated for northern fowl mite control by dipping. *Poult. Sci.*

61:443–446.

196. Ivey, M.C., G.W. Ivie, J.A. Devaney, et al. 1984. Residues of carbaryl and two of its metabolites in eggs of laying hens treated with Sevin® for northern fowl mite control by dipping. *Poult. Sci.* 63:61–65.

197. Ivie, G.W., and H.W. Dorough (eds.). 1977. *Fate of Pesticides in Large Animals.* New York: Academic Press.

198. Izat, A.L., J.M. Kopek, and J.D. McGinnis. 1991. Research note: Incidence, number, and serotypes of *Salmonella* on frozen broiler chickens at retail. *Poult. Sci.* 70:1438–1440.

199. Jackson, G.J. 1990. Public health and research perspectives on the microbial contamination of foods. *J. Anim. Sci.* 68:884–891.

200. Jackson, S.G. 1991. *Bacillus cereus. J. Assoc. Off. Anal. Chem.* 74:704–706.

201. Jacobs, L., and M.L. Melton. 1966. Toxoplasmosis in chickens. *J. Parasitol.* 52:1158–1162.

202. Jacobus, C.H., M.F. Holick, Q. Shao, et al. 1992. Hypervitaminosis D associated with drinking milk. *N. Engl. J. Med.* 326:1174–1177.

203. James, L.F., and W.J. Hartley. 1977. Effects of milk from animals fed locoweed on kittens, calves, and lambs. *Am. J. Vet. Res.* 38:1263–1265.

204. Jan, J., and M. Adamic. 1991. Polychlorinated biphenyl residues in foods from a contaminated region of Yugoslavia. *Food Addit. Contam.* 8:505–512.

205. Janda, J.M. 1991. A lethal leviathan–*Vibrio vulnificus. West. J. Med.* 155:421–422.

206. Janda, J.M., C. Powers, R.G. Bryant, et al. 1988. Current perspectives on the epidemiology and pathogenesis of clinically significant *Vibrio* spp. *Clin. Microbiol. Rev.* 1:245–267.

207. Jarvis, W.R. 1992. *Yersinia enterocolitica:* A new or unrecognized nosocomial pathogen? *Infect. Control Hosp. Epidemiol.* 13:137–138.

208. Johnson, A.E. 1976. Changes in calves and rats consuming milk from cows fed chronic lethal doses of *Senecio jacobaea* (tansy rag-wort). *Am. J. Vet. Res.* 37:107–110.

209. Joint FAO/WHO Codex Alimentarius Commission. 1977. *Recommended International Code of Hygiene Practice for Egg Products.* Rome: Food and Agriculture Organization.

210. Jones, D. 1990. Foodborne listeriosis. *Lancet* 336:1171–1174.

211. Jones, F.T., R.C. Axtell, D.V. Rives, et al. 1991. A survey of *Campylobacter jejuni* contamination in modern broiler production and processing systems. *J. Food Prot.* 54:259–262, 266.

212. Joseph, C.A., and S.R. Palmer. 1989. Outbreaks of salmonella infection in hospitals in England and Wales, 1978–87. *Br. Med. J.* 298:1161–1164.

213. Juranovic, L.R., and D.L. Park. 1991. Foodborne toxins of marine origin: Ciguatera. *Rev. Environ. Contam. Toxicol.* 117:51–94.

214. Kahunyo, J.M., A. Froslie, and C.K. Maitai. 1988. Organochlorine pesticide residues in chicken eggs: A survey. *J. Toxicol. Environ. Health* 24:543–550.

215. Kampelmacher E.H., and L.M. van Noorle Jansen. 1969. Isolation of *Listeria monocytogenes* from faeces of clinically healthy humans and animals. *Zbl. Bakt. Parasit. Infekt. Hyg.* 21:354–359.

216. Kapperud, G. 1991. *Yersinia enterocolitica* in food hygiene. *Int. J. Food Microbiol.* 12:53–65.

217. Karabiber, N., and F. Aktas. 1991. Foodborne giardiasis. *Lancet* 337:376–377.

218. Katz, S.E., C.A. Fassbender, and J.J. Dowling, Jr. 1973. Oxytetracycline residues in tissue, organs, and eggs of poultry fed supplemented rations. *J.A.O.A.C.* 56:77–81.

219. Kaysner, C.A. 1981. Incidence of *Vibrio alginolyticus* and bacteria of sanitary significance in the Bering Sea. *Appl. Environ. Microbiol.* 41:1279–1282.

220. Kefenie, H., and G. Bero. 1992. Trichinosis from wild boar meat in Gojjam, North-West Ethiopia. *Trop. Geogr. Med.* 44:278–280.

221. Kello, D. 1990. Epidemiological aspects in food safety. *Food Addit. Contam.* 7(Suppl. 1):S5–S11.

222. Khakhria, R., D. Duck, and H. Lior. 1991. Distribution of *Salmonella enteritidis* phage types in Canada. *Epidemiol. Infect.* 106:25–32.

223. Khamboonruang, C. 1991. The present status of trichinellosis in Thailand. *Southeast Asian J. Trop. Med. Public Health* 22 (Suppl.):312–315.

224. Khan, S.U., and T.S. Foster. 1976. Residues of atrazine (2-chloro-4-ethylamino-6-isopropylamino-s-triazine) and its metabolites in chicken tissues. *J. Agric. Food Chem.* 24:768–771.

225. King, D.F. 1939. *The Detection of Infertile Eggs and Its Application to Hatchery Management.* Circ. 82. Auburn: Alabama Polytechnic Institute.

226. Kingsbury, J.M. 1964. *Poisonous Plants of the United States and Canada.* Englewood Cliffs, N.J.: Prentice-Hall.

227. Koenig, K.L., J. Mueller, and T. Rose. 1991. *Vibrio vulnificus*—hazard on the half shell. *West. J. Med.* 155:400–403.

228. Kohl, S. 1978. *Yersinia enterocolitica*: A significant "new" pathogen. *Hosp. Pract.* 13:81–85.

229. Kosters, J. 1969. [Can infertile, incubated eggs be used for food?] *Fleischwirtschaft* 49:224–226.

230. Kotula, A.W., and N.O. Stern. 1984. The importance of *Campylobacter jejuni* to the meat industry: A review. *J. Sci.* 58(6):1561–1566.

231. Kovalyov, G.K. 1989. On human tuberculosis due to *M. bovis:* a review. *J. Hyg. Epidemiol. Microbiol. Immunol.* 33:199–206.

232. Kraft, A.A., G.S. Torrey, J.C. Ayres, et al. 1967. Factors influencing bacterial contamination of commercially produced liquid egg. *Poult. Sci.* 46:1204–1210.

233. Kreuzer, R. (ed.). 1972. *Fish Inspection and Quality Control.* London: Fishing News (Books).

234. Krogh, P. (ed.). 1987. *Mycotoxins in Food.* New York: Academic Press.

235. Kumazawa, N.H., E. Nakagaki, Y. Yonekawa, et al. 1991. Ecological cycle of thermostable direct hemolysin-producing strains of *Vibrio parahaemolyticus* in a brackish-water area with special reference to molluscs and attached microalgae. *J. Vet. Med. Sci.* 53:263–267.

236. Kumpulainen, J., and P. Koivistoinen. 1977. Fluorine in foods. *Residue Rev.* 68:37–57.

237. Laitinen, O., M. Leirisalo, and E. Allander. 1975. Rheumatic fever and *Yersinia* arthritis. Criteria and diagnostic problems in a changing disease pattern. *Scand. J. Rheumatol.* 4:145–157.

238. Lamabadusuriya, S.P., C. Perera, I.V. Devasiri, et al. 1992. An outbreak of salmonellosis following consumption of monkey meat. *J. Trop. Med. Hyg.* 95:292–295.

239. Lamm, S.H., and J.F. Rosen. 1974. Lead contamination in milks fed to infants: 1972–1973. *Pediatrics* 53:137–141; Comments 142–146.

240. Larsen, J.L., and P. Willeberg. 1984. The impact of terrestrial and estuarial factors on the density of environmental bacteria (vibrionaceae) and faecal coliforms in coastal water. *Zbl. Bakt. Hyg., I. Abt. Orig. B* 179:308–323.

241. Larsen, J.L., A.F. Farid, and I. Dalsgaard. 1981. A comprehensive study of environmental and human pathogenic *Vibrio alginolyticus* strains. *Zbl. Bakt. Hyg., Abt. Orig. A* 251:213–222.

242. ———. 1981. Occurrence of *Vibrio parahaemolyticus* and *Vibrio alginolyticus* in marine and estuarine bathing areas in Danish coast. *Zbl. Bakt. Hyg., Abt. Orig. B* 173:338–345.

243. Larson, B.L., and K.E. Ebner. 1960. Strontium-90 and milk, considerations on

SR-90 and the role of milk in our diet. *J. Dairy Sci.* 43:119–124.

244. Lee, L.A., A.R. Gerber, D.R. Lonsway, et al. 1990. *Yersinia enterocolitica* O:3 infections in infants and children, associated with the household preparation of chitterlings. *N. Engl. J. Med.* 322:984–987.

245. Lee, L.A., J. Taylor, G.P. Carter, et al. 1991. *Yersinia enterocolitica* O:3: An emerging cause of pediatric gastroenteritis in the United States. *J. Infect. Dis.* 163:660–663.

246. Leo, M.A., and C.S. Lieber. 1988. Hypervitaminosis A: A liver lover's lament. *Hepatology* 8:412–417.

247. Leonard, B.J. (ed.). 1978. *Toxicological Aspects of Food Safety.* New York: Springer-Verlag.

248. Lessof, M.H. (ed.). 1983. *Clinical Reactions to Food.* New York: John Wiley and Sons.

249. ———. 1992. *Food Intolerance.* London: Chapman and Hall.

250. Levine, W.C., J.F. Smart, D.L. Archer, et al. 1991. Foodborne disease outbreaks in nursing homes, 1975 through 1987. *J. Am. Med. Assoc.* 266:2105–2109.

251. Levine, W.C., P.M. Griffin, C.H. Woernle, et al. 1993. *Vibrio* infections on the Gulf Coast: Results of first year of regional surveillance. *J. Infect. Dis.* 167:479–483.

252. Lewus, C.B., A. Kaiser, and T.J. Montville. 1991. Inhibition of food-borne bacterial pathogens by bacteriocins from lactic acid bacteria isolated from meat. *Appl. Environ. Microbiol.* 57:1683–1688.

253. Libby, J.A. (ed.). 1975. *Meat Hygiene.* 4th ed. Philadelphia: Lea and Febiger.

254. Liston, J. 1990. Microbial hazards of seafood consumption. *Food Technol.,* Dec., 56–62.

255. Lossos, I.S., I. Felsenstein, R. Breuer, et al. 1992. Food-borne outbreak of group A ß-hemolytic streptococcal pharyngitis. *Arch. Intern. Med.* 152:853–855.

256. Luby, S.P., J.L. Jones, and J.M. Horan. 1993. A large salmonellosis outbreak associated with a frequently penalized restaurant. *Epidemiol. Infect.* 110:31–39.

257. Lucas, J. 1974. *Our Polluted Food: A Survey of the Risks.* New York: John Wiley and Sons.

258. Lund, B.M. 1990. Foodborne disease due to *Bacillus* and *Clostridium* species. *Lancet* 336:982–986.

259. MacDonald, K.L., and M.T. Osterholm. The emergence of *Escherichia coli* O157:H7 infection in the United States. *J. Am. Med. Assoc.* 269:2264–2266.

260. Maguire, H., J. Cowden, M. Jacob, et al. 1992. An outbreak of *Salmonella dublin* infection in England and Wales associated with a soft unpasteurized cows' milk cheese. *Epidemiol. Infect.* 109:389–396.

261. Maguire, H.C.F., A.A. Codd, V.E. Mackay, et al. 1993. A large outbreak of human salmonellosis traced to a local pig farm. *Epidemiol. Infect.* 110:239–246.

262. Majeed, K.N., and I.C. MacRae. 1991. Experimental evidence for toxin production by *Aeromonas hydrophila* and *Aeromonas sobria* in a meat extract at low temperatures. *Int. J. Food Microbiol.* 12:181–188.

263. Majeed, K.N., A.F. Egan, and I.C. MacRae. 1989. Incidence of aeromonads in samples from an abattoir processing lambs. *J. Appl. Bacteriol.* 67:597–604.

264. Majeed, K.N., A.F. Egan, and I.C. MacRae. 1990. Production of exotoxins by *Aeromonas* spp. at 5°C. *J. Appl. Bacteriol.* 69:332–337.

265. Marth, E.H., and B.E. Ellickson. 1959. Problems created by the presence of antibiotics in milk and milk products. *J. Milk Food Technol.* 22:266–272.

266. Matthysse, J.G. 1974. Insecticides used on dairy cattle and in dairy barns: Toxicity to man and cattle, hazards to the consumer and the environment. *J. Milk Food Technol.* 37:255–264.

267. Mauskopf, J.A., and M.T. French. 1991. Estimating the value of avoiding morbidity and mortality from foodborne illnesses. *Risk Anal.* 11:619–631.

268. Maxwell, E.L., B.C. Mayall, S.R. Pearson, et al. 1991. A case of *Vibrio vulnificus* septicaemia acquired in Victoria. *Med. J. Aust.* 154:214–215.

269. McCullough, N.B., and C.W. Eisele. 1951. Experimental human salmonellosis. IV. Pathogenicity of strains of *Salmonella pullorum* obtained from spray-dried whole egg. *J. Infect. Dis.* 88:259–265.

270. McLauchlin, J., S.M. Hall, S.K. Velani, et al. 1991. Human listeriosis and pate: A possible association. *Br. Med. J.* 303:773–775.

271. Meehan, P.J., T. Atkeson, D.E. Kepner, et al. 1992. A foodborne outbreak of gastroenteritis involving two different pathogens. *Am. J. Epidemiol.* 136:611–616.

272. Melconian, A.K., Y. Brun, and J. Fleurette. 1983. Enterotoxin production, phage typing and serotyping of *Staphylococcus aureus* strains isolated from clinical materials. *J. Hyg.* 91:235–242.

273. Melvin, D.M., M.M. Brooke, and E.H. Sadun. 1959. *Life Charts: Common Intestinal Helminths of Man.* Atlanta, Ga.: Communicable Disease Center.

274. Meredith, W.E., H.H. Weiser, and A.R. Winter. 1965. Chlortetracycline and oxytetracycline residues in poultry tissues and eggs. *Appl. Microbiol.* 13:86–88.

275. Merino, S., S. Camprubi, and J.M. Tomas. 1993. Detection of *Aeromonas hydrophila* in food with an enzyme-linked immunosorbent assay. *J. Appl. Bacteriol.* 74:149–154.

276. Mes, J., D.E. Coffin, and D. Campbell. 1974. Polychlorinated biphenyl and organochlorine pesticide residues in Canadian chicken eggs. *Pestic. Monit. J.* 8:8–11.

277. Mes, J., W.H. Newsome, and H.B.S. Conacher. 1989. Determination of some specific isomers of polychlorinated biphenyl congeners in fatty foods of the Canadian diet. *Food Addit. Contam.* 6:365–375.

278. ——. 1991. Levels of specific polychlorinated biphenyl congeners in fatty foods from five Canadian cities between 1986 and 1988. *Food Addit. Contam.* 8:351–361.

279. Miller, A.J., J.L. Smith, and G.A. Somkuti. 1990. *Topics in Industrial Microbiology: Foodborne Listeriosis.* Amsterdam: Elsevier.

280. Miller, K. 1987. *Toxicological Aspects of Food.* London: Elsevier Applied Science.

281. Misbah, S.A., J.B. Peiris, and T.M.S. Atukorala. 1984. Ingestion of shark liver associated with pseudotumour cerebri due to acute hypervitaminosis A. *J. Neurol. Neurosurg. Psychiatry* 47:216–220.

282. Mitchell, E., M. O'Mahony, D. Lynch, et al. 1989. Large outbreak of food poisoning caused by *Salmonella typhimurium* definitive type 49 in mayonnaise. *Br. Med. J.* 298:99–101.

283. Mitchell, R.B., F.C. Garlock, and R.H. Broh-Kahn. 1946. An outbreak of gastroenteritis presumably caused by *Salmonella pullorum*. *J. Infect. Dis.* 79:57–62.

284. Mohd, M.G. 1989. Brucellosis in the Gezira area, Central Sudan. *J. Trop. Med. Hyg.* 92:86–88.

285. Mol, H. 1975. *Antibiotics and Milk.* Rotterdam: A.A. Balkema.

286. Molitoris, E., S.W. Joseph, M.I. Krichevsky, et al. 1985. Characterization and distribution of *Vibrio alginolyticus* and *Vibrio parahaemolyticus* isolated in Indonesia. *Appl. Environ. Microbiol.* 50:1388–1394.

287. Morgan, M.R.A., and G.R. Fenwick. 1990. Natural foodborne toxicants. *Lancet* 336:1492–1495.

288. Morris, J.G., and R.E. Black. 1985. Cholera and other vibrioses in the United States. *N. Engl. J. Med.* 312:343–350.

289. Morris, J.G., J.D. Snyder, R. Wilson, et al. 1983. Infant botulism in the United States: An epidemiologic study of cases occurring outside of California. *Am. J. Public Health* 73:1385–1388.

290. Morse, L.J., and A.D. Rubenstein. 1967. A food-borne institutional outbreak of enteritis due to *Salmonella blockley*. *J. Am. Med. Assoc.* 202:939–940.

291. Mossel, D.A.A. 1982. *Microbiology of Foods.* 3d ed. Utrecht: University of Utrecht, Faculty of Veterinary Medicine.

292. Mozar, H.N., D.G. Bal, and S.A. Farag. 1989. Human cancer and the food chain: An alternative etiologic perspective. *Nutr. Cancer* 12:29–42.

293. Myers, J.A., and J.H. Steele. 1969. *Bovine Tuberculosis in Man and Animals.* St. Louis: Green.

294. Nagata, T., M.Saeki, T. Ida, et al. 1992. Sulfadimethoxine and Sulfa-monomethoxine residue studies in chicken tissues and eggs. In *Analysis of Antibiotic Drug Residues in Food Products of Animal Origin,* ed. V.K. Agarwal. New York: Plenum Press, 173–185.

295. Naguib, M.M., and M.A. Nour. 1981. Viability of *Salmonella typhi, Salmonella paratyphi* and *Salmonella typhimurium* in white cheeses. 2. Domiati cheese. *Mikrobi-ologija* 18:29–37.

296. Nakano, H., Y. Yoshikuni, H. Hashimoto, et al. 1992. Detection of *Clostridium botulinum* in natural sweetening. *Int. J. Food Microbiol.* 16:117–121.

297. Nassar, A., and A. Abou-Elala. 1993. Tuberculosis in slaughtered animals in As-siut Province (Upper Egypt): A foodborne infection. *Proc. 11th Int. Symp. WAVFH,* 139–146.

298. Ndon, J.A., S.M. Udo, and W.B. Wehrenberg. 1992. *Vibrio*-associated gas-troenteritis in the Lower Cross-River Basin of Nigeria. *J. Clin. Microbiol.* 30:2730–2732.

299. Nelson, N. (chrmn.). 1975. *Principles for Evaluating Chemicals in the Envi-ronment.* Washington, D.C.: National Academy of Sciences.

300. Nerkar, D.P., and N.F. Lewis. 1982. Radicidation for elimination of salmonel-lae in frog legs. *J. Food Prot.* 45:820–823.

301. Newell, D.G. (ed.). 1982. *Campylobacter: Epidemiology, Pathogenesis and Biochemistry.* Lancaster, United Kingdom: MTP Press, Ltd.

302. Newell, K.W. 1955. Outbreaks of paratyphoid B fever associated with imported frozen egg. I. Epidemiology. *J. Appl. Bacteriol.* 18:462–477.

303. Nocera, D., M. Altwegg, G.M. Lucchini, et al. 1993. Characterization of *Liste-ria* strains from a foodborne listeriosis outbreak by DNA gene restriction patterns com-pared to four other typing methods. *Eur. J. Clin. Microbiol. Infect. Dis.* 12:162–169.

304. Noegrohati, S., P. Sardjoko, K. Untung, et al. 1992. Impact of DDT spraying on the residue levels in soil, chicken, fish-pond water, carp, and human milk samples from malaria infested villages in Central Java. *Toxicol. Environ. Chem.* 34:237–251.

305. Notermans, S., and A. Hoogenboom-Verdegaal. 1992. Existing and emerging foodborne diseases. *Int. J. Food Microbiol.* 15:197–205.

306. Okolo, M.I.O. 1992. Tuberculosis in apparently healthy milch cows. *Microbios* 69:105–111.

307. Olaison, L., and I. Ljungstrom. 1992. An outbreak of trichinosis in Lebanon. *Trans. R. Soc. Trop. Med. Hyg.* 86:658–660.

308. Olsvik, O., Y. Wasteson, A. Lund, et al. 1991. Pathogenic *Escherichia coli* found in food. *Int. J. Food Microbiol.* 12:103–113.

309. O'Mahony, M., E. Mitchell, R.J. Gilbert, et al. 1990. An outbreak of foodborne botulism associated with contaminated hazelnut yoghurt. *Epidemiol. Infect.* 104:389–395.

310. Onah, D.N., S.N. Chiejina, and C.O. Emehelu. 1989. Epidemiology of echinococcosis/hydatidosis in Anambra State, Nigeria. *Ann. Trop. Med. Parasitol.* 83:387–393.

311. Ostroff, S.M., P.M. Griffin, R.V. Tauxe, et al. 1990. A statewide outbreak of *Es-cherichia coli* O157:H7 infections in Washington State. *Am. J. Epidemiol.* 132:239–247.

312. Osweiler, G.D., T.L. Carson, W.B. Buck, et al. 1985. *Clinical and Diagnostic Veterinary Toxicology.* 3d ed. Dubuque, Iowa: Kendall-Hunt.

313. Palumbo, S.A. 1988. The growth of *Aeromonas hydrophila* K144 in ground

pork at 5°C. *Int. J. Food Microbiol.* 7:41–48.

314. Palumbo, S.A., M.M. Bencivengo, F.D. Corral, et al. 1989. Characterization of the *Aeromonas hydrophila* group isolated from retail foods of animal origin. *J. Clin. Microbiol.* 27:854–859.

315. Palumbo, S.A., J.L. Smith, B.S. Marmer, et al. 1993. Thermal destruction of *Listeria monocytogenes* during liver sausage processing. *Food Microbiol.* 10:243–247.

316. Pamukcu, A.M., E. Erturk, S. Yalciner, et al. 1978. Carcinogenic and mutagenic activities of milk from cows fed bracken fern (*Pteridium aquilinum*). *Cancer Res.* 38:1556–1560.

317. Pande, P.G., R.R. Shukla, and P.C.Sekariah. 1961. *Toxoplasma* from the eggs of the domestic fowl (*Gallus gallus*). *Science* 133:648.

318. Park, S.D., H.S. Shon, and N.J. Joh. 1991. *Vibrio vulnificus* septicemia in Korea: Clinical and epidemiologic findings in seventy patients. *J. Am. Acad. Dermatol.* 24:397–403.

319. Pearson, L.J., and E.H. Marth. 1990. *Listeria monocytogenes*—threat to a safe food supply: A review. *J. Dairy Sci.* 73:912–928.

320. Pether, J.Y., and E.O. Caul. 1983. An outbreak of foodborne gastroenteritis in two hospitals associated with a Norwalk-like virus. *J. Hyg.* 91(2):343–350.

321. Phillips, I. 1991. Epidemic potential and pathogenicity in outbreaks of infection with EMRSA and EMREC. *J. Hosp. Infect.* 18(Suppl. A):197–201.

322. Pinner, R.W., A. Schuchat, B. Swaminathan, et al. 1992. Role of foods in sporadic listeriosis. II. Microbiologic and epidemiologic investigation. *J. Am. Med. Assoc.* 267:2046–2050.

323. Pomeroy, B.S., and R. Fenstermacher. 1944. *Salmonella* infections in turkeys. *Am. J. Vet. Res.* 5:282–288.

324. Pop, A., A. Oprisan, A. Pop, et al. 1989. Toxoplasmosis prevalence parasitologically evaluated in meat animals. *Arch. Roum. Pathol. Exp. Microbiol.* 48:373–378.

325. Poppe, C., R.J. Irwin, C.M. Forsberg, et al. 1991. The prevalence of *Salmonella enteritidis* and other *Salmonella* spp. among Canadian registered commercial layer flocks. *Epidemiol. Infect.* 106:259–270.

326. Poskanzer, D.C., and A.L. Herbst. 1977. Epidemiology of vaginal adenosis and adenocarcinoma associated with exposure to stilbestrol in utero. *Cancer* 39:1892–1895.

327. Potter, M.E. 1992. The changing face of foodborne disease. *J. Am. Vet. Med. Assoc.* 201:250–253.

328. Pozio, E., O. Cappelli, L. Marchesi, et al. 1988. Third outbreak of trichinellosis caused by consumption of horse meat in Italy. *Ann. Parasitol. Hum. Comp.* 63:48–53.

329. Pozio, E., G. La Rosa, K.D. Murrell, et al. 1992. Taxonomic revision of the genus *Trichinella. J. Parasitol.* 78:654–659.

330. Pozio, E., P. Varese, M.A.G. Morales, et al. 1993. Comparison of human trichinellosis caused by *Trichinella spiralis* and by *Trichinella britovi. Am. J. Trop. Med. Hyg.* 48:568–575.

331. Prelusky, D.B., R.M.G. Hamilton, and H.L. Trenholm. 1989. Transmission of residues to eggs following long-term administration of ^{14}C-labelled deoxynivalenol to laying hens. *Poult. Sci.* 68:744–748.

332. Press, E., and L. Yeager. 1962. Food "poisoning" due to sodium nicotinate. *Am. J. Public Health* 52:1720–1728.

333. Pulce, C., D. Lamalson, G. Keck, et al. 1991. Collective human food poisonings by clenbuterol residues in veal liver. *Vet. Hum. Toxicol.* 33:480–481.

334. Radeleff, R.D. 1970. *Veterinary Toxicology.* 2d ed. Philadelphia: Lea and Febiger.

335. Read, S.C., R.C. Clarke, A. Martin, et al. 1992. Polymerase chain reaction for detection of verocytotoxigenic *Escherichia coli* isolated from animal and food sources.

Mol. Cell. Probes 6:153–161.

336. Renwick, S.A., R.J. Irwin, R.C. Clarke, et al. 1992. Epidemiological associations between characteristics of registered broiler chicken flocks in Canada and the *Salmonella* culture status of floor litter and drinking water. *Can. Vet. J.* 33:449–458.

337. Richardson, G.H. (ed.). 1985. *Standard Methods for the Examination of Dairy Products.* 15th ed. Washington, D.C.: American Public Health Association.

338. Riemann, H., and F.L. Bryan (eds.). 1979. *Food-borne Infections and Intoxications.* 2d ed. New York: Academic Press.

339. Riemann, H.P., M.E. Meyer, J.H. Theis, et al. 1975. Toxoplasmosis in an infant fed unpasteurized goat milk. *J. Pediatr.* 87:573–576.

340. Riordan, T., T.J. Humphrey, and A. Fowles. 1993. A point source outbreak of campylobacter infection related to bird-pecked milk. *Epidemiol. Infect.* 110:261–265.

341. Roach, R.L., and D.G. Sienko. 1992. *Clostridium perfringens* outbreak associated with minestrone soup. *Am. J. Epidemiol.* 136:1288–1291.

342. Robens, J.F., and J.L. Richard. 1992. Aflatoxins in animal and human health. *Rev. Environ. Contam. Toxicol.* 127:69–94.

343. Robert, R., G. Grollier, F. Malin, et al. 1991. Isolation of *Vibrio alginolyticus* from blood cultures in a leukaemic patient after consumption of oysters. *Eur. J. Clin. Microbiol. Infect. Dis.* 10:987–988.

344. Robertson, L. 1961. *Brucella* organisms in milk. *R. Soc. Health J.* 81:46–50.

345. Robinson, P. 1988. *Mycobacterium bovis* as an occupational hazard in abattoir workers. *Aust. N. Z. J. Med.* 18:701–703.

346. Rodricks, J.V., C.W. Helleltine, and M.A. Mehlman (eds.). 1977. *Mycotoxins in Human and Animal Health.* Park Forest South, Ill.: Pathotox Publishers.

347. Rodriguez, D.C., R.V. Tauxe, and B. Rowe. 1990. International increase in *Salmonella enteritidis:* A new pandemic? *Epidemiol. Infect.* 105:21–27.

348. Rodriguez-Osorio, M., V. Gomez-Garcia, J. Rodriguez-Perez, et al. 1990. Sero-epidemiological studies on five outbreaks of trichinellosis in southern Spain. *Ann. Trop. Med. Parasitol.* 84:181–184.

349. Romalde, J.L., J.L. Barja, and A.E. Toranzo. 1990. Vibrios associated with red tides caused by *Mesodinium rubrum. Appl. Environ. Microbiol.* 56:3615–3619.

350. Rommel, M. 1989. Recent advances in the knowledge of the biology of the cyst-forming coccidia. *Angew. Parasitol.* 30:173–183.

351. Roueche, B. 1953. *Eleven Blue Men.* New York: Berkley Publishing.

352. Rowe, P.C., E. Orrbine, H. Lior, et al. 1993. Diarrhoea in close contacts as a risk factor for childhood haemolytic uraemic syndrome. *Epidemiol. Infect.* 110:9–16.

353. Royal, L., and I. McCoubrey. 1989. International spread of disease by air travel. *Am. Fam. Physician* 40:129–136.

354. Rubenstein, A.D., R.F. Feemster, and H.M. Smith. 1944. Salmonellosis as a public health problem in wartime. *Am. J. Public Health* 34:841–853.

355. St. Louis, M.E., D.L. Morse, M.E. Potter, et al. 1988. The emergence of grade A eggs as a major source of *Salmonella enteritidis* infections: New implications for the control of salmonellosis. *J. Am. Med. Assoc.* 259:2103–2107.

356. Salmon, R.L., S.R. Palmer, C.D. Ribeiro, et al. 1991. How is the source of food poisoning outbreaks established? The example of three consecutive *Salmonella enteritidis* PT4 outbreaks linked to eggs. *J. Epidemiol. Community Health* 45:266–269.

357. Sang, F.C., M.E. Hugh-Jones, and H.V. Hagstad. 1987. Viability of *Vibrio cholerae* O1 on frog legs under frozen and refrigerated conditions and low dose radiation treatment. *J. Food Prot.* 50:662–664.

358. Sarles, J., C. Scheiner, M. Sarran, et al. 1990. Hepatic hypervitaminosis A: A familial observation. *J. Pediatr. Gastroenterol. Nutr.* 10:71–76.

359. Sarti, E., P.M. Schantz, A. Plancarte, et al. 1992. Prevalence and risk factors for *Taenia solium* taeniasis and cysticercosis in humans and pigs in a village in Morelos, Mexico. *Am. J. Trop. Med. Hyg.* 46:677–685.

360. Schandevyl, P., E. Van Dyck, and P. Piot. 1984. Halophilic *Vibrio* species from seafood in Senegal. *Appl. Environ. Microbiol.* 48:236–238.

361. Schantz, P.M., and J. McAuley. 1991. Current status of food-borne parasitic zoonoses in the United States. *Southeast Asian J. Trop. Med. Public Health* 22 (Suppl.):65–71.

362. Scheuplein, R.J. 1992. Perspectives on toxicological risk—an example: Food-borne carcinogenic risk. *Crit. Rev. Food Sci. Nutr.* 32:105–121.

363. Schiemann, D.A., and S. Toma. 1978. Isolation of *Yersinia enterocolitica* from raw milk. *Appl. Environ. Microbiol.* 35:54–58.

364. Schlech, W.F., III. 1991. Listeriosis: Epidemiology, virulence and the significance of contaminated foodstuffs. *J. Hosp. Infect.* 19:211–224.

365. ———. 1992. Expanding the horizons of foodborne listeriosis. *J. Am. Med. Assoc.* 267:2081–2082.

366. Schnoor, J.L. (ed.). 1992. *Fate of Pesticides and Chemicals in the Environment.* New York: John Wiley and Sons.

367. Schroeder, J.P., J.G. Wallace, M.B. Cates, et al. 1985. An infection by *Vibrio alginolyticus* in an Atlantic bottlenose dolphin. *J. Wildl. Dis.* 21:437–438.

368. Schuchat, A., B. Swaminathan, and C.V. Broome. 1991. Epidemiology of human listeriosis. *Clin. Microbiol. Rev.* 4:169–183.

369. Schuchat, A., K.A. Deaver, J.D. Wenger, et al. 1992. Role of foods in sporadic listeriosis, I. Case-control study of dietary risk factors. *J. Am. Med. Assoc.* 267:2041–2045.

370. Schwartz, B., D. Hexter, C.V. Broome, et al. 1989. Investigation of an outbreak of listeriosis: New hypotheses for the etiology of epidemic *Listeria monocytogenes* infections. *J. Infect. Dis.* 159:680–685.

371. Shaffer, N., R.B. Wainwright, J.P. Middaugh, et al. 1990. Botulism among Alaska natives: The role of changing food preparation and consumption practices. *West. J. Med.* 153:390–393.

372. Shandera, W.X., C.O. Tacket, and P.A. Blake. 1983. Food poisoning due to *Clostridium perfringens* in the United States. *J. Infect. Dis.* 147:167–170.

373. Shane, S.M., and M.S. Montrose. 1985. The occurrence and significance of *Campylobacter jejuni* in man and animals. *Vet. Res. Commun.* 9:167–198.

374. Sharp, J.C.M. 1991. Meat and milkborne infections. *Proc. Soc. Vet. Epidemiol. Prev. Med.,* ed. M.V. Thrusfield, April 17, 1991, 34–45.

375. Shelef, L.A. 1989. Listeriosis and its transmission by food. *Prog. Food Nutr. Sci.* 13:363–382.

376. Shinagawa, K. 1990. Analytical methods for *Bacillus cereus* and other *Bacillus* species. *Int. J. Food Microbiol.* 10:125–141.

377. Shrivastava, K.P. 1978. The occurrence of salmonellas in raw frozen frog legs. *J. Appl. Bacteriol.* 45:407–410.

378. Sibley, L.D., and J.C. Boothroyd. 1992. Virulent strains of *Toxoplasma gondii* comprise a single clonal lineage. *Nature* 359:82–85.

379. Singh, M., Y.E. Hian, and C. Lay-Hoon. 1991. Current status of food-borne parasitic zoonoses in Singapore. *Southeast Asian J. Trop. Med. Public Health* 22 (Suppl.):27–30.

380. Skirrow, M.B. 1990. *Campylobacter. Lancet* 336:921–923.

381. Sklan, D. 1987. Vitamin A in human nutrition. *Prog. Food Nutr. Sci.* 11:39–55.

382. Slaten, D.D., R.I. Oropeza, and S.B. Werner. 1992. An outbreak of *Bacillus cereus* food poisoning—are caterers supervised sufficiently? *Public Health Rep.* 107:477–480.

383. Slater, P.E., D.G. Addiss, A. Cohen, et al. 1989. Foodborne botulism: An international outbreak. *Int. J. Epidemiol.* 18:693–696.

384. Small, R.G., and J.C.M. Sharp. 1979. A milk-borne outbreak due to *Salmonella dublin. J. Hyg.* (Camb.) 82:95–100.

385. Smith, C.T., F.R. Shaw, D.L. Anderson, et al. 1965. Ronnel residues in eggs of poultry. *J. Econ. Entomol.* 58:1160–1161.

386. Smith, M.G. 1992. Destruction of bacteria on fresh meat by hot water. *Epidemiol. Infect.* 109:491–496.

387. Snapper, I. 1944. Salmonellosis caused by the ingestion of ducks' eggs. *Am. J. Digest. Dis.* 11:8–10.

388. Sockett, P.N. 1991. The economic implications of human salmonella infection. *J. Appl. Bacteriol.* 71:289–295.

389. Sockett, P.N., J.M. Cowden, S. LeBaigue, et al. 1993. Food borne disease surveillance in England and Wales: 1989–1991. *CDR Rev.* 3:R159–R173.

390. Solly, S.R.B., V. Shanks, R.T. Steele, et al. 1976. Effects of polychlorinated biphenyls on poultry. I. Residues in tissues and eggs. *N. Z. J. Agric. Res.* 19:225–229.

391. Soule, C., J. Dupouy-Camet, P. Georges, et al. 1989. Experimental trichinellosis in horses: Biological and parasitological evaluation. *Vet. Parasitol.* 31:19–36.

392. Spink, W.W. 1956. *The Nature of Brucellosis.* Minneapolis: University of Minnesota Press.

393. Stevens, A., C. Joseph, J. Bruce, et al. 1989. A large outbreak of *Salmonella enteritidis* phage type 4 associated with eggs from overseas. *Epidemiol. Infect.* 103:425–433.

394. Stiles, G.W. 1945. Brucellosis in goats: Recovery of *Brucella melitensis* from cheese manufactured from unpasteurized goats' milk. *Rocky Mt. Med. J.* 42:18–25.

395. Stiles, M.E. 1979. Survival of *Salmonella* during cooking of eggs. *J. Can. Diet. Assoc.* 40:155–158.

396. Stone, W., and F.W. Smith. 1973. Infection of mammalian hosts by milk-borne nematode larvae: A review. *Exp. Parasitol.* 34:306–313.

397. Svabic-Vlahovic, M., D. Pantic, M. Pavicic, et al. 1988. Transmission of *Listeria monocytogenes* from mother's milk to her baby and to puppies. *Lancet* 2(8621):1201.

398. Swaddiwudhipong, W., P. Kunasol, O. Sangwanloy, et al. 1989. Foodborne disease outbreaks of chemical etiology in Thailand, 1981–1987. *Southeast Asian J. Trop. Med. Public Health* 20:125–132.

399. Swerdlow, D.L., E.D. Mintz, M. Rodriguez, et al. 1992. Waterborne transmission of epidemic cholera in Trujillo, Peru: Lessons for a continent at risk. *Lancet* 340:28–32.

400. Swerdlow, D.L., B.A. Woodruff, R.C. Brady, et al. 1992. A waterborne outbreak in Missouri of *Escherichia coli* O157:H7 associated with bloody diarrhea and death. *Ann. Intern. Med.* 117:812–819.

401. Tacket, C.O., J.P. Narain, R. Sattin, et al. 1984. A multistate outbreak of infections caused by *Yersinia enterocolitica* transmitted by pasteurized milk. *J. Am. Med. Assoc.* 251:483–486.

402. Tamplin, M.L., and G.M. Capers. 1992. Persistence of *Vibrio vulnificus* in tissues of Gulf Coast oysters, *Crassostrea virginica,* exposed to seawater disinfected with UV light. *Appl. Environ. Microbiol.* 58:1506–1510.

403. Tarlo, S.M., and G.L. Sussman. 1993. Asthma and anaphylactoid reactions to food additives. *Can. Fam. Physician* 39:1119–1123.

404. Tauxe, R.V. 1991. *Salmonella:* A postmodern pathogen. *J. Food Prot.* 54:563–568.

405. Taylor, D.N., and P. Echeverria. 1986. Etiology and epidemiology of travelers' diarrhea in Asia. *Rev. Infect. Dis.* 8(Suppl. 2):S136–S141.

406. Taylor, J.L., D.M. Dwyer, C. Groves, et al. 1993. Simultaneous outbreak of *Salmonella enteritidis* and *Salmonella schwarzengrund* in a nursing home: Association of *S. enteritidis* with bacteremia and hospitalization. *J. Infect. Dis.* 167:781–782.

407. Taylor, P.R., W.M. Wernstein, and J.H. Bryner. 1979. *Camplyobacter fetus* infection in humans: Association with raw milk. *Am. J. Med.* 66:779–783.

408. Telzak, E.E., E.P. Bell, D.A. Kautter, et al. 1990. An international outbreak of

type E botulism due to uneviscerated fish. *J. Infect. Dis.* 161:340–342.

409. Telzak, E.E., M.S. Zweig Greenberg, L.D. Budnick, et al. 1991. Diabetes mellitus—a newly described risk factor for infection from *Salmonella enteritidis*. *J. Infect. Dis.* 164:538–541.

410. Terr, A.I. 1989. Clinical ecology. *Ann. Intern. Med.* 111:168–178.

411. Thomas, G.W. 1978. *Salmonella paratyphi* B in cattle. *Vet. Rec.* 103:512.

412. Thomas, G.W., and J.F. Harbourne. 1972. *Salmonella paratyphi* B infection in dairy cows. Part II. Investigation of an active carrier. *Vet. Rec.* 91:148–150.

413. Thompkin, R.B. 1980. Botulism from meat and poultry products—a historical perspective. *Food Technol.* 34:229–236, 257.

414. Thompson, E.M., G.J. Mountney, and G.W. Ware. 1967. Methoxychlor residues in chicken eggs. *J. Econ. Entomol.* 60:235–237.

415. Todd, E.C.D. 1989. Foodborne and waterborne disease in Canada—1984 annual summary. *J. Food Prot.* 52:503–511.

416. ——. 1989. Costs of acute bacterial foodborne disease in Canada and the United States. *Int. J. Food Microbiol.* 9:313–326.

417. ——. 1990. Epidemiology of foodborne illness: North America. *Lancet* 336:788–790.

418. ——. 1992. Foodborne disease in Canada—a 10 year summary from 1975 to 1984. *J. Food Prot.* 55:123–132.

419. ——. 1993. Domoic acid and amnesic shellfish poisoning—a review. *J. Food Prot.* 56:69–83.

420. Todd, L.S., J.C. Hardy, M.F. Stringer, et al. 1989. Toxin production by strains of *Aeromonas hydrophila* grown in laboratory media and prawn puree. *Int. J. Food Microbiol.* 9:145–156.

421. Tranter, H.S. 1990. Foodborne staphylococcal illness. *Lancet* 336:1044–1046.

422. Trott, D., P. Seneviratna, and J. Robertson. 1991. *Listeria* in cooked chicken, pate and mixed smallgoods. *Aust. Vet. J.* 68:249–250.

423. Trott, D.J., I.D. Robertson, and D.J. Hampson. 1993. Genetic characterisation of isolates of *Listeria monocytogenes* from man, animals and food. *J. Med. Microbiol.* 38:122–128.

424. Tullett, S.G. (ed.). 1991. *Avian Incubation.* London: Butterworth-Heinemann.

425. United Nations. 1972. *A Review of the Technological Efficacy of Some Antioxidants and Synergists.* WHO Food Addit. Ser. 3. Geneva: World Health Organization.

426. —— 1972. *Evaluation of Mercury, Lead, Cadmium and the Food Additives Amaranth, Diethylpyrocarbonate, and Octyl Gallate.* WHO Food Addit. Ser. 4. Geneva: World Health Organization.

427. ——. 1974. *Fish and Shellfish Hygiene.* WHO Tech. Rep. Ser. 550. Geneva: World Health Organization.

428. United Nations. World Health Organization. 1971. *Evaluation of Some Pesticide Residues in Food.* WHO Pestic. Residues Ser. 1. Geneva: World Health Organization.

429. ——. 1974. *Food-borne Disease: Methods of Sampling and Examination in Surveillance Programs. Report of a WHO Study Group.* WHO Tech. Rep. Ser. 543. Geneva: World Health Organization.

430. U.S. Army Medical Service. 1962. *Inspection of Waterfoods.* Washington, D.C.: U.S. Government Printing Office.

431. Utsalo, S.J., F.O. Eko, and O.E. Antia-Obong. 1991. Cholera and *Vibrio parahaemolyticus* diarrhoea endemicity in Calabar, Nigeria. *West Afr. J. Med.* 10:175–180.

432. van de Giessen, A.W., J.B. Dufrenne, W.S. Ritmeester, et al. 1992. The identification of *Salmonella enteritidis*-infected poultry flocks associated with an outbreak of human salmonellosis. *Epidemiol. Infect.* 109:405–411.

433. van Egmond, H.P. (ed.). 1989. *Mycotoxins in Dairy Products.* London: Elsevier Applied Science.

434. Vugia, D.J., B. Mishu, M. Smith, et al. 1993. *Salmonella enteritidis* outbreak in

a restaurant chain: The continuing challenges of prevention. *Epidemiol. Infect.* 110:49–64.

435. Waldron, A.C., and E.C. Naber. 1974. Importance of feed as an unavoidable source of pesticide contamination in poultry meat and eggs. *Poult. Sci.* 53:1428–1435.

436. Wasti, S.S., and F.R. Shaw. 1971. Residues of encapsulated Rabon in tissues and eggs of poultry. *J. Econ. Entomol.* 64:224–225.

437. Weber, J.T., R.G. Hibbs, Jr., A. Darwish, et al. 1993. A massive outbreak of type E botulism associated with traditional salted fish in Cairo. *J. Infect. Dis.* 167:451–454.

438. Wesley, I.V., and F. Ashton. 1991. Restriction enzyme analysis of *Listeria monocytogenes* strains associated with food-borne epidemics. *Appl. Environ. Microbiol.* 57:969–975.

439. Wharton, M., T.L. Chorba, R.L. Vogt, et al. 1990. Case definitions for public health surveillance. *MMWR* 39(RR-13):1–43.

440. Wieneke, A.A., D. Roberts, and R.J. Gilbert. 1993. Staphylococcal food poisoning in the United Kingdom, 1969–90. *Epidemiol. Infect.* 110:519–531.

441. Wiesenthal, A.M., M. Ressman, S.A. Caston, et al. 1985. Toxic shock syndrome. I. Clinical exclusion of other syndromes by strict and screening definitions. *Am. J. Epidemiol.* 122:847–856.

442. Williams, J.E., and A.D. Whittemore. 1967. A method for studying microbial penetration through the outer structures of the avian egg. *Avian Dis.* 11:467–490.

443. Wise, K.A., and P.J. Newton. 1992. A fatal case of *Vibrio vulnificus* septicemia. *Pathology* 24:121–122.

444. Wogan, G.N. 1992. Aflatoxins as risk factors for hepatocellular carcinoma in humans. *Cancer Res.* 52(Suppl.): 2114S–2118S.

445. Wong, H.-C., S.-H. Ting, and W.-R. Shieh. 1992. Incidence of toxigenic vibrios in foods available in Taiwan. *J. Appl. Bacteriol.* 73:197–202.

446. Wood, R.C., K.L. MacDonald, and M.T. Osterholm. 1992. *Campylobacter* enteritis outbreaks associated with drinking raw milk during youth activities. *J. Am. Med. Assoc.* 268:3228–3230.

447. Woodruff, B.A., P.M. Griffin, L.M. McCroskey, et al. 1992. Clinical and laboratory comparison of botulism from toxin types A, B, and E in the United States, 1975–1988. *J. Infect. Dis.* 166:1281–1286.

448. Woodruff, C.W. 1976. Milk intolerances. *Nutr. Rev.* 34:33–37.

449. Xu, Z.-Y., Z.-H. Li, J.-X. Wang, et al. 1992. Ecology and prevention of a shellfish-associated hepatitis A epidemic in Shanghai, China. *Vaccine* 10(Suppl. 1):S67–S68.

450. Yamaguchi, T. 1991. Present status of trichinellosis in Japan. *Southeast Asian J. Trop. Med. Public Health* 22 (Suppl.):295–301.

451. Zanen, H.C., and H.W.B. Engel. 1975. Porcine streptococci causing meningitis and septicaemia in man. *Lancet* 1:1286–1288.

452. Zimmerman, M.R. 1990. The paleopathology of the liver. *Ann. Clin. Lab. Sci.* 20:301–306.

453. Zottola, E.A., and L.B. Smith. 1991. Pathogens in cheese. *Food Microbiol.* 8:171–182.

454. Bean, N.H. 1990. *Morb. Mortal. Wkly. Rep.* 39 (55-1): 15-57.

455. *U.S. Department of Agriculture Yearbook 1942.* Washington, D.C.: U.S. Government Printing Office.

4

Consumer Protection

Objectives

- Describe the practical aspects involved in applying the principles of foodborne disease prevention.

- Describe the problems of unacceptable drug and chemical residues in meat and milk, including control methods used.

- Describe the procedures utilized in investigating outbreaks of foodborne disease.

- Describe the purposes and benefits of meat, poultry, and milk inspection.

- Describe the evolution of meat, poultry, and milk inspection in Australia, Canada, the United Kingdom, and the United States.

- Describe antemortem inspection and why it is necessary.

- Describe and contrast the effectiveness, benefits, and disadvantages of various (humane) stunning methods.

- Describe the economic and food hygiene considerations involved in postmortem examination.

- Describe the procedures for control of rejected (condemned) and inedible material.

- Identify conditions that make a carcass or product acceptable for animal food but not for human food.

- Describe the conditions (generalized or localized) found at postmortem examination that may require rejection of the carcass or offal and the criteria used in the decision (disposition) for each, including those rejected for aesthetic reasons.

- Describe the essential features of meat and poultry product labels, including warning statements, and how they should inform consumers of quality standards.

- Describe safety and quality standards for milk and milk products and governmental responsibility for quality control of the dairy industry.

Preventing Foodborne Disease[51,54,56,77,105,125,126,173,174,176,184,207]

Factors Essential for Occurrence of Foodborne Disease[159,160,161,162,238]

Several factors are essential for foodborne microbial or parasitic disease to occur. They are (1) microbial or parasitic pathogens present (they may have been present in the meat or milk preharvest), (2) a source of postharvest contamination (knife, cutting board, hands, mouth, etc.), (3) a medium in which bacterial pathogens can grow, (4) suitable environmental conditions (time, temperature, moisture, etc.), and (5) consumption of a sufficient quantity of the contaminated food.

Preharvest Prevention[102,143,144,146,168,169,193,208,236]

If the food animal is a primary reservoir host of a microbial or parasitic pathogen, then eradication of the agent from this reservoir is of major importance in the prevention of foodborne exposure. Similarly, prevention of the accumulation of chemical residues must be achieved while the animal is still alive by control of production activities.

Postharvest Prevention[220]

Four facets of postharvest food processing are essential for preventing foodborne disease. These are (1) **preventing contamination** of foods during each step of processing (beginning at harvest of meat or milk), especially when handling foods already cooked; (2) **inhibiting growth** of organisms in foods already contaminated; (3) **destroying pathogens** to prevent foodborne disease; and (4) **preventing adulteration** by chemical additives or contaminants. Because chemical adulteration may occur both preharvest and postharvest, and involves characteristics distinct from organic agents, it is presented in a separate section to follow.[199]

Preventing Microbial Contamination

Personal Hygiene. Postharvest contamination of food can be prevented or at least minimized by practicing good personal hygiene at all points along the food chain.

Equipment. Preventing contamination of food products includes using properly cleaned equipment. A three-step process (involving washing with cleansing compounds and water, rinsing, and sanitizing effectively) cleans utensils and equipment. The rinsing procedure removes excess cleaning compounds. Sanitizing may be achieved by immersion in hot water (76.7°C [170°F]) for at least 30 s or by use of a chemical rinse. Three factors, which vary depending on the chemical used, are important in the chemical method: (1) immersion time, (2) temperature, and (3) concentration of active ingredients.

Spread of microorganisms from contaminated food to other foods can be prevented if separate utensils are used for each. If this is not possible, the equipment should be cleaned thoroughly and disinfected between uses. Of prime consideration is cross-contamination between raw and cooked foods.

Ingredients. Unwholesome ingredients can contaminate a prepared food product. Only inspected foods with undamaged containers should be used.

Inhibiting Growth of Microorganisms. Foods may be divided into three categories: (1) those that **support bacterial growth** (propagative vehicles), (2) those in which **bacteria survive** but do not multiply, and (3) those that **actually kill** pathogens. The growth of microorganisms can be inhibited or retarded by controlling temperature, pH, and the water activity of the food.

Low Temperatures. Most pathogens are mesophilic organisms that grow best at temperatures between 4.4°C (40°F) and 60°C (140°F). Foods should be stored at temperatures outside of this range. Cooked food should be cooled as rapidly as possible, although this may be difficult to achieve when large quantities of meat are cooked.

After food is chilled, it should be refrigerated at 4.4°C (40°F) or less. Clostridial spores and staphylococcal toxins are not destroyed by normal cooking, so preventing contamination and inhibiting their growth is necessary to effectively prevent outbreaks of foodborne disease.

Precooling ingredients helps inhibit microbial growth during preparation. In sausage production, meat processors often add ice instead of water to cool the mixture.

High Temperatures. Hot foods held for a period before serving should be maintained at temperatures in excess of 60°C (140°F) at the center.

Water Activity. The amount of water available to microorganisms is termed the water activity (Aw) and is defined as the ratio of water vapor pressure of the food (p) to that of pure water (po) at the same temperature.

$$Aw = p/po$$

Aw level should not be confused with water content because foods with the same water content may have different Aw levels. Fresh meat, fish, and milk usually have an Aw level of .98 or higher; evaporated milk and cooked sausage will be between .98 and .93; dried beef, sausage, sweetened condensed milk, and cheese will be between .93 and .85; whereas dried milk and dried eggs will be below .60. Even though microorganisms may survive for some time at unfavorable Aw levels, knowledge of optimum Aw levels is useful in determining potential hazards. Salmonellae and *Escherichia coli,* for example, have an optimum Aw level for growth of .95, whereas *Staphylococcus aureus* and many mycotoxin-producing fungi can grow at an Aw level of approximately .86. Microorganisms do not multiply, but can remain viable, at levels below .60.

Time.   Although food should be consumed as soon as possible after preparation, this is not always practical. Generally, foods held between 4.4°C (40°F) and 60°C (140°F) should be eaten within 4 h.

Destroying Pathogens.   Vegetative forms of microorganisms are killed by thorough cooking, whereas spores and some toxins may not be destroyed. Because adequate heat penetration is more difficult to achieve when large quantities of food are cooked, food should be cooked in small quantities when possible. Some foods may be treated with irradiation, freezing, or pasteurization to kill specific microbial or parasitic pathogens (see Chapter 2 and later in this chapter).

Chemical Adulteration[32,53,123]

The industrial revolution had far-reaching effects, among them a dramatic change in techniques for producing food, particularly those of animal origin. Technology took the responsibility for the purity of food out of the kitchen and relegated it to the increasingly complex food chain described in Chapter 2. Concomitant with these technological advances was an increased potential for the adulteration of foods of animal origin with toxic materials. A food is **adulterated** whenever it contains any chemical that may render it injurious to health or otherwise unfit for human food. Meat and poultry may become adulterated with unacceptable concentrations of chemicals during production of the animals before slaughter or during processing, packaging, or storage of the meat after slaughter.

Sources of Chemical Adulteration.   During their growth, animals may be intentionally or inadvertently exposed to antibiotics, sulfonamides, growth-promoting substances, mycotoxins, pesticides, toxic metals, and radionuclides that may persist in the body as **residues**. Some chemicals may be added on purpose during processing or packaging, and these are considered as **additives**. These chemicals should not be confused with **feed additives** incorporated in animal rations. Other chemicals, termed *contaminants,* may occur unintentionally during processing, packaging, or storage, such as the results from leaching of chemical constituents from the packaging materials. All of the above chemicals may cause adulteration if not used properly. Residues are of preharvest origin; additives and contaminants of postharvest.

Standard Measures of Acceptable Residue Concentrations in Food.[42,65] Human exposure to foodborne toxic chemicals can be assessed either by direct measurement of human tissues or fluids or by food analysis. Results obtained by the former method, however, also include chemicals accumulated from other sources, such as air and water, and may be the result of long-term accumulation in the tissues. Direct measurement, on the other hand, provides definitive evidence of absorption, although not necessarily through ingested food in the gut.

Health risks from residues in food depend on both the toxicity of the chemical and the likelihood of exposure, i.e., consuming it in a given food. The likelihood of toxic chemicals in foods of animal origin varies according to the likelihood of exposure of the food animal to the chemical. Average daily intake of the food is an important factor in determining exposure-related risk, especially for milkborne residues and particularly in infants inasmuch as milk represents a high portion of their diet.

International standards have been developed for more than 2,000 compounds by the joint Food Agriculture Organization/World Health Organization (FAO/WHO) Codex Alimentarius Commission (CAC) and are referred to as *maximum residue limits* (*MRLs*). In Australia MRLs are established by the National Registration Authority and the Department of Community Services and Health and become part of the Food Standards Code. In Canada the setting of MRLs is done by Health Canada under the authority of the Food and Drug Act and Regulations. The guidelines are set out in the act, and specific provisions are published in Division 15 of the Food and Drug Regulations. MRLs applied in the United Kingdom are set by the Committee on Veterinary Products, DG III, European Commission, Brussels. In the United States, the maximum acceptable residue concentrations in foods are established by the Environmental Protection Agency (EPA) for pesticides and the Food and Drug Administration (FDA) for animal drugs and unavoidable environmental contaminants. The concentrations, designated in ppm or ppb, are referred to as *tolerances.*

Inasmuch as most data used to establish these limits are derived from animal studies, an "uncertainty factor" may be added to provide a margin of safety beyond the level demonstrated to cause no observable effect. Usually, the uncertainty factor is 100- to 1,000-fold for short- and long-term hazards other than for carcinogens. For carcinogens, a level of safety is calculated that represents an estimated dietary exposure that is 95% confident that, during a 40-yr lifetime, a risk of one in a million of developing cancer will not be exceeded.

Residues in foods of animal origin may be derived from several sources: industrial and agricultural chemicals and veterinary therapeutic agents. Those in the first two categories occasionally cause clinical problems, usually manifesting as herd problems. The latter, the therapeutic agents, are of constant concern to veterinarians involved in the production of food animals. Origins of chemical residues follow:

1. **Agricultural and industrial chemicals**.[65,73,101,108,147,157,167,180, 189,221] Although these chemicals may be responsible for occasional clinical episodes in food animals, they are more important as causes of adulteration. Therefore, veterinarians need to be aware of how agricultural and industrial chemicals may contaminate animal feed and water sources if they are to alert their food-animal producer clients to these dangers. For example, when heptachlor-treated seed grains were used as feed for dairy cattle, poultry, and swine, extensive losses resulted from condemnation of adulterated meat and milk. Heptachlor persists for long periods in body fat. Polybrominated and polychlorinated biphenyls may also cause clinical signs in animals and persist in fat as residues. Some chlorinated hydrocarbon pesticides, such as DDT, persist in the environment for many years, and forage harvested from land treated long ago may still be contaminated.

2. **Therapeutic and growth-promoting agents**.[33,47,52,59,72,92,111,134,185,192,212,232,241,245] Therapeutic and growth-promoting agents, including anabolic agents, antibiotics, parasiticides, pesticides, and sulfonamides, may be administered orally, parenterally, or topically to food animals and may persist as residues. Most residue problems result from misuse of approved drugs, particularly the use of either an inappropriate dose or route of administration, use in an unintended species or age group, or failure to observe appropriate withdrawal times. The **withdrawal time (period)** is the interval from last medication until the concentration of the drug in the tissues has decreased sufficiently so that the animal can be slaughtered or the milk can be used without the presence of violative residues.

   A few drugs are prohibited from use in **any** food animal. Use of diethylstilbestrol (DES) will produce a 1,000-lb beef animal 30 d sooner than if not used and will save about 500 lb of feed, an increase of 10–12% in feed efficiency. Concern about the causal association between sex hormones and neoplasia, however, resulted in the use of DES in food animals being banned in most countries by the early 1980s. The use of chloramphenicol was banned because of its association with aplastic anemia. The ß-agonist clenbuterol is approved in some countries for use in treatment of respiratory conditions in horses. It has also been used illegally to produce leaner carcasses in food animals.

Violative concentrations of residues are those that exceed (i.e., are in violation of) the legal tolerance in products to be sold for human consumption. The presence of these concentrations per se is **not illegal**, as long as the product is not sold for human food. On the other hand, the presence of any amount of a banned compound, such as DES, is illegal regardless of the intended use of the product.

Utilization of antibiotics in feed to promote growth started in the 1940s. In the 1960s, concerns about antibiotic resistance were voiced in several countries, with subsequent recommendations to eliminate or at least severely curtail the use of medicated feeds. The principal reasons for their concern were (1) the prevalence of multiresistant, R-plasmid-bearing pathogenic, and nonpathogenic bacteria in animals was increasing and was related to the use of antibiotics and sulfonamide drugs, especially in growth promotant and subtherapeutic amounts, and (2) there was also an increase in the prevalence of antibiotic- and sulfonamide-resistant bacteria in humans.

Nitrates and Nitrites.[62] These chemicals are **food additives** used in meat and meat products to protect against botulism, develop cured meat flavor and color, and retard rancidity during storage. Nitrites are the active salts. Nitrates convert to nitrites with time, which continue the desired action. In recent years, the occurrence of nitrosamines, which are produced by the interaction of nitrites and secondary or tertiary amines in meat, has been a subject of concern. Dimethylnitrosamine (DMN), the simplest member of the nitrosamines, is a hepatotoxic agent. In addition, among 130 nitroso compounds tested, more than 80% were found to be carcinogenic. On the other hand, elimination of nitrites may result in an increase in the number of cases of botulism among consumers. Government agencies must consider the positive effects of nitrites as well as their hazards. In an effort to balance these effects, the U.S. Department of Agriculture (USDA) banned inclusion of nitrates in seasoning premixes and placed an upper limit of 120 ppm sodium nitrite in bacon.

Preventing Violative Residues
Preventing Residues from Therapeutic Agents.[17,60,104,107,115,119,120,121,141,148,149,] [154,175,242] Guidelines listing drugs with government approval for use in food animals and appropriate withdrawal times are widely available. Lactating dairy cows present the greatest problem inasmuch as only a handful of drugs are approved for use without some withholding of milk. Drug manufacturers furnish information about withdrawal times for medications they manufacture and sell. It is sometimes worthwhile to be reminded of the obvious—reading the label and adhering to stated guidelines will avoid most potential problems. Drugs such as sulfamethazine persist in the environment, and this factor must be considered as well in preventing residues.

Veterinary Therapy. A decision regarding therapy of any food animal must first be justified on a cost-benefit basis that takes into consideration the value of the animal when healthy, the cost of therapy, and the salvage value of the unhealthy animal (Fig. 4.1). Food-animal practitioners must be familiar with the pharmacology of drugs they use for therapy or in preventive medicine programs, particularly, the required withdrawal times. Drugs with a short biologic half-life, such as penicillin, usually create very few residue problems. Those with a longer half-life, such as streptomycin, are more likely to be associated with violative concentrations in meat or milk. Veterinarians may be held legally responsible for utilizing therapeutic agents that are not allowed in food-producing animals or are used in such a manner that illegal concentrations of medications are found in carcasses of food animals.

Identification of treated animals and the maintenance of a permanent record that documents drugs used, route, and date of administration and recommended withdrawal periods will avoid possible legal judgments against the practitioner. This is particularly true when using "extra-label" (unapproved) drug therapy. Extra-label use involves a careful decision by a veterinarian to use a product in a species for which it has not been specifically approved. Some drugs, such as chloramphenicol, have been **prohibited** from use in food animals, and any administration would be illegal.

Advice to Clients.   Livestock and poultry producers are recognizing that they must consider themselves food producers rather than merely feeders of livestock or poultry. Two of the major responsibilities of the veterinary profession are (1) to advise the food-animal producer how to eliminate financial loss as the result of carcass adulteration resulting from unlawful residues and (2) to protect the consumer from exposure to food containing unlawful concentrations of residues. Veterinarians may be held responsible for failing to advise clients of potential residue problems. Many producers, for example, do not realize that calves can have measurable residue concentrations when they are nursing dams that are treated with antibiotics. Even pooled colostrum fed to calves may produce residues. To be most effective in advising food-animal producer clients, veterinarians should be prepared to develop an efficient program of residue prevention. This involves examining **all** critical points in the client's production enterprise wherein residue problems may evolve. Elements to consider include the following:

Animals: source and history, identification, medications, withdrawal periods.

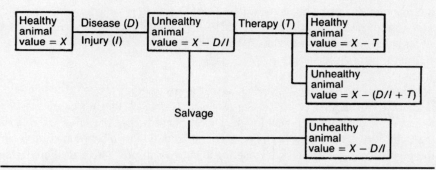

Fig. 4.1.   Decision algorithm for food-animal therapy.

Feed: source, possible pesticide or other contamination, separation of medicated from nonmedicated, water supply.

Management: trained personnel, records for feed and medication.

Environment: isolation pens; storage of drugs and other chemicals (pesticides, etc.), including proper labeling, mixers, and other feed-handling equipment; bioaccumulation by recycling in manure.

Quality control: testing milk and/or urine, samples of purchased feed if of uncertain origin.

Show Animals. Market cattle, hogs, or lambs slaughtered after exhibition at livestock shows represent a minute portion of the meat consumed in any country. These animals are of concern to veterinarians, however, because veterinarians often provide general health certification as well as perform various tests for specific communicable diseases of animals as a prerequisite to exhibition. Occasionally, the exhibitors are uninformed youth (e.g., 4-H or Future Farmers of America) or overly competitive persons who use illegal drugs or fail to observe appropriate withdrawal periods. The education of exhibitors before the show should ensure that no misunderstanding exists. This, combined with appropriate testing at entry, is important to prevent the occurrence of violative residues at slaughter. The adverse publicity to the livestock industry can be serious when residues are detected in a single show animal.

On-Farm Testing. There are two components to on-farm testing. First, and often overlooked, is the collection of specimens from animals—feed, water, or bedding sent to the laboratory to be tested for organochlorines or other residue-producing chemicals for which on-site tests are not available. Second, there are on-site tests available commercially that veterinarians and producers can use on the farm to detect antibiotic and sulfonamide residues in animals (in their milk, serum, tissues, or urine) and feed. These on-site tests are often referred to as *rapid, screening,* or *cowside tests*. The sensitivity (usually in ppb) of the test to detect antibiotics or sulfonamides depends on the test mechanism (e.g., microbial inhibition, ELISA, latex agglutination), the drug and its metabolites, and the specimen tested (e.g., milk, serum, or urine). Typically, the tests are simple to perform under farm conditions and provide results within a few hours. The results are usually qualitative; i.e., the test is designed to react (have a positive result) whenever the residue exceeds a given concentration rather than provide a precise measurement. This minimum detectable concentration is generally intended to be less than the tolerance. Problems arise when official (confirmatory) tests used by governmental agencies provide positive results at lower concentrations than the on-farm test, particularly for drugs in milk for which no tolerance exists (i.e., the tolerance is zero). The sulfa-on-site (SOS) test for sulfonamides in urine has been especially useful in testing live hogs on the farm and in the abattoir to prevent those with violative concentrations of sulfamethazine (SMZ) from being slaughtered. The results of the urine test correlate well with the likelihood of violative concentrations in tissues. Once hogs are removed from a contaminated environment, SMZ is eliminated within a few days.

Quality Assurance Programs.   Many producer groups within animal agriculture are initiating quality assurance programs to promote residue prevention. These industry-driven programs generally encourage the establishment of a valid veterinarian-client-patient relationship so that drugs and pesticides will be used most effectively to promote animal health and well-being as well as to ensure preharvest food safety. Animal industry groups have recognized that effective producer education is a critical element in the initiation of their quality assurance programs as well as in the overall success of the programs. In the United States, the Pork Quality Assurance Program, developed by the National Pork Producers Council, and the Milk and Dairy Beef Residue Prevention Protocol, developed by the National Milk Producers Federation and the American Veterinary Medical Association, are two excellent producer-industry examples. Both programs have incorporated the following 10 critical control points:

1. Establish an efficient and effective herd health management plan.
2. Establish a valid veterinarian-client-patient relationship.
3. Store all drugs correctly.
4. Use only government-approved over-the-counter or prescription drugs with a veterinarian's guidance.
5. Administer all drugs properly and identify all treated animals.
6. Be sure that all drugs have labels that comply with legal requirements and follow label instructions for use of feed additives.
7. Maintain and use proper treatment records on all treated animals.
8. Use drug residue tests when appropriate.
9. Implement employee and family awareness of proper drug use.
10. Complete the quality assurance program checklist annually.

Control of Residues in Food by Governmental Agencies.[3,15,19,34,38,41,145,206, 209,210,229]   The legal authority of governmental agencies in food safety usually begins with postharvest handling of food. Therefore, residue tests are generally performed to measure concentrations in the commodity to be eaten, e.g., milk or meat, rather than the live animal. This is in contrast to routine blood or urine testing in race horses, when the presence of any amount of a prohibited substance is illegal. To be most effective in consumer protection, endpoint testing by governmental agencies must be designed to support industry-based on-farm residue prevention programs. An increasingly important role of these agencies is to present the results of residue prevention efforts in forms acceptable to both the scientific community and the consuming public.

Milk.   In addition to standard tests for butterfat, protein content, and somatic cell count, milk is now routinely examined for the presence of antibiotic residues upon receipt in milk-processing plants. The tests currently in use utilize inhibition of growth of antibiotic susceptible organisms, such as *Bacillus subtilis* or *B. stearothermophilus* that can detect concentrations of penicillin as low as 0.004 IU/ml. Test kits are available for use by milk truck drivers as a screening procedure at the farm. Microbial inhibition tests may give false-positive results associated with the presence of naturally occurring antibiotics in feed, e.g., penicillin produced by *Penicillium* spp. on citrus pulp.

In the United States, governmental testing for residues (primarily antibiotics and sulfonamides) in milk is performed by various state health or agriculture agencies. Uniformity in standards used is largely achieved through the adoption by the National Conference on Interstate Milk Shipments (an organization with representatives from all the states and the FDA) of a Recommended Milk Ordinance and Code. A recent revision of the code has included a requirement that any dairy producer found shipping milk with violative residues must participate in the industry-sponsored Milk and Dairy Beef Residue Prevention Protocol.

Meat. Detection of meat adulterated with chemical residues depends on two factors: (1) the professional expertise of the individual veterinarian, and (2) the extent of the governmental residue program. Veterinarians working in an inspection program must know the antemortem signs associated with exposure to naturally occurring toxic substances, drugs, pesticides, feed additives, and environmental contaminants as well as the postmortem lesions produced by them. They must be familiar with food-animal production practices and be able, therefore, to anticipate potential problems such as are seen among cull dairy cows and newborn or "bob" veal calves. (In the United States, meat from calves slaughtered during the first 3 wk of life [150 lb, 68.2 kg, or less] is called "bob" veal.) Finally, they must be able to collect limited case histories to help identify an adulterant and its source.

In the United States, the USDA is capable of testing tissues from slaughtered animals for more than 400 drugs, pesticides, herbicides, industrial chemicals, and heavy metals. The tissue tested may be fat, kidney, liver, or muscle, depending on the affinity of the compound. In-plant screening tests are available for detection of antibiotic and sulfonamide residues. If the screening test result is positive, the product is retained until confirmatory laboratory test results are available. This governmental residue program consists of two parts: **individual enforcement testing** and **population sampling**. **Individual enforcement testing** involves the assaying of specimens selected by the in-plant veterinarians individually from animals or lots based on herd history or clinical signs observable before or during slaughter. Individual enforcement testing is used principally in problem populations to prevent residues from entering the human food chain. **Population sampling** programs involve **monitoring**, **exploratory studies**, and **surveillance**. **Monitoring** is a population-based sampling used to detect a threshold level of residues in food animal slaughter classes (i.e., steers, boars, lambs, etc.). Monitoring provides information about residue violations in a slaughter class for compounds with established residue limits and for which appropriate laboratory methods are available. **Exploratory studies** focus on residues for which there are no established residue limits and are conducted to gather information about the occurrence (including frequency) and amounts of such residues in meat animals at slaughter. **Surveillance** is designed to measure the extent of a residue problem after the problem has been detected by monitoring. The purpose of surveillance is to characterize the residue problem by distinguishing the components of the problem population, measuring the magnitude of the problem, and evaluating control efforts.

Although available, international standards for acceptable residue concentrations in meat are not used universally. Exporters must, therefore, meet the standards acceptable to the importing country, and this can prove a major problem for the international meat trade.

Foodborne Disease Prevention Systems[100]

The type of foodborne disease prevention system will depend on the extent of food safety controls in place throughout food production, processing, and distribution. If there is no control before the product reaches the consumer, then a **passive control** system exists that depends entirely on appropriate food safety action by the consumer. If the safety of the end product is checked, as in traditional endpoint inspection of meat and poultry, then an **end product control** system exists. **Process control** exists when procedures are monitored at various points in the production process, as in milk inspection involving both the dairy farm and the processing plant. An **integrated quality control (IQC)** system is a system of process control expanded to involve every facet of the food chain from farm to table, including production, processing, distribution, and preparation. *Longitudinally integrated safety assurance* (LISA) is another term used to describe such a system.

In 1987, the International Organization for Standardization issued the ISO 9000 series of international generic standards for quality management and quality assurance. Most countries have adopted the ISO 9000 standards, which may be applied to manufacturing and service industries of all types, including food production and handling. An industry site may be registered as meeting the related quality standards after providing the necessary specific documentation and completing a successful audit of the quality system.

Risk Analysis[11,96,120,181]
Overview of Risk Analysis.   Risk analysis is a formal process that is being adopted by governments throughout the world to help evaluate existing programs in food safety and to develop new programs. Risk analysis has been used

to formulate public policy on concerns ranging from the setting of speed limits, to fish consumption advisories based on monitoring of waterborne pollutants, to the evaluation of new food manufacturing processes.

Formal risk analysis is the way of the future for the prevention of foodborne disease. Risk analysis is becoming increasingly important as the trend toward globalization of trade continues. The North American Free Trade Agreement (NAFTA) and the General Agreement on Tariff and Trade (GATT) provide for the removal of artificial trade barriers for food and agricultural products. While these agreements have as their goal the facilitation of international trade, they must also ensure the safety of the people and livestock industries of importing countries that, in the past, have not been exposed to the global marketplace.

The application of risk analysis to food safety will help make the best use of available scientific information about the risks of foodborne disease, will formalize the decision-making process about specific steps in food production, processing, and inspection, will assist in resource allocation, and will help communication between all parties involved, including the consumer. By identifying problem areas in the food production cycle, risk analysis will make government programs more efficient by allocating economic resources to areas of greatest need.

Risk analysis is a complex process that can be broken down into three parts: risk assessment, risk management, and risk communication.

Risk Assessment.   Risk assessment is the process of identifying hazards and describing risks. Describing risks includes estimating the likelihood (probability) and potential severity (magnitude) of damage resulting from exposure to harmful agents or situations. Hazards in foods of animal origin include all the microbial, chemical, and physical agents of foodborne illness. For example, in the case of setting speed limits, describing risk involves the gathering of information about the relationship between increased speed and injury. Since the exact probability of an event is not always known, risk assessment must also take into account uncertainty in the estimates of risk. Quantitative computer programs have been developed to help model the uncertainty of risk characterization. Qualitative models are also possible. In risk assessment regarding foodborne disease the emphasis is on biology, but economic information may also be brought into the decision-making process.

Risk Management.   Risk management is the development of policy regarding risk and is developed from information provided by a variety of sources, including risk assessment data. The question of who benefits from the risk and who is bearing the risk comes into play when decisions about acceptable risk are made. This involves value judgements about the risk, and any individuals who will be affected should be consulted.

Risk Communication.   Risk assessment and risk management are dependent on communication between policy makers and the stakeholders (producers, processors, and consumers) for information relevant to the process. Communication also includes education about how risks are assessed and how they are managed. In application, communication is the most important part of risk analysis.

Risk analysis is an iterative process. Public input is often essential. In setting speed limits, for example, it is well-known that increased speed results in more injuries and deaths. But the public is willing to accept this increased risk in order to more quickly reach the places to which it is traveling. (As a result, many states have raised the maximum speed limit from 55 mph to 65 mph.) It is not possible to eliminate all risks from every situation. For example, even though utilization of risk analysis may reduce the risk of foodborne illness from eating raw shellfish, some risk will always be present. It cannot be completely eliminated.

Hazard Analysis Critical Control Points (HACCP) (see next section) is a natural realization of the risk analysis approach to food safety. Risk analysis will eventually lead to HACCP programs covering all aspects of food safety, from the farm to the preparation of food at home.

Hazard Analysis Critical Control Points [2,14,26,28,29,30,31,44,45,150,155,195,203,219,233,243]

Hazard Analysis Critical Control Points (HACCP) is a system of process control aimed at ensuring food safety by anticipating foodborne hazards and instituting appropriate controls to prevent their occurrence. HACCP was designed in the early 1970s initially to control postharvest processing to ensure safe food for astronauts. The food-processing industry quickly adopted the principles and expanded their application. Today, the system is also applied to preharvest production of food. The following are definitions used in HACCP:

Hazard analysis is the identification of sensitive ingredients or areas in the production or processing of food that represent critical points that must be monitored to ensure food safety. **Critical control points** (**CCPs**) are those areas in the food chain where the loss of control can result in an unacceptable risk to food safety. A **hazard** is an unacceptable contamination with and/or production or persistence in food of a pathogenic organism or toxic chemical. **Severity** is the seriousness of a hazard. **Risk** is an estimate of the likelihood of occurrence of a hazard. **Criteria** are specified limits or characteristics of a physical, chemical, or biological nature. **Monitoring** is surveillance to determine if the procedure instituted at each CCP meets the established criteria, is functioning properly, and is under control. **Verification** is the use of supplemental procedures to ensure that the HACCP system is in place and functioning properly.

The principal steps of the HACCP system of control are as follows:

1. Assess hazards and risks associated with growing, harvesting, raw materials

and ingredients, processing, manufacturing, distribution, marketing, preparation, and consumption of the food.
2. Determine CCPs required to control the identified hazards.
3. Establish the critical limits (criteria) that must be met at each identified CCP.
4. Establish procedures to monitor CCPs.
5. Establish corrective action to be taken when there is a deviation identified by monitoring a CCP.
6. Establish effective record keeping systems that provide documentation of the HACCP plan.
7. Establish procedures for verification that the HACCP system is working correctly.

To assess hazards, a schematic or flow chart of the entire production, processing, and distribution system is first developed. For beef, this would begin with production on the farm or in the feedlot, progressing through the slaughtering plant, and ending at the delivery point of the finished product (retail market). Next, points in the flow chart that are potential trouble spots (CCPs) are identified. For example, common CCPs in food service systems are depicted in Table 4.1. The emphasis regarding "trouble" under the HACCP system is on hazards to human health, not production efficiency or aesthetics. The selection of these potential control points is based on a thorough knowledge of the hazards involved (physical, chemical, or microbial) and experience with the production system in use. The latter is important because, even though basic procedures are similar, each stage in production, processing, and delivery has different trouble spots as a result of differing physical facilities, equipment, source of raw material, and personnel.

Two categories of CCP are used, based upon the level of confidence that hazards can be prevented. Action at a CCP[1] will ensure effective control of the hazard, whereas action at a CCP[2] will minimize risk but cannot absolutely guarantee control of the hazard. For example, adequate cooking can ensure destruction of salmonellae in meat (a CCP[1]) whereas hygienic practices at slaughter can

Table 4.1. Common critical control points in food service systems

System	Receipt	Formulation	Handling of raw ingredients	Cooking	Hot-holding	Cooling	Handling of cooked products	Reheating
Cook/serve				x			x[a]	
Prepare/serve cold	x	x[a]	x			x		
Cook/hold hot				x	x		x	
Cook/chill				x		x	x	x
Cook/freeze				x		x	x	x
Assemble/serve	x						x	x

Source: 30.
[a]Sometimes a critical control point.

reduce, but not eliminate, contamination of raw meat (a CCP[2]). Flow charts illustrating CCPs for milk processing, livestock slaughter, and poultry meat production are presented in Figs. 4.2–4.4.

Appropriate surveillance (monitoring) procedures are selected. For each CCP, well-defined criteria are needed to determine if the observations are within acceptable limits. The procedures and criteria depend upon the particular hazard; e.g., a hazard associated with employee activity would be best observed visually, and the criteria would focus on activities having an impact on hygiene. Records of production and purchase are important in monitoring feed-associated hazards. Organoleptic procedures are appropriate for detecting potential hazards associated with carcasses and finished products. Testing feed ingredients, tissue samples, or environmental surfaces may be essential to detect chemical and micro-

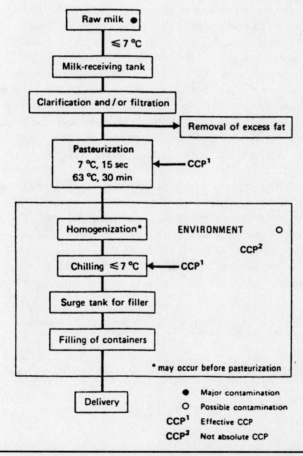

Fig. 4.2.   Sources of contamination and critical control points during the processing of milk. Source: 247.

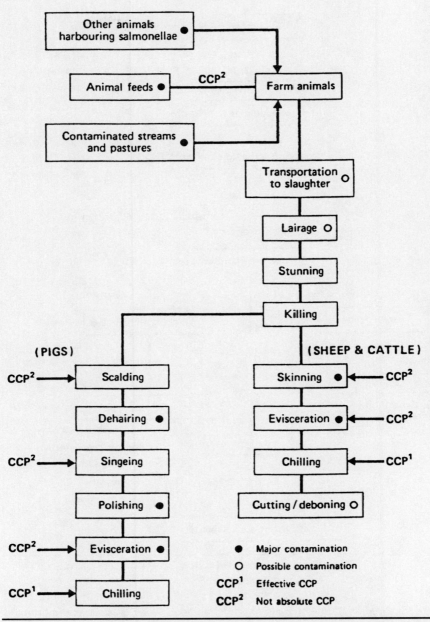

Fig. 4.3. Sources of contamination and critical control points before and during the slaughter of pigs, sheep, and cattle. Source: 247.

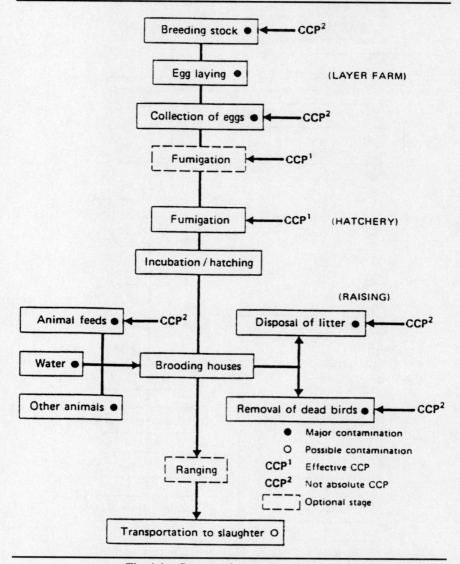

Fig. 4.4.   Sources of contamination and critical control points in poultry husbandry. Source: 247.

biologic contamination. Timing and frequency of surveillance is as important as point-of-sampling. Equipment swabs, for instance, may reveal nearly sterile surfaces at 8:00 a.m. but unacceptable levels of contamination at 11:00 a.m. The frequency of tissue sampling must be acceptable statistically. An abattoir slaughtering 1,000 cows a day will require more sampling than one slaughtering 100 cows a day, but not necessarily 10 times more. Statistical tables (or a statistician) will need to be consulted.

A method for evaluating the effectiveness of the surveillance system must be determined. This will normally include an ongoing review of recorded data, particularly of the results of observations and testing at appropriate CCPs.

Safe Food Handling at Home[12,66,69,110,213,226]

Most consumers live in urban areas, and the ingredients they obtain for the foods they prepare at home are produced elsewhere, whereas rural and suburban residents may produce some of their own food. All need to be able to handle (store and prepare) food safely in the kitchen. Those who prepare food at home for persons with lowered resistance to disease must take special care to avoid foodborne hazards. People who produce their own eggs, meat, or milk should be prepared to bring a safe and sanitary product from the field into the kitchen.

Based on epidemiologic evidence, some foods of animal origin are considered to have a high potential to be vehicles of foodborne pathogens. The reasons for this include: (1) high susceptibility to microbial contamination, (2) considerable opportunity for survival of the contaminants, (3) high likelihood that growth of the contaminants will occur at some point before serving, and (4) reasonable likelihood that the food may be subjected to mistreatment just before serving.

The World Health Organization (WHO) Golden Rules for safe food preparation are

1. Choose foods processed for safety.
2. Cook food thoroughly.
3. Eat cooked foods immediately.
4. Store cooked foods carefully.
5. Reheat cooked foods thoroughly.
6. Avoid contact between raw foods and cooked foods.
7. Wash hands repeatedly.
8. Keep all kitchen surfaces meticulously clean.
9. Protect foods from insects, rodents, and other animals.
10. Use pure water.

Examples of foods processed for safety are pasteurized milk and irradiated poultry. Foods purchased cold should be selected last and promptly taken home to the refrigerator. Fresh meat and poultry should not be stored refrigerated for more than 7 d (1–2 d for raw ground meat) whereas eggs may be kept for as long as 3 wk. All parts of cooked food must reach at least 70°C (184°F). Thaw frozen food before cooking in a microwave (then cook immediately) or in a refrigerator, but not at room temperature. Do not hold food at room temperature for more than 2 h. If food must be cooked earlier or leftovers are to be stored, keep either cold or hot (<4°C [<40°F] to >60°C [>140°F]). Foods for infants or others with lowered resistance should never be stored. In warm climates where refrigeration may be limited or nonexistent, many people customarily prepare no more than enough food to be eaten during the meal thereby avoiding storage problems. Thoroughly reheat food to at least 70°C (184°F). Home-canned foods (especially meats and low-acid vegetables) should be heated before **tasting** inasmuch as botulinum toxin in the juice is tasteless and potentially lethal. Modified atmosphere (e.g., vacuum) packaging of raw meat and fish extends their shelf life but may increase the risk of botulism.

Cross-contamination between raw foods containing pathogens and cooked foods can reintroduce pathogens into cooked foods. Raw foods of animal origin, e.g., fish, meat, or poultry, can contaminate cutting boards and other surfaces where food is handled. These surfaces should be washed thoroughly after contact with raw product before cooked foods are placed on them. Washing hands and utensils, as well as surfaces, using clean cloths or towels after **each** contact with raw food is also important. Cooked foods should be stored in clean, sealed containers to protect them from contamination. Raw fish, meats, or poultry should be stored on a plate or in a container that will prevent contamination of other foods with their juices.

Water is essential for food preparation and may be overlooked as a source of foodborne pathogens. Ice made from contaminated water has been the culprit on occasion. In rural areas, fecal contamination of well water is not uncommon. If in doubt, to make it safe for drinking and food preparation, bring water to a rolling boil for at least 1 min because this will kill vegetative enteric pathogens.

The health of the person preparing the food is critical to the success of safe food handling. Good habits such as regular hand washing are essential. Individuals with diarrhea, with pustules on their hands, or who are coughing should not be preparing food.

Use of microwave ovens has become a common means of defrosting and cooking food in many countries. Microwave cooking, however, does not heat food in the same way as traditional methods and, unless care is taken, cold spots may occur thereby allowing pathogens to survive. The following rules are recommended to ensure safe microwave cooking of meat:

- Debone large pieces, as bone can shield the meat from thorough cooking.
- Arrange meat uniformly in a covered dish and add a little liquid as steam aids in killing bacteria.
- Rotate meat for even cooking.
- Do not cook whole, stuffed poultry in the microwave.
- Never partially cook meat—if combined with conventional heat, complete cooking immediately.
- Use a temperature probe or meat thermometer to verify the meat has reached a safe temperature.
- Observe the standing time in the recipe (it is needed to complete the cooking process).

Cows, does, and ewes are kept worldwide for production of milk to be consumed at home. Although the animals may be tested for the presence of infections that cause such milkborne zoonotic diseases as brucellosis and tuberculosis, raw milk may not be pathogen-free. All milk to be consumed at home as fluid milk or a dairy product (e.g., butter, cheese, and yogurt) should first be pasteurized. For many persons who hand milk one or two animals to provide milk for home use, the adoption of sanitary milking procedures, including cleansing the udder, is essential. The steps in cleaning the milk bucket and strainer are

1. Rinse with cold water immediately after use.
2. Wash thoroughly with a brush, hot water, and an efficient (soapless) cleaner.
3. Rinse with hot water.
4. Clean with an approved sanitizing agent.
5. Store between milkings so the equipment remains dry and is protected from contamination.

Electric home pasteurizers with agitators, suitable for small volumes (1–2 gal) of milk, are available. These pasteurizers heat milk for 30 min at 63°C (145°F). Milk in jars can be heated in a water bath canner or a double boiler at 73°C (163°F) for 1 min. Regardless of the method used, the milk should be cooled immediately in a cold running water bath and then stored in the refrigerator.

Many publications provide illustrated descriptions of the equipment required for home (on-farm) slaughter of livestock and poultry. However, the guidance they provide regarding the sanitary aspects of slaughter is frequently insufficient. For example, straw placed on an earthen floor is sometimes recommended for the site where cattle are to be slaughtered. Although such preparations may be considered for emergency slaughter, they should not be recommended under normal circumstances. Careful preparation is needed, beginning with the site, and it may be useful to encourage several neighbors to develop a site for collective use. The site should have a supply of potable water, a solid floor (e.g., concrete) that can be cleaned, adequate lighting if slaughtering is to be done at night, as is often the case in the tropics, and protection from dust and insects. If poultry or swine are to be slaughtered, a means of heating water for scalding will be needed. In warm climates, slaughtering should be done during the coolest time of the day, e.g., night or early morning, whereas time of day is not a factor in cool weather (<4°C [<39°F]). Adequate refrigerator space, which may be a problem for large beef carcasses, is needed for chilling of the carcass during warm weather. Chilling can be done outside during cool weather. Allow at least 24 h after slaughter for complete rigor mortis before freezing unless electrical stimulation of the carcass is undertaken. Hanging indoors initially may be necessary in cold climates to avoid premature freezing. Hot water should be available to rinse tools frequently during slaughter. Time should be allotted for thorough cleanup between animals. A pit should be dug beforehand to bury the discarded (nonoffal) internal organs deep enough to prevent scavenging by dogs and other carnivores. Removal of the hide and internal organs without excessive contamination of the carcass requires skill gained through practice. A novice owner who intends to perform on-farm slaughter should obtain some prior practical instruction or at least observe actual slaughter by a skilled person.

On-farm slaughter of a healthy animal under sanitary conditions presents no unusual risk of foodborne disease. Questions regarding the safety of the meat may arise, however, whenever conditions such as epithelioma or prolapse occur. Also, the owner may be concerned about the safety of meat from an animal that recovered from an illness after therapy, even though the recommended withdrawal period for the administered drugs has elapsed. Whenever a client slaughters an animal, it is an excellent opportunity for the veterinarian to learn about the condition of the flock or herd comparable to performing a necropsy without the associated economic loss. It is also an important time for the veterinarian to provide sound advice and peace of mind to the client regarding food that will be eaten at home. In addition to advice on the safety of meat from animals with various conditions, the veterinarian can establish an arrangement with the client that he or she be contacted in advance whenever an animal is to be slaughtered. The veterinarian can thereby be present to observe the condition of the animal. If this is not feasible, the client should be encouraged to save tissues that do not appear normal for the veterinarian to examine later. Frequently, when conditions such as liver abscess or parasite cyst in the mesentery occur, they will be noticed by the owner at slaughter. These conditions will not always be recognized as abnormal, however, so some prior guidance by the veterinarian regarding what to look for may be useful.

Satisfactory freezing and storage of meat at home depends on the freezer capacity. The freezing temperature should be $-23°C$ ($-10°F$) or lower, and the storage temperature for frozen foods should be no higher than $-18°C$ ($0°F$). The freezing compartment of household refrigerators is usually satisfactory for storage but is not cold enough for satisfactory freezing. Chilled meat within a week after slaughter should be frozen in quantities of 2.5 kg(5.5 lb) or less, well wrapped in freezer paper with a label that includes the date. When large quantities are to be frozen, such as 200 kg(440 lb) of meat from a calf, it may be best to use a commercial freezer and then store the frozen meat in the home freezer. Under some conditions, beef can be stored up to 36 mo without rancidity developing. Generally, however, to avoid excessive deterioration (rancidity), it is not recommended to store frozen meat for more than 12 mo (3–6 mo for ground meat and pork products).

Investigation of Outbreaks of Foodborne Disease[27,170,194]

Value of Investigating Outbreaks of Foodborne Disease

Over the years, significant advances have been made in food technology to produce a wholesome product. Despite these advances, foodborne diseases remain an important problem.

Investigation of foodborne disease outbreaks is an essential part of a total surveillance program. By accumulating data about outbreaks, three important objectives can be achieved. First, **controlling the disease** is possible by rapid identification of contaminated food sources, correcting faulty processing or preparation practices, and identifying and treating infected people. Second, **identification of etiologic** agents causing outbreaks is necessary. The accumulation of information assists in making correlations among several outbreaks in which a causative agent was not identified by indicating events common to all. With this information, probable causes can be proposed. In subsequent outbreaks, efforts can be made to increase the probability of identifying suspected agents. Third, **administrative guidance** based on surveillance data can be offered. Public health agencies are thus better informed when attempting to implement disease control programs.

Outbreak investigations often involve correlating disease data from several places. When there are only a few individuals affected with rather mild signs and symptoms, the disease is often not regarded as foodborne. But when information on outbreaks is reported to a single agency, factors common to all outbreaks may be identified and thus may allow the cause to be determined.

Procedure for Investigating Outbreaks of Foodborne Disease

Investigating outbreaks of foodborne disease involves applying principles of epidemiology to determine which food was the source for an outbreak and why. The eleven essential steps in an investigation are as follows:

1. Establishing the existence of an outbreak.
2. Verifying the diagnosis.
3. Planning a detailed investigation.
4. Organizing interviews or a survey (e.g., a case control study).
5. Formulating a hypothesis.
6. Conducting an investigation.
7. Analyzing the data.
8. Testing the hypothesis.
9. Formulating a conclusion.
10. Implementing controls.
11. Writing a report.

Establishing the Existence of an Outbreak.   In Chapter 3, a definition of an outbreak of human foodborne disease is presented. That cases result from a common source is understood. In veterinary medicine, a similarly precise definition of a common source (e.g., feed) outbreak does not exist. In addition, reporting is not sufficiently adequate to regularly identify cases in animals on different premises that have become infected from a common source. As a consequence, the outbreak will go unrecognized without a thorough investigation of the problem.

Verifying the Diagnosis. The first step in investigating an outbreak suspected of being foodborne is to verify the diagnosis. Assuming that this is correct (i.e., it meets appropriate diagnostic criteria), it must be decided whether or not the agent or circumstances suggest a foodborne origin. The investigator looks for similarities among the patients' symptoms, signs, and incubation periods (time of onset of clinical illness) and attempts to identify any foods that all patients have consumed recently. With this information, it is possible to determine the probability that a foodborne outbreak exists.

The investigator is given a certain amount of information at the time a report is filed, but this is usually inadequate to determine the source of the causative agent. Also, some cases may not have been suspected of being foodborne at the time they were reported.

Planning a Detailed Investigation. Before the investigator conducts interviews and collects laboratory specimens, a detailed plan should be prepared that will give pertinent information.

Organizing Interviews or a Survey. Interviews with patients or clients are an important part of conducting an investigation. Appropriate questions can give important clues as to why some exposed individuals had signs and symptoms and others did not. Likewise, poorly phrased questions may be misleading and produce erroneous answers. Responses to closed questions (those with "yes" or "no" answers) can often reflect patient or interviewer bias; an open question, and a careful listener, may gather more useful information. The first question that the investigator wants answered is, Where did the outbreak occur? A food can be contaminated through improper handling and/or unsanitary preparation, so the potential for food to become a source of a disease agent exists wherever food is produced, processed, prepared, or served.

Certain foods have over the years been found to be associated more frequently with foodborne outbreaks. The investigator should ascertain whether these potentially hazardous foods were consumed when reviewing attack-rate calculations.

Identification of the hazardous food results from the cooperative efforts of the investigator and the laboratory. The laboratory relies on the investigator to submit specimens of foods most likely to be involved for investigation. On the other hand, the laboratory's analyses of patients' specimens, as well as food samples, are essential if the agent and its source are to be identified.

Formulating a Hypothesis. After the investigator has collected preliminary data about the circumstances surrounding an outbreak, a hypothesis is formulated. For example, it may be suggested at this time that the agent is either infectious or noninfectious.

To identify accurately the source of the problem, the investigator should ask the following questions: (1) Is the agent responsible for the outbreak associated with raw foods? (2) Did the outbreak occur because of faulty handling, processing, or preparation practices? (3) Was an improper environment maintained for holding the food(s) after preparation?

Conducting an Investigation

Unsafe Sources. One of the first questions to answer is whether or not the food was obtained from a source that may be unsafe, e.g., raw milk and milk products, shellfish from polluted waters, and cracked or checked eggs. Animal food sources include feed from a mill with a record of contamination, hay baled in a field containing toxic plants, and contaminated carcass meats used in pet food.

Ingredients of a food product may be contaminated by improper storage practices. Storing chemicals such as pesticides near foods provides a potential opportunity for contamination. Leaky pipes (especially sewer drains), insects, and rodents are also sources of contamination.

Many disease agents are spread by foods of animal origin. These may be infectious microorganisms or noninfectious agents. In the slaughtering process, agents may be transmitted readily to other carcasses when contaminated equipment is used or when they are allowed to touch. Chemicals, either in the environment or iatrogenically introduced, also are found in foods of animal origin.

Some food products require the use of uncooked ingredients. Food handlers should be advised not to use food from damaged containers, such as cracked eggs or dented cans. Numerous outbreaks have occurred following addition of cracked or checked eggs to milk in preparation of eggnog, hollandaise sauce, or ice cream.

Faulty Handling and Preparation. Foodborne outbreaks that occur as a result of unhygienic or improper food preparation can be attributed to either equipment or food handlers. In reality, the outbreak is nearly always the result of human error, whether contamination is from equipment or worker.

In reviewing food preparation practices, the investigator should determine if the outbreak has been caused by cross-contamination during preparation. Raw foods, e.g., fish, poultry, eggs, and meat, are obvious sources of microorganisms in the kitchen. If the same utensils (knife, cutting board, etc.) are used for raw products and then later used for prepared foods, without first being cleaned and disinfected, contamination will occur.

Certain disease agents may be spread by personnel preparing foods, particularly enterotoxigenic staphylococci.

Improper Environment. The factors needed for growth of microorganisms in prepared food are (1) a sufficient nutrient (the **food** itself), (2) a proper environment for growth (**moisture** and **appropriate temperatures**), and (3) **time** for replication so that sufficient numbers are present to cause illness.

Some people, and especially those in food service establishments, routinely prepare food well in advance of when it is to be served. If the temperature at which the food is held is between 4.4°C (40°F) and 60°C (140°F), growth of thermoduric bacteria will occur.

Improper refrigeration is the one factor that is associated frequently with foodborne disease outbreaks. The investigator should evaluate closely the cold chain during production, processing, and distribution.

Analyzing the Data.   After the investigator has conducted interviews, observed food-handling areas and practices, and obtained laboratory results, the data can be analyzed to suggest a probable source of the agents that caused the outbreak. When several food items are involved, it may be necessary to perform an **attack-rate analysis**.

To make this analysis, the interview data regarding consumption of foods and beverages obtained from individuals involved in an outbreak are summarized in tabular form. For each food, those who ate it and those who did not are tabulated. Each of these major categories is further subdivided into the numbers of those who became ill and those who did not.

Attack rates are then calculated for each food in relation to each group. (For an example, see Table 4.2.)

To identify the food most likely to be the vehicle involved, the attack rate for persons eating each food is compared with that for persons not eating it, using the standard procedure for determining relative risk.

$$\text{Relative risk} = \frac{\text{Attack rate among those eating item}}{\text{Attack rate among those \textbf{not} eating item}}$$

The food with the greatest relative risk is considered the food involved. In the attack-rate table (Table 4.2), the greatest relative risk existed for those who ate the roast beef.

$$\text{Relative risk} = \frac{73.2}{8.3} = 8.82$$

Table 4.2.   Sample attack-rate table

| Food or beverage | Persons who ate specific food | | | | Persons who did not eat specific food | | | |
	Sick	Not sick	Total	Attack-rate, %	Sick	Not sick	Total	Attack-rate, %
Roast beef	104	38	142	73.2	1	11	12	8.3
Potatoes, gravy	87	35	122	71.3	8	24	32	25.0
Green beans	76	30	106	71.7	25	30	55	45.4
Bread	55	20	75	73.3	50	45	95	52.6
Butter	55	20	75	73.3	50	45	95	52.6
Salad	3	7	10	30.0	98	57	155	63.2
Dessert	25	17	42	59.5	76	46	122	62.3
Coffee	60	40	100	60.0	39	21	60	65.0
Milk	13	7	20	65.0	85	55	140	60.7

Calculating the greatest difference between attack rates is another method of analyzing the data to identify possible food vehicles. When there is a causal relationship, this calculation is referred to as *attributable risk*. Attributable risk is an indicator of the cases that **would not occur** if the food had not been eaten.

Attributable risk = Attack rate among those eating item − attack rate among those **not** eating item

The attributable risks for foods listed in Table 4.2 are as follows: roast beef, 64.9; potatoes and gravy, 46.3; green beans, 26.3; bread, 20.7; butter, 20.7; salad, −33.2; dessert, −2.8; coffee, −5.0; and milk, 4.3.

Testing the Hypothesis. After analyzing the data, the investigator must test the hypothesis. Analysis may be possible with available data (attack-rate calculations), or it may be necessary to conduct further laboratory tests or reexamine food-handling practices.

Formulating a Conclusion. The investigator then attempts to assemble facts into a logical sequence to formulate a conclusion about factors that led to the outbreak.

Implementing Controls. When a faulty step in the food chain is identified, it must be corrected. Often this is a matter of properly educating and supervising food handlers. If a specific food item is identified as the probable source of illness, further use of that food must be stopped until analyzed.

Writing a Report. The final step in conducting an investigation involves writing a report. Disseminating reports of an outbreak is a valuable aid to other investigators who may at some time encounter a similar outbreak.

Inspection of Food Products[16,39,40,63,70,71,81,82,91,127,131,139,140,163,171,177,188,202,216,223,225,231]

Purposes and Benefits of Meat Inspection

Although the inspection of livestock to be used for food has been referred to traditionally as *meat inspection,* the process usually begins with observation of the live animal. Herein the term *meat inspection* will be used in the traditional sense, referring to the inspection of livestock slaughtered for food.

Purposes of Meat Inspection. There are four major purposes of meat and poultry inspection to provide consumer protection.

Eliminating Diseased Meat.   The most important purpose of an inspection program is to prevent entry of diseased meat into the food chain. This is more effective than trying to separate unwholesome meat and poultry, or meat products, from wholesome products at the time of retailing.

Preventing Sale of Objectionable Meat.   A second reason for inspection is to prevent objectionable, undesirable, or aesthetically unacceptable meat from being sold. Thus, all meat that is rejected is not necessarily diseased.

Maintaining Strict Hygiene.[23,55,95,130,152,153,191,200,211,235]   Clean equipment and environment are essential to provide high-quality (including enhanced safety from foodborne disease) meat and meat products for the consumer. Note that the objective here is prevention!

Ensuring Proper Labeling.   Proper labeling is important to prevent adulteration and misrepresentation of products. Certain essential information must be included on the label, thereby providing consumers with criteria for evaluating the product. Proper labeling helps prevent processors from taking advantage of consumers.

Benefits of Meat Inspection.[239,240]   An effective meat inspection program offers several benefits to the community from the producer to the consumer.

Control of Animal Diseases.   Recognition of diseases by veterinarians inspecting animals at slaughter assists disease surveillance and control. Cooperation between the meat inspection program and other programs involved in on-farm animal disease investigation is an important part of animal disease control and eradication. An animal with lesions of tuberculosis or taeniasis (cysticercosis) detected at slaughter, for example, may be traced back to the farm of origin. Individual identification of the slaughter animals (or lots) with relevant information regarding the farm of origin is critical to efficient traceback because epidemiologic investigations can then more precisely identify the source of the problem.

Information Regarding Causes of Condemnation.   The term *condemnation* is widely used to identify the crucial action taken by the inspector to prevent sale of objectionable meat. (In the United Kingdom, this action is referred to as *rejection* inasmuch as condemnation requires court action.) Veterinarians in private practice who advise food-animal producers should be aware of the most common causes of condemnation. For example, the mean annual data for condemnations in the United States during 1986–1990 are presented in Tables 4.3–4.5. The Food Safety Inspection Service (FSIS) combines the results of antemortem and postmortem inspection and publishes only the total. This can be misleading, however. For instance, many cows are condemned on antemortem inspection for epithelioma (cancer eye). However, as there is no separate reporting of antemortem inspection results, all epitheliomas are listed under postmortem conditions. Some conditions, such as pyrexia, can only be determined on antemortem inspection and therefore can be distinguished as such (Table 4.4). From these data, it is evident that considerable losses result from disease and mishandling (Table 4.5). These data also indicate the considerable opportunity for improvement from the point of view of both animal welfare and economics.

Prevention of Foodborne Disease.   Several disease agents may be transmitted through the food chain. Animal by-products from abattoirs are used in feed for livestock as well as food for companion animals. Inspection helps to prevent transmission of those agents that infect animals and people.

Increased Marketability of Products.   Uniformity of meat inspection standards increases marketability by reducing trade barriers between localities resulting from differences in criteria for an acceptable product. National and international standards for shipment effectively eliminate local trade barriers.

Consumer Confidence.   Establishing consumer confidence that the product is safe and wholesome is an important benefit of meat and poultry inspection.

Improving Production Efficiency.   In Denmark and Sweden, computer-based systems of data gathering have been used for nearly 3 decades in swine slaughter. Data on weight, grade, yield, and price received are recorded for individual pigs because this is the basis for payment to the producer. This information is then used in the evaluation and selection of breeding animals. The condition of the animal at slaughter is also recorded, and this information is available to the owner and animal health agencies. Whenever any disease is noted, all concerned are alerted immediately through this system. An up-to-date and accurate estimate of disease status can be provided for individual farms as well as surveillance on the progress of disease control throughout the country.

Evolution of Meat Inspection in Australia[7,10,68,80,84,98,109,129,199,218,244]
Historical Issues in Veterinary Public Health.   The geographic isolation of
Australasia and its relatively late settlement were important factors in limiting the introduction of many livestock diseases of public health importance into Australia and New Zealand. This fortunate position was maintained by the erection of quarantine barriers (restricting the importation of new genetic stock to the few countries with well-established veterinary forces where local disease status was reliably known) and by improved facilities for excluding exotic pathogens in imported semen.

Infections introduced early after settlement were insidious endemic diseases (such as tuberculosis and brucellosis) in carrier stock and anthrax through the importation of unsterilized bone dust or meal from Southeast Asia. Anthrax cap-

Table 4.3. Number of animals condemned on postmortem inspection, annual mean, FY 1986-1990

Condition	Cattle	Calves	Sheep/ lambs	Goats	Swine	Equine
Degenerative and dropsical						
Emaciation	3,749	1,033	2,440	211	403	58
Miscellaneous degenerative	4,008	80	49	7	553	15
Total	7,757	1,113	2,489	218	956	73
Infectious diseases						
Actinomycosis, actinobacillosis	908	2	6	—	11	—
Caseous lymphadenitis	—a	—	4,424	135	—	—
Coccidioidal granuloma	14	—	1	—	4	—
Swine erysipelas	—	—	—	—	3,627	—
Tuberculosis (nonreactor)	100	—	9	—	3,290	—
Tuberculosis (reactor)	21	—	—	—	—	—
Miscellaneous infectious	135	27	17	1	124	4
Total	1,178	29	4,457	136	7,056	4
Inflammatory diseases						
Arthritis	1,726	2,923	1,635	14	11,258	4
Eosinophilic myositis	3,183	6	265	—	16	—
Mastitis	836	1	4	—	47	—
Metritis	1,650	1	8	—	319	2
Nephritis, pyelitis	3,530	163	173	5	1,956	7
Pericarditis	5,124	86	57	2	1,322	3
Peritonitis	5,928	1,468	172	8	8,791	27
Pneumonia	11,273	2,777	3,774	81	10,616	147
Uremia	992	27	653	2	809	5
Miscellaneous inflammatory	2,352	364	39	6	677	17
Total	36,594	7,452	6,780	118	35,811	212
Neoplasms						
Carcinoma	4,344	27	20	2	351	52
Epithelioma	18,773	5	—	—	18	3
Malignant lymphoma	13,684	95	53	3	1,261	21
Sarcoma	282	5	25	—	85	4
Miscellaneous neoplasms	500	18	28	2	1,160	35
Total	37,583	150	126	7	2,875	115
Parasitic conditions						
Cysticercosis	128	1	339	1	9	—
Myiasis	5	5	—	—	3	—
Miscellaneous parasitic	245	3	1,339	2	551	1
Total	377	9	1,678	3	563	1
Septic conditions						
Abscess, pyemia	11,698	491	804	67	25,120	34
Septicemia	12,324	3,295	599	42	8,024	81
Toxemia	6,267	600	314	22	3,172	19
Total	30,289	4,386	1,717	131	36,316	134
Other conditions						
Contamination	2,079	188	397	30	7,597	7
Icterus	787	5,747	933	16	8,350	2
Injuries	4,312	788	173	12	4,391	49
Pigmentary conditions	191	19	6	2	356	227
Residue	166	2,887	26	—	195	2
General miscellaneous	443	89	57	3	4,098	4
Other reportable diseases	14	11	1	1	17	—
Sexual odor	—	—	—	—	90	—
Total	7,992	9,729	1,593	64	25,094	291

Source: 246.

Note: In some instances, figure may represent combined antemortem and postmortem inspections (e.g., epithelioma and abscesses).

aMean annual condemnation frequency <0.5 animal.

Table 4.4. **Number of animals condemned on antemortem inspection, annual mean, FY 1986-1990**

Condition	Cattle	Calves	Sheep/lambs	Goats	Swine	Equine
CNS disorders	269	79	29	6	335	2
Dead	10,226	16,697	3,316	591	60,768	105
Moribund	2,628	1,363	231	15	1,381	19
Pyrexia	397	36	25	1	218	2
Tetanus	29	13	11	—[a]	30	1
Total	13,549	18,188	3,612	613	62,732	129

Source: 246.

Note: Condemnations for these conditions were the result of only antemortem inspection. Conditions that resulted in some antemortem condemnations and some postmortem condemnations (e.g., epithelioma) are not included.

[a]Mean annual condemnation frequency <0.5 animal.

Table 4.5. Condemnation percentages and rates, annual mean, FY 1986-1990

Condition group	Cattle	Calves	Sheep/lambs	Goats	Swine	Equine
			Percentage			
Antemortem	10.0	44.3	16.1	47.5	36.6	13.4
Postmortem						
Degenerative and dropsical	5.7	2.7	11.1	16.9	0.6	7.6
Infectious diseases	0.9	>0.1	19.8	10.5	4.1	0.4
Inflammatory diseases	27.0	18.2	30.2	9.1	20.9	22.1
Neoplasms	27.8	0.4	0.6	0.5	1.7	12.0
Parasitic conditions	0.3	>0.1	7.5	0.2	0.3	0.1
Septic conditions	22.4	10.7	7.6	10.2	21.2	14.0
Other conditions	5.9	23.7	7.1	5.0	14.6	30.3
Total	100.0	100.0	100.0	99.9	100.0	99.9
			Mean annual rate			
Mean number condemned	135,319	41,056	22,452	1,290	17,1403	959
Mean number inspected	32,966,117	2,519,240	4,996,382	200,228	79,077,502	275,083
Mean condemnation rate (/1000 inspected)	4.1	16.3	4.5	6.4	2.2	3.5

Source: 246.

tured the attention of the public more than any other disease. It was introduced in 1847 and by the 1880s had spread throughout many areas of southeastern Australia via traveling stock by producers topdressing pasture with unsterilized bone dust and feeding contaminated bone meal. It is recorded that in 1885 on one farm of 220,000 sheep, 5,000 cattle, and 600 horses 22% of the sheep and 10% of the cattle and horses died from anthrax.

Such catastrophic episodes attracted both government attention and the interest of Louis Pasteur. While Pasteur's staff was in Australia to compete for a reward of £25,000 to exterminate rabbits, it was asked to test his anthrax vaccine in 1888. The staff's tests and trials were successful, but the need for better quality control and marketing made it necessary to produce anthrax vaccine in Australia. For a while a Parisian syndicate funded the "Pasteur Institute in Australia" to produce anthrax vaccine, but eventually Pasteur's two-dose vaccine was replaced by a single-dose vaccine developed locally by McGarvie Smith. This single-dose attenuated spore vaccine was produced commercially in 1897, well before the first successful spore vaccines were developed in Europe.

Ceasing to use unsterilized bone dust as fertilizer, and vaccination of stock on farms where anthrax was endemic, substantially reduced its prevalence, and it now is reported annually from very few farms. No cases of human anthrax from eating infected meat have occurred in Australia because of the requirement that all livestock must be walked onto the killing floor before being slaughtered.

Development of Export Trade in Meat.   Meat inspection procedures adopted in Australasia were derived from systems developed in Great Britain. The impetus for developing a formal meat inspection approach arose after the building of the world's first freezing works in the port of Sydney in 1861. This technology eventually enabled frozen meat to be exported from Australasia to Europe, but continuing acceptance of the product required that inspection procedures in the colonies were comparable to those in place in Britain. Cargoes of frozen beef were successfully imported from Australia in 1880 and of frozen mutton, lamb, and pork from New Zealand in 1882.

At the time of Australian federation, control of meat inspection passed from colonial governments to the states, and any variations in procedures between states reflected the regional prevalence of different diseases. For example, there was a need for more-intensive searching for *Cysticercus bovis* lesions in carcasses of stock grazing pastures irrigated with treated human sewage of metropolitan Melbourne compared with less rigorous inspection of carcasses of stock raised extensively in semiarid rural areas where there were few people.

In New Zealand, the genesis for development of its meat inspection service was the passing of the **Slaughtering and Inspection Act of 1890**. This legislation assisted government to assess the prevalence of many diseases, leading to introduction of improved measures for control of endemic diseases.

As evidence of the imperfect meat inspection system of the time, some consignments of forequarter beef exported from Australia in the early 1900s were rejected in the United Kingdom because of residual onchocerca nodules in beef briskets. While these were of no public health significance, they were aesthetically displeasing to consumers, and new procedures were adopted to excise the brisket from forequarter meat and to examine the stifle joint and connective tissue of carcasses routinely for nodules. Because of the potential threat to trade, in 1911 the Commonwealth government invoked its constitutional powers and installed a veterinary officer to supervise inspection of export beef in Queensland. By 1916 the entire meat inspection service for export meat had been expanded to all states and was firmly under federal control.

Just prior to the start of the twentieth century the Australian export market for frozen bone-in beef faced substantial competition from Argentina, which was able to exploit the export of chilled meat to service the European market. Chilled meat was preferred by consumers because it maintained its flavor, texture, color, and weight. Unlike frozen meat, it was immediately available for retail sale, and it did not "drip" moisture (due to thawing). "Drip" was shown to be reduced by fast-freezing rates and when the pH of beef muscle was 5.4 to 6.8. Conventional chilling was unsuitable for exporting Australian meat to the northern hemisphere because of the microbial growth and oxidation that affected the product during the long sea voyages. Some chilled meat was successfully landed in the United Kingdom by flooding meat holds with formaldehyde gas (1 h daily), but this was soon discontinued because of concerns about public health. In 1932 workers at the University of Cambridge found such spoilage could be inhibited when chilled meat was held in atmospheres of carbon dioxide. This discovery was the catalyst for definitive work to determine how temperature, water content, atmospheric CO_2 concentration, and initial microbial load influenced the storage life of chilled beef. In 1933, an experimental consignment of chilled meat was successfully sent by sea from Australia to the United Kingdom, and within 5 yr about one-third of its beef exports were in the form of chilled product.

During the 1950s to 1970s considerable attention was directed at investigating how carcasses became contaminated with salmonellae and developing slaughtering and carcass-dressing protocols to reduce the contamination of meat with fecal bacteria. The recognition that the stress of transportation (travel, food and water deprivation, crowding, and exposure to climatic extremes) led to buildup of salmonellae in the rumen (and cecum of pigs) provided impetus for developing improved animal welfare measures for transporting and handling stock prior to slaughter and for more-strict processes to minimize the opportunity for meat to become infected as carcasses were skinned, dressed, trimmed, inspected, packaged, and refrigerated. Preslaughter stress of stock was also found to be important in adversely affecting tenderness.

The early meat inspection procedures relied particularly upon visual inspection of the lymphatic system and incision of lymph nodes to detect abnormalities. This system was based on the predicted distribution and frequency of tubercular lesions in cattle herds with uncontrolled tuberculosis. As bovine tuberculosis became eradicated from the national herd in southern Australia, the practice of routinely incising lymph nodes was substantially curtailed to reduce cross-infection between carcasses from contaminated knives. New protocols were also developed for both export and domestic meat to minimize transfer of bacterial contamination from hides, scabbards, knives, hooks, steels, cutting boards, tables, aprons, processing machinery, and hands of operators. Awareness developed that the variable microbial load on carcasses at different stages of processing reflected various work practices in different abattoirs. The new procedures were particularly important while inspecting pork because of the high carrier rate of salmonellae in the lymph nodes of pigs. As potentially all cattle hides can be expected to be contaminated with salmonellae, skinning knives are also a common vehicle for transmitting contamination. Immersing knives in hot water (82°C [180°F]) for short periods was found to be sufficient to kill the majority of salmonellae present and to limit cross-contamination. It is becoming increasingly common practice to use one knife while a second knife used on the previous carcass lies immersed in hot water. Surface contamination of carcasses has replaced diseased organs as the most significant hazard to human health; it can be reduced by showering carcasses in a cabinet with hot water (akin to surface "pasteurization").

Other measures introduced to minimize bacterial contamination included designing machinery and workstations for easy cleaning; the erection of barriers between areas where raw and processed meats were handled; enforced control of rodents and insects; and education of staff by emphasizing the importance of maintaining a clean work environment and good personal hygiene.

An important innovation in the 1930s was the development of a chain and rail system for dressing sheep carcasses. This technology was modified over many years and is now used in all export abattoirs, permitting highly effective hygienic dressing and inspection of ovine carcasses. Periodic bans have been placed by the United States on the importation of Australian mutton because of caseous lymphadenitis (CLA) in sheep carcasses. The most recent was in 1970 when the inspection procedures for mutton were modified to remove lesions or reject carcasses where the disease was systemic. CLA occurs infrequently in New Zealand, and any affected case there is subjected to detailed inspection.

When Australia first began exporting frozen beef in quantity to the United States in the late 1950s, processing methods in abattoirs changed from supplying 90% bone-in beef to 90% boneless product to satisfy that market. This required the development of new wrapping methods using polyethylene to prevent shrinkage of product and loss of flavor and packaging in corrugated fiberboard cartons. The principle involved wrapping meat in high-grade plastic film that had low permeability to CO_2, O_2, and water vapor. After sealing the product under vacuum, sufficient CO_2 diffused from the meat to maintain an environment that restricted bacterial growth. Good abattoir hygiene to minimize bacterial load on meat and avoidance of high pH beef were essential to prevent deterioration of the packaged product; one difficulty encountered was H_2S production by lactobacilli.

Meat inspection procedures changed further in the 1960s to meet U.S. import regulations, under the newly enacted U.S. Pure Foods Act of 1965. This legislation provided rules and standards for slaughtering of stock and processing and storing of meat not only within the United States but also for countries exporting meat to the United States. It included the employment of veterinarians at all export abattoirs to undertake inspection of stock prior to slaughter and to supervise the work of meat inspectors. Regular visits were made by USDA veterinarians to Australian abattoirs to ensure that graded improvements were made to export meat to meet USDA standards. As a result, Australia has established high standards of hygiene at its abattoirs, with mechanical skinning of carcasses, improved dressing procedures, and sophisticated quality control for holding meat under refrigeration for extended periods. Primal cuts of meat are now packaged to specifications of overseas customers and transported chilled in airtight bags of multilayer polyethylene and plastic to prevent spoilage (for 8 to 10 wk). Such packaging has also helped satisfy local needs, permitting increased shelf life of special cuts of meat used in the restaurant and tourist industries widely dispersed throughout Australia.

Conditions for bulk transport of meat to Northern Hemisphere destinations have improved substantially by the use of temperature-controlled sealed containers (each containing 16 t of chilled or frozen meat), which are equipped with instruments for monitoring temperature over the complete voyage. The seals remain intact until the product is inspected at its destination, along with appropriate certification. Although expensive, air consignments of beef and lamb carcasses and cuts are regularly dispatched overseas and can be on sale in supermarkets in Japan and North America within 72 h of being packaged in Australian abattoirs.

Foodborne transmission of disease from animal products is rare in Australasia, with most food-poisoning episodes occurring from handling practices in the retail trade. Two recent exceptions were a serious episode of salmonellosis in Sweden in 1977 that was traced to the ingestion of raw meat from Australia, and multiple cases of endemic salmonellosis in 1980 derived from salami, which led to increased use of pure cultures of lactobacilli in making fermented sausage. However, processed meats remain under scrutiny because of the occasional finding of *Listeria* in pate and smallgoods and a recent episode of enterohemorrhagic *Escherichia coli* O111 food poisoning from eating contaminated sausage.

Current Diagnostic Services.   The commonwealth government is responsible for quarantine and provides advice, coordination, and sometimes finance for eradication of disease. Each state is controlled by a State Government Veterinary Officer who is responsible for monitoring and improving the health of livestock in each state, supervising veterinary, diagnostic, and extension services, and administering regulatory legislation. The trend to deregulation has led to one state government contracting out diagnostic services, including testing for chemical residues, to private enterprise.

Current Red Meat Inspection Services.[8,46,97,151,165,166]   The responsibility for meat inspection is shared between state and federal government authorities. Inspection of all meat for export is controlled by the Australian Quarantine Inspection Service (AQIS), a division within the Commonwealth Department of Primary Industries and Energy (DPIE). Responsibility for inspection of domestic red meat product lies with AQIS by contract arrangements with two state governments, and with two state governments independently, and in at least one other state it has been contracted to private enterprise. AQIS is also responsible for inspection of white meat product in one state. All export abattoirs are reviewed monthly by senior staff.

Australia is in danger of squandering its relatively disease-free status through deciding to decrease import barriers because of international agreements (the General Agreement on Tariffs and Trade [GATT], now the World Trade Organization [WTO]) to free up trade. Increased sophistication of procedures to transfer genetic material via ovum transplant, and improved sensitivity and specificity of diagnostic tests, induce a false sense of security in legislators, but no matter how good the risk management strategies put in place, there must be no complacency as the introduction of new and old diseases is inevitable.

In 1991, AQIS employed 175 veterinary officers and 1,650 meat inspectors in the supervision of 85% of all red meat produced in Australia. Numbers have declined slightly since then. At each establishment the Veterinary Officer-in-Charge is responsible for ensuring compliance with regulations; supervising production of a safe wholesome product true to labeling; making disposition judgments on carcasses; directly supervising abattoir hygiene, construction and equipment, security, and documentation; protecting animal welfare including authority for emergency killing; and ensuring that all testing policies and programs are operating efficiently (e.g., microbiological and residue testing, hormone growth promotant exclusion, disposal of effluent, etc.).

Inspectors working on meat-processing chains are responsible to the Veterinary Officer in Charge. Their duties include safeguarding public health by identifying and condemning unsafe and unwholesome meat, controlling zoonoses, ensuring correct operation of plant and refrigeration, ensuring cleanliness and hygiene, inspecting carcasses, and suspending operations in the event of failure to meet standards.

Industry is accepting increasing responsibility for control of its own product through gradual introduction of quality assurance programs using **Hazard Analysis Critical Control Points** principles. HACCP strategy focuses more upon identifying risk factors on farms and during processing, and then eliminating or reducing those hazards, rather than upon inspection procedures to remove potential risks to public health. The supply of protective clothing and equipment, better lighting, revised work practices, greater awareness of the importance of high standards of personal cleanliness, and educational campaigns have all helped to bring a more hygienic and safe product to the consumer. Included in the HACCP approach is inspection of stock immediately preceding slaughter; this was initially introduced to ensure that sick animals were not slaughtered, thereby limiting potential cross-contamination between carcasses of sick and healthy animals. This policy has had the effect of increasing veterinary involvement in inspection procedures and permits the rehabilitation of sick stock before they are slaughtered. Acceptance by industry that horned cattle induce more bruising during transport than do polled or dehorned cattle is another example of HACCP principles. Polled beef cattle are forming an increased proportion of the national herd, and other cattle are commonly dehorned.

Monitoring for the presence of pesticide and herbicide residues in tissues, urine, and feces of slaughtered animals and tracing affected stock back to their herd of origin can lead to altered management strategies on farms that will ensure that stock intended for human consumption does not have residues at the time of slaughter. For example, the introduction of buffalo fly traps through which cattle can be trained to walk to obtain water or supplementary feed has led to the phasing out of chemical control of the buffalo fly on many farms.

In 1987 U.S. authorities detected violative amounts of organochloride residues in Australian beef. Trade ceased until Australia assured importing countries that no export meat contained violative levels of pesticide. Trade resumed only after extensive testing (of abattoir product and of stock on individual farms) was able to show that the residue violations occurred in about 0.4% of animals on few farms. As a result Australia has now in place a monitoring system at abattoirs that enables stock identified with violative levels of antibacterial, pesticide, or herbicide residues to be traced to their herd of origin. This sophisticated system is being used to provide information to producers to aid in endemic disease control on farms (e.g., specific pneumonia in pigs or CLA in sheep) and has the capacity to provide information to assist in the eradication of endemic and exotic disease and to determine payment of premiums for disease-free stock. As part of this approach farmers are counseled about the use and misuse of adventitious substances, and standards and prohibitions have been developed for livestock feed additives.

Incidents of meat substitution detected in 1981 highlighted the need for more stringent quality control of meat and meat products at abattoirs. While only a few unscrupulous people were involved in the fraud, an enquiry (by the Woodward Royal Commission) the following year recommended substantial changes to existing meat export regulations to prevent processors and exporters operating outside the law; greatly increased penalties were introduced for noncompliance with the legislation. These new rules included stringent security measures in boning rooms, chillers, and stores such as sealing of cartons, chambers, and meat transport vehicles and the introduction of meat transfer certificates. Many new tests were developed in Australia (agar gel diffusion, isoelectric focusing, ELISA, and radioimmunoassays) that were more reliable than precipitin tests to detect natural host antigens in raw and cooked food products, and random tests for species of origin are now continually performed on packaged product.

Deer and other game, crocodiles, buffalo, and horses (export trade only) raised for human consumption are emerging as important industries, and similar principles are applied to ensure their meat is fit for eating. Exceptional care has to be taken to prevent contamination of crocodile meat with salmonellae.

In New Zealand all meat produced for sale must be examined by inspectors of the Ministry of Agriculture and Fisheries (MAF). This is done by MAF Quality Meat Services, who employ more than 100 veterinarians and some 1,000 trained meat inspectors. Some 95% of meat is produced in abattoirs with export licenses and the Supervising Veterinarian has similar responsibilities to the Australian Veterinary Officer in Charge in ensuring compliance with regulations.

Inspection procedures and judgments for international trade of meat are now based on European and North American disease regulations, which differ in some respects. For cattle, such regulations are based upon historical attitudes to the detection of tuberculosis lesions (see above). For sheep, the above judgments are particularly harsh for New Zealand lamb, which is free of all Office International des Epizooties (OIE) List A diseases, and enforcement of these regulations increases the cost of processing. In such circumstances, less rigorous inspection adds little or no risk to the search for lesions. Reduction in the number of lymph nodes incised and other modifications have been made to meat inspection procedures for product destined for the domestic market in Australasia to effect financial savings and increased efficiency without compromising public health safety.

It is becoming increasingly evident that in countries where disease recognition systems are effective, inspection procedures for the export meat trade should be based upon the prevalence of defects or diseases in each country and not on an internationally common procedure.

Future Technology.[234]   Meat industry processing is labor intensive with Australia's costs being at the top of the international range. Australian scientists have thus looked at the development of automated technology to reduce processing costs, to improve efficiency, to decrease carcass contamination, and to improve occupational safety. One recently designed prototype consists of a series of modules with complementary functions. Using computers and photo-optic sensors, the technology facilitates procedures such as exsanguination, removal of hides and horns, cleavage of the pubic symphysis and xiphoid cartilage, removal of hide, excision of viscera, and division of carcasses into longitudinal halves. Modules can be used as stand-alone or in an integrated system linking almost the entire process from slaughter to finishing. Commercial development of individual modules is imminent and should become part of the management of all new abattoirs and slaughterhouses; development of the fully integrated automated system awaits further refinement.

Occupational Health and Safety.   The most important occupational zoonotic diseases of meat workers have been bovine brucellosis, Q fever, leptospirosis, and skin infections. Brucellosis was widespread in cattle at the start of the twentieth century. Initial control programs were limited to voluntary participation using strain 19 vaccination. From 1970 a national eradication policy was adopted based on compulsory immunization of cattle. As herd immunity increased, compulsory immunization was gradually replaced by a strategy of testing all slaughtered cattle for *Brucella* antibody, tracing infected cattle to their herd of origin, and quarantining positive herds for more-intensive control measures. Since Australia was declared free of bovine brucellosis in 1989, all the new cases of endemic human brucellosis have been due to *Brucella suis* from contact with carcasses of feral pigs, as part of an export trade in feral pork. Q fever remains important but is now effectively prevented in many abattoirs by immunizing nonsensitized staff (Q-vax®). Human cases of leptospirosis are predominantly due to serovars **pomona** (from pigs and calves) and **hardjo** (from cattle).

Live Sheep Export Trade.   With increased sophistication in shipping, some Middle East countries have sought to change their reliance on the importation of meat to the transport of live sheep for slaughter at local abattoirs according to regional customs. The export trade in live sheep for Australia has grown from 2×10^6 head in 1975 to 7.8 $\times 10^6$ in 1987. Death rates during voyages (largely due to depression of appetite, salmonellosis, and trauma) have been reduced to 2% or less as a result of provision for improved ventilation, temperature control, and optimal spacing in pens and by encouraging sheep to become accustomed to shipboard diets before transportation; shy feeders can then be culled.

Poultry Inspection.   Almost all poultry meat is produced for the domestic market. Interest in quality control of poultry meats developed post-World War II with the realization that surface contamination of carcasses with salmonellae (and later *Campylobacter* spp.) was significant with some systems of slaughtering and processing (chain slaughter, dressing, and spin-chilling procedures). While salmonellae readily attach to poultry skin, scalding and plucking were also demonstrated to facilitate the attachment of salmonellae to exposed collagen fibers when skin was removed from carcasses during processing.

The catalyst for developing the poultry meat industry in Australia came with the control of pullorum disease, importation of new genotypes of table poultry, and intensification of broiler production with "all-in, all-out" management systems concentrated among very few commercial producers. These systems (and the relative infancy of turkey production) were **not** conducive to the emergence of important zoonotic diseases, e.g., psittacosis. Hence poultry meat inspection was slow to be regulated. However, the need to prevent significant contamination of carcasses with *Salmonella* spp. and *Campylobacter* spp. has led to the introduction of procedures for inspection of domestic poultry meat, the responsibility for which is shared between the state departments of agriculture and AQIS. An important recent development in Australia has been the widespread dissemination of the nonpathogenic *S. sophia* among poultry flocks, which appears to have resulted in decreased prevalence in the isolation of virulent salmonellae. The reason for this fortuitous position is not known.

Evolution of Meat Inspection in Canada[4,35]

The history of red meat and poultry inspection in Canada is similar to that in the United States but is complicated by the fact that the infrastructure of government has changed dramatically during the same time period. In spite of this, inspection programs have been in place since the beginning of the nineteenth century.

Red Meat Inspection.   The following is a summary of significant events in Canadian red meat and poultry inspection:

1805   Upper and Lower Canada Legislature passed laws regulating the slaughter and processing of beef and pork.

1800s   Certain municipalities began regulating the slaughter and processing of meat animals for human consumption.

1850s   The Bureau of Agriculture, which was initially responsible for disease control in livestock, and then also meat inspection, came into existence.

1867   British North America Act created a federal state in Canada, consisting of the provinces of New Brunswick, Nova Scotia, Ontario, and Quebec and later all 10 Canadian provinces.

1869   Animal Disease Control Act was passed by the new federal government for regulating animal health.

1873   Federal Beef and Pork Inspection legislation passed.

1907   The Inspection Service that has since become the Animal Inspection Directorate was inaugurated by the Meat and Canned Foods Act passed in 1907. Inspectors were initially trained at the Chicago Veterinary College, but shortly thereafter they were trained in Canada. The Meat and Canned Foods Act regulated meat destined for interprovincial and export markets.

Poultry Inspection.   Poultry inspection came under the same federal legislation and jurisdiction as red meat with the passage of the first Canada Agricultural Products Standards (CAPS) Act in 1955. This was replaced by the Canada Agricultural Products Act, still known as CAPS, in 1980. The CAPS Act is a general act that controls inspection for wholesomeness and quality and marketing of agricultural products. These are defined, under Chapter 27 of the act, as "an animal, plant, or an animal or plant product, or a product, including any food or drink, wholly or partly derived from an animal or plant." Because it also controls marketing, it sets the stage for supply management of agricultural commodities, specifically, dairy products and most poultry products, including production of eggs, broilers, turkeys, and broiler chicks. Regulations are published and updated periodically for the implementation of the act. The Meat Inspection Act, first adopted in 1955, elaborates and extends the provisions of the CAPS Act to red meat and poultry. It excludes caribou and other game species that are important to indigenous northern peoples.

Evolution of Meat Inspection in the United Kingdom[25,50,79,94,118,186]
Food Law.   Food law developed, initially during the Middle Ages, to control the adulteration of food. Many of the laws introduced at that time served to protect the "honest trader" as much as the consumer. This situation remains today because much of legislative action taken by the state has been in response to pressure from commercial interests that require controls to maintain fair trading.

In contrast, the consumer movement, which made its appearance during the 1960s, and which has developed strongly since, has had surprisingly little impact on food law.

The origins of modern food and drugs law in England and Wales can be traced to the Adulteration of Food and Drink Act of 1860 and the Pharmacy Act of 1868. Between the years 1875 and 1928, the principles of food law were established. It became an offense (1) to mix injurious ingredients with any food to be sold for human consumption and (2) to sell any food not of the nature or substance or quality demanded by the purchaser. This means that consumers do not have to prove something is wrong because it is their right to assume that any food sold is both safe and wholesome. In order to assist the enforcement of food legislation, the local authorities must appoint (1) sampling officers, who have powers to procure samples for investigation, and (2) public analysts.

The United Kingdom is a member of the European Union (EU). This obliges the government to harmonize national food law with that of the EU. Most of EU law is introduced in the form of directives, and these are immediately binding on all member states. The time it takes for national governments to enforce these directives is variable, and this can cause disagreements between them and ill feeling among those involved in both national and intracommunity trade.

Food law in the United Kingdom was updated and revised substantially by the passing of the Food Safety Act of 1990, which replaced the Food Act of 1984. Its title is something of a misnomer, however, because the act covers, in addition to safety, such topics as advertising, labeling, and the composition of food. The 1990 act covers all stages of the food chain, and it has improved and strengthened the powers of local authorities to enter premises and to inspect, to sample, and to seize suspect foods. The act also introduced the concept of "due diligence"; i.e., it may be an offense if it can be proved that not all reasonable precautions were taken to ensure that food is safe and wholesome. The ministers also have powers to issue emergency control orders when there is an important risk of injury to human health. The Food Safety Act of 1990, as did its predecessors, empowers ministers to enact supplementary legislation. This usually takes the form of regulations that deal specifically with different aspects of food production and processing. The three most important in the veterinary context are the Fresh Meat (Hygiene and Inspection) Regulations of 1992, the Poultry Meat, Farmed Game Bird Meat, and Rabbit Meat (Hygiene and Inspection) Regulations of 1994, and the Dairy Products (Hygiene) Regulations of 1994.

The implementation and enforcement of food legislation is the responsibility of local authorities. The Local Government Act of 1972 reduced the number of food authorities in England and Wales from 360 to 78, and county councils, both metropolitan and nonmetropolitan, became the food authorities under the Food Act of 1984. However, the discharge of these responsibilities is usually devolved by the county councils to the environmental health departments of district councils. This decentralization has advantages because much food is still produced and processed locally and local knowledge of the people and the premises involved is extremely valuable for ensuring that safety standards are maintained. The large number of district councils involved makes it inevitable that there will be occasional problems of uniform enforcement. In order to attempt to reduce this, the Association of County Councils and the Association of Metropolitan Authorities set up, in 1976, a committee known as Local Authorities Coordinating Body on Food and Trading Standards (LACOTS). Among other things, LACOTS assists with resolving differences in the interpretation of the law and specific problems of enforcement.

Within the Ministry of Agriculture, there is the Food Safety Directorate, which is responsible for all matters of food safety. The directorate is divided into four parts: the Food Safety Group, the Food Science Group, the Animal Health and Veterinary Group (which includes the Animal Health Zoonoses Division and Meat Hygiene Division), and the Agricultural Inputs, Plant Protection, and Emergencies Group (includes the Pesticide Safety Division and Veterinary Medicines Directorate).

In order to ensure that the food legislation is appropriate and takes account of recent developments, there are nine committees, comprising both independent experts and officials, which advise government ministers on food and drugs matters. These include the Committee on Microbiological Safety of Food, the Steering Group of Food Surveillance, the Veterinary Products Committee, and committees concerned with toxicity, mutagenicity, and carcinogenicity of chemicals in foods.

Welfare of Animals at Slaughter.   The welfare of farm mammals at slaughter is covered by the Slaughterhouses Act of 1974. This act requires that all slaughterhouses and slaughterhouse workers are licensed and that all animals killed in slaughterhouses are either killed instantaneously or rendered insensible to pain until death supervenes. Exceptions are made for ritual slaughter. The detailed requirements are embodied in the Slaughter of Animals (Humane Conditions) Regulations of 1990. These regulations deal with the construction and equipment of lairages (holding pens) and slaughterhouses, the confinement and treatment of animals awaiting slaughter, and the stunning and slaughter of animals. There are additional provisions for horses. The corresponding welfare legislation for birds includes the Slaughter of Poultry Act of 1967 and Slaughter of Poultry (Humane Conditions) Regulations of 1984. In addition, the Ministry of Agriculture, Fisheries, and Food has published codes of practice on (1) the welfare of red meat animals at slaughter and (2) the construction and layout of red meat slaughterhouses in relation to animal welfare. These codes are not legally binding in themselves, but the failure to take account of their provisions would be taken into account in any prosecution taken under the welfare laws.

Red Meat Inspection.   The Public Health Act of 1875 empowered medical officers of health and inspectors of nuisances, the forerunners of today's environmental health officers, to seize unsound meat and also unsound animals. This legislation was subsequently used to prevent the meat of tuberculous animals from being used for human consumption. In some parts of the country, the seizures were so numerous that farmers and meat traders complained to the government in no uncertain terms!

In 1922, the minister of health issued a memorandum that recommended a procedure for the examination of carcasses for meat inspectors. The Public Health (Meat) Regulations were introduced in 1924. These required butchers to notify the times of slaughter and to retain carcasses for a short period for inspection. These developments did not result in meat inspection being enforced by all local authorities, however. Little changed during the next 20 yr, although the number of licensed slaughterhouses, excluding bacon factories, was reduced to fewer than 500 during World War II, and this made it, for the first time, a practical proposition to examine all meat. This did not come about, however, and after meat rationing ceased during the early 1950s, the number of licensed slaughterhouses increased dramatically. It was not until 1966 that meat inspection eventually became compulsory in England and Wales when the provisions of the Meat Inspection Regulations of 1963 came fully into force. At this time, there were about 1,800 slaughterhouses in England and Wales, and the district councils became responsible for meat inspection. Separate legislation existed for Scotland, and the situation was altogether more satisfactory because there had been much tighter legislative control of slaughterhouse development during the late nineteenth and early twentieth centuries.

The United Kingdom joined the European Community, now the EU, in 1973, and this resulted in a two-tier system of inspection. Slaughterhouses that complied with Directive 64/433/EEC were licensed by the Ministry of Agriculture, Fisheries, and Food so that they could participate in intracommunity trade whereas others continued to be licensed by the local authorities and could only sell their meat for domestic consumption. The standards of hygiene and the ante- and postmortem inspection of meat, in EEC-approved slaughterhouses, was the responsibility of veterinarians, most of whom were private practitioners and worked part-time as official veterinary surgeons. In both categories of slaughterhouse, the inspection of the meat was carried out by authorized meat inspectors.

Several pieces of legislation were introduced after 1973, including the Slaughterhouses (Hygiene) Regulations of 1977, the Meat Inspection Regulations of 1987, and the Fresh Meat Export (Hygiene and Inspection) Regulations of 1987. These were all revoked in 1992 with the introduction of the Fresh Meat (Hygiene and Inspection) Regulations of 1992. This legislation put into effect parts of several EU directives concerned with red meat and applies to England, Scotland, and Wales, thereby unifying the legislation in the three countries for the first time. The regulations apply to slaughterhouses killing cattle, pigs, sheep, farmed game-handling and -processing facilities, cutting premises, and cold stores. There are separate rules for slaughterhouses with a low throughput, whereas butchers who supply only the final consumer are exempt from the regulations. The standards of hygiene and inspection, and the certification of meat, remain the responsibility of the official veterinary surgeon.

The Fresh Meat (Hygiene and Inspection) Regulations of 1992 require that all licensed premises, that is, (1) slaughterhouses, (2) premises in which fresh meat, not destined for the final consumer, is cut up or stored, and (3) farmed game-handling and -processing facilities, are inspected from time to time. The following categories of officials (as of August 1994) may be involved with inspections:

1. **Ministry of Agriculture, Fisheries, and Food (MAFF) Veterinary Officers (VOs).** The VOs inspect premises periodically, usually at agreed times. They will usually complete a hygiene assessment form and will discuss the findings with the abattoir management and the OVS. On some occasions, the VO may be accompanied by a regional meat hygiene advisor, who is also a veterinarian and an MAFF employee. Exceptionally, a VO may recommend that a license should be revoked.
2. **Official Veterinary Surgeons (OVSs).** The OVSs are appointed by the local authority either on a full- or part-time basis. They are responsible for ensuring that the overall hygiene standards are maintained and that animals and carcasses are properly handled and inspected.
3. **Authorized Meat Inspectors (AMIs).** AMIs are appointed by local authorities and work under the supervision of the OVS. They do much of the physical meat inspection and health marking and may assist with antemortem inspection.
4. **Environmental Health Officers (EHOs).** EHOs are employed by the local authority, and they are responsible for enforcing the Food Safety Act of 1990 and associated legislation, including the Fresh Meat (Hygiene and Inspection) Regulations of 1992.
5. **Official Inspector of the European Commission.** Each year, community veterinary inspectors, accompanied by MAFF veterinary staff, visit a small number of licensed premises in each member state.

The inspectors may (1) collect samples and inspect records (VOs, OVSs, AMIs, and EHOs), (2) detain or seize meat that is a danger to human health (OVSs, VOs, EHOs, and AMIs), and (3) recommend carcasses or fresh meat be regarded as unfit for human consumption (OVSs, VOs, EHOs, and AMIs). In addition, authorized OVSs and EHOs may (1) require that breaches of the law are remedied, (2) issue improvement notices, and (3) prohibit the use of premises or equipment if there is a risk of injury to health. The VOs may also (1) recommend the revocation of the license and (2) give directions to the OVSs to ensure that the regulations are properly complied with.

Inspectors should give reasons in writing for any action they recommend. They should provide a clear distinction between what must be done to comply with the law and what should be done because it is good practice. If the abattoir proprietor disagrees with the opinion of either the inspector or his or her superior officer after informal discussion, then the proprietor may seek advice of the LACOTS, or the proprietor may appeal to a magistrate's court.

Further developments were scheduled in 1995 with the formation of a national Meat Hygiene Service (MHS). The MHS took over from local authorities and employs the OVSs and AMIs. This means that local authorities are no longer involved directly with meat hygiene although they will continue to have responsibility for hygiene in retail outlets, restaurants, etc.

Poultry Meat Inspection. There was no organized statutory inspection of poultry before the United Kingdom joined the EU in 1973. The poultry meat directive 71/118/EEC was approved during 1973 and was given effect by the Poultry Meat (Hygiene) Regulations of 1976. This applied to all poultry slaughtered in England, Scotland, and Wales except that sold directly by the producer to the consumer, so-called farm gate sales. All licensed processing plants were of European Economic Community (EEC) standard and could, therefore, participate in intracommunity trade. The ante- and postmortem inspection was carried out by an Official Veterinary Surgeon (OVS) assisted by trained poultry meat inspectors (PMIs). The problems of complying with this legislation were enormous. The industry had to invest substantial amounts of money to bring the processing plants up to the standard required while the district councils, which had no previous experience of inspecting poultry meat as they had with red meat, had to provide an inspection service to cater to more than 350 million birds per annum.

The Poultry Meat (Hygiene) Regulations of 1976 remained in force, with amendments, until 1994 when they were replaced by the Poultry Meat, Farmed Game Bird Meat, and Rabbit Meat (Hygiene and Inspection) Regulations of 1994, which gave effect to several EU directives. These 1994 regulations are more comprehensive than those of 1976, as their title implies, and they introduced some new features. These include the (1) introduction of a new system of preslaughter assessment that includes a declaration on health status by the flock owner and the supervising veterinary surgeon and (2) provision for a new category of inspector, the plant inspection assistant (PIA). Plant inspection assistants will be employees of the processing plant and will be trained by the OVS to carry out certain postmortem duties under his or her supervision. The regulations **do not** apply to (1) premises where fresh meat is prepared for direct sale to the final consumer, e.g., butcher shops, (2) premises with an annual throughput of <10,000 birds or rabbits, (3) fresh meat intended for exhibitions, and (4) fresh meat not intended for human consumption.

Residue Testing. The United Kingdom has a national sampling and surveillance scheme that screens specimens of animal origin for residues of "veterinary" compounds. The Animals, Meat, and Meat Products (Examination for Residues and Maximum Residue Limits) Regulations of 1991 gives effect to the EU Directive 86/469/EEC. Specimens are taken from live cattle, horses, sheep, pigs, and their carcasses selected at random from farms and at abattoirs by members of the State Veterinary Service of the MAFF. During 1994, about 52,000 tests were performed on about 45,000 specimens at MAFF laboratories. Of these, about 80% were screened for antimicrobial agents. The remainder were assayed for hormones, pesticides, ß-agonists, heavy metals, tranquilizers, and gestagens. The two most notable residue problems were cadmium in the kidneys of horses and antibiotics and sulfonamides in the kidneys of pigs. The proportion with concentrations exceeding the maximum residue limit (MRL) during the first 6 mo of 1994 being 100%, 2.5%, and 0.3%, respectively. When a positive (exceeding the MRL) specimen is found, the farmer is advised of the steps that need to be taken to avoid residues entering the food chain. Prosecutions are only considered if there is evidence of serious disregard of the regulations.

In addition to the statutory National Surveillance Scheme (NSS), the Veterinary Medicines Directorate, MAFF, undertakes other surveillance programs that cover products and animal species not included in the NSS. Action is taken, as above, if the concentration exceeds the MRL, or if one is not set, the limit of quantification (LOQ). The principal residue problems identified by these non-statutory sampling programs have been the presence of coccidiostats and sulfonamides in the tissues of poultry.

Extensive sampling of animals, animal products, and the environment may also be undertaken by MAFF following a serious pollution problem. In recent years, these have included the extensive contamination of cattle feed with lead and the Chernobyl disaster. In both these instances, animals on farms were placed under restriction for a considerable period of time until the concentrations of the lead and radionuclides had decreased to acceptable levels in the animals and the environment.

Evolution of Federal Meat Inspection in the United States[197,214]
Chronology of Legislation

1906 Meat Inspection Act. Although the first U.S. federal legislation on meat inspection was passed in 1890, the basis for the current program is the Meat Inspection Act of 1906. This act provided for (1) antemortem and postmortem inspection of each animal, (2) prevention of diseased and unwholesome meat from entering the food chain, (3) reinspection during processing, (4) continuous inspection, (5) factual and informative labeling, (6) approval of floor plans and specifications, and (7) control of meat shipped interstate, imported from foreign countries, or sold to federal agencies. This program was administered by the Bureau of Animal Industry, USDA.

1957 Poultry Products Inspection Act. In the 1920s, the USDA entered into agreements with certain state and local agencies to inspect poultry, either live or dressed. The 1957 federal Poultry Products Inspection Act extended inspection to poultry similar to that established for red meat in 1906. The law provided for inspection of poultry destined for interstate commerce and for poultry from foreign sources. A major stimulus for mandatory poultry inspection was the increase in size of the poultry meat industry since the 1940s. In the late 1920s, "voluntary" inspection for wholesomeness was begun by some packers because of regulations promulgated by some large city health departments that prohibited sale of uninspected poultry. Elsewhere, poultry was still sold as live birds to be killed and inspected by the homemaker, restaurant cook, or butcher, or it was marketed as "New York–dressed" poultry (a trade term for a bird that had its feathers and blood removed but not its internal organs).

1958 Humane Slaughter Act.   This act applied to all plants selling meat items to the federal government. Before 1958, a knocking hammer commonly was used to stun cattle, the effectiveness of which depended entirely on the skill of the individual slaughterer. Although most cattle were stunned with the first blow, some required several attempts, especially a novice.

1967 Wholesome Meat Act.   This act closed loopholes in earlier laws that permitted sale of uninspected meat to the public. Its provisions are as follows:

1. **Inspection of all meat sold**. All meat must be inspected, whether for sale in intrastate or interstate commerce or in export to foreign markets. All meat for intrastate sale must be inspected in a program at least equal to the standards of the federal inspection act.
2. **Federal assistance to states**. A cooperative agreement could be developed between state and federal governments in which a state would be subsidized for up to half the cost of the state-operated program.
3. **Designating plants as health hazards**. The secretary of agriculture was given authority to designate a plant a health hazard and to close the plant if necessary, e.g., a plant handling dead, dying, diseased, or disabled animals or one without potable water, proper sewage disposal, or adequate sanitation.
4. **Inspection of meat establishments in territories** (e.g., Guam).
5. **Inspection of boner and cutter plants**. Plants referred to as "boners" and "cutters" divide whole carcasses into the primal cuts such as ribs, loin, shoulder, and round before further distribution. Before this act, plants receiving carcasses from federally inspected abattoirs could process and distribute the meat in interstate commerce as federally inspected on the basis of inspection in the previous plant.
6. **Inspection of sausage-packing plants**. Before this act, plants that merely packaged sausage could apply for a certificate of exemption from federal inspection.
7. **Authority over renderers, transporters, warehouses, and animal-food manufacturers**. The 1967 act required registration and periodic inspection of plants rendering carcasses of animals into inedible products; transporters such as truckers, airlines, and railroads; warehouses where meat may be stored; and manufacturers of food for animals such as pets, mink, and fish. Emphasis is placed on preventing unwholesome meat from entering the food chain from any of these sources.

8. **Increased inspection of imported products**. A country that wants to export products to the United States is now required (a) to have its inspection law approved by the USDA as at least equal in stringency to the U.S. law and (b) to allow a USDA veterinarian to visit the country and review its inspection procedures for compliance with the law. Plants in the country are then approved individually. No country with endemic foot-and-mouth disease may ship uncooked meat to the United States.

There are still two exemptions from the 1967 Wholesome Meat Act. First, **custom plants** that slaughter animals for the owner could be exempt as long as all meat is returned to the owner for personal use. These plants were subject to sanitation and equipment inspections. Second, exemptions were allowed for retail dealers or stores that did not slaughter and at least 75% of whose sales were to retail customers. The retail store could cut up, slice, and trim carcasses into retail cuts; grind and freeze meat products; and cure, smoke, and cook products for retail customers only.

The Curtis Amendment made an important change in the 1967 act. Because many small custom plants remained that both slaughtered animals and had retail counters, for which inspection would require expensive changes in their plants, the act was amended to allow them to stay in business on a legitimate basis. These operators must stamp all custom slaughtered meat "not for sale." All meat sold at the retail counter must be derived from properly inspected carcasses.

1968 Wholesome Poultry Products Act (WPPA). The USDA inspects all poultry in states that either never developed a state poultry inspection program or discontinued their program after the 1968 act was passed. Federal inspection is an advantage in states with large broiler or turkey industries because most of these products enter interstate commerce. The 1968 act permits two exemptions from inspection of poultry. The first allows farmers who raise up to 250 turkeys or 1,000 chickens annually, or combinations thereof, to sell uninspected dressed birds directly to consumers, but not for resale. The second exemption allows small processors to handle up to 5,000 turkeys or 20,000 chickens annually. The processing plant must still pass equipment and sanitation inspection, but each bird does not have to be inspected. Poultry inspected under a state "equal to" program are not eligible for interstate commerce.

1978 Humane Methods of Slaughter Act. This act extended the 1958 law to include all federally inspected plants. The law regulates handling procedures in the abattoir up to and including the stunning process. Ritual slaughter and the handling of livestock in preparation for ritual slaughter are exempt from the act.

Antemortem Inspection[6,21,74,112,113,115,116,135,138,142,215]
Terminology. The following terms are used in this section.

1. *Antemortem.* This term, literally meaning "before death," refers to the inspection of an animal before it is slaughtered.
2. *Suspect animal.* A *suspect animal* is one possibly affected by a condition or disease that requires rejection of the carcass, either wholly or in part, if found at slaughter.
3. *Condemned animal.* A *condemned animal* is one that may not go to slaughter because it is judged unfit at antemortem inspection. The veterinarian bases the decision on clinical signs.
4. *Subject to inspection.* Abattoirs may establish conditions before purchasing an animal. In this situation, the seller (the farmer) and the buyer (the abattoir operator) agree on payment for that portion of the animal that passes inspection; thus the animal is purchased subject to passing inspection.
5. *4D.* This is abattoir jargon for animals that are dead, dying, diseased, or disabled.
6. *Disposition.* This refers to the ultimate handling of a carcass, or its parts, as a result of decisions reached during inspection.

Basis for Antemortem Inspection. Antemortem inspection is the beginning of traditional endpoint inspection performed at the abattoir. As HACCP-based process control programs are introduced in meat production, we can anticipate antemortem inspection originating on the farm with preharvest health evaluations and progressing to the final evaluation at the abattoir. Culled livestock and spent hens, as well as any animals with a history of illness, will continue to require greater scrutiny than young, healthy meat animals.

There are several reasons for antemortem inspection. The most important is to remove animals with conditions that cannot be detected at postmortem inspection. Several central nervous system (CNS) diseases, for example, cannot be readily detected at necropsy. Prior observation of the live animal is useful in making sound postmortem decisions. The knowledge of the veterinarian during antemortem inspection is also a deterrent to farmers who may send ill or injured animals to slaughter, an important animal welfare consideration. Antemortem inspection also prevents unnecessary contamination of personnel and equipment inside the abattoir by diseased animals. This is important from an economic point of view, as it avoids unnecessary abattoir shutdowns and cleanup. Cooperative programs with other animal disease control agencies are an integral component of antemortem inspection. The antemortem inspector may be the first medically trained person to observe the animal clinically, and so the inspector can play a vital part in some disease eradication programs. Some clinically ill animals that would probably be condemned on postmortem inspection can be salvaged with proper treatment, if this is permitted by the veterinary inspector. Swine erysipelas, for example, responds readily to antibiotic treatment.

Antemortem Inspection Procedure.[187] It is imperative that antemortem inspection is done on the day of slaughter. It should include, ideally, observation of the animal at rest and in motion. Lameness and locomotor abnormalities are best observed while the animal is moving whereas some abnormalities, such as dyspnea and shivering, are difficult to evaluate unless the animal is resting. The clinical evaluation may include determining the temperature, pulse, and respiratory rates. Additional tests may be required in some circumstances. Acute phase reactants (APRs) such as C-reactive protein and haptoglobin are nonimmunoglobulin plasma proteins produced during the acute phase of inflammation and tissue injury. An on-site test for APRs has been suggested to complement detection of fever and other clinical signs as evidence of disease.

Adequate facilities and equipment for antemortem inspection are required. These include proper lighting and holding facilities that are well designed and constructed because the inspector may need to observe an animal closely in order to make a decision about the disposition. It is important to keep the facilities clean, because animal discharges, indicative of disease, may then be observed on the floor.

Antemortem Dispositions of Abnormal Animals.   Handling of abnormal animals involves three considerations (Fig. 4.5.). (1) Animals with a localized lesion, not indicative of a generalized condition, may go to slaughter. (2) Animals may be slaughtered as "suspects." A tag is attached to the ear of these animals (or they may be tattooed) so that they can be identified in the abattoir and subjected to detailed postmortem inspection. (3) Any portion of the carcass that is condemned must be routed so that it does not enter the food chain. An animal that is condemned at antemortem inspection must never enter the abattoir.

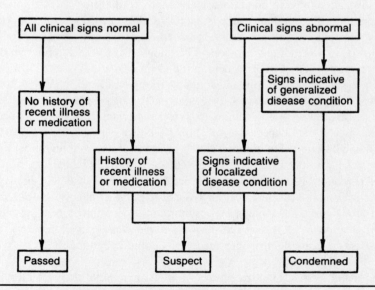

Fig. 4.5.   Antemortem inspection and dispositions.

Animals may become disabled because of injury during loading, transport, or unloading. Many of these injuries can be prevented if animals are handled by trained personnel using properly designed equipment, which should always be available when needed. Ill or injured animals should not be sent for slaughter but should be kept on the farm for treatment or euthanasia because admission of such animals to the abattoir creates both disease and animal welfare problems. In some countries, the term *emergency slaughter* is used for animals other than those that are injured for which the term *salvage slaughter* is used. (See "Disabled Animals" below.) Laboratory examination may be required as a basis for disposition of all animals presented for emergency slaughter. Immediate salvage slaughter of injured animals is both humane and economically sound as meat may be salvaged for either human or animal consumption.

Dispositions for Commonly Encountered Conditions

Scirrhous Cords. Swine with scirrhous cords may go to slaughter because the localized lesion and associated tissues will be rejected at postmortem inspection.

Multiple Abscesses. Multiple abscesses usually indicate a generalized condition for which the animal must be condemned because it is unfit for food.

Epithelioma. Epithelioma (cancer eye) is seen commonly in Herefords. Cattle with enucleated eyes are usually classified as suspects. If the neoplasia is extensive and a secondary infection is present, or if the animal has become cachectic, condemnation is the logical disposition. On the other hand, if the tumor is less advanced, the animal may be tagged a suspect and sent to slaughter.

Actinomycosis. Actinomycosis (lumpy jaw) is an infectious disease caused by *Actinomyces bovis* that forms abscesses in the mandibular region. Because this is usually a localized condition, the animal may be sent to slaughter as a suspect.

Disabled Animals. Disabled or nonambulatory animals are animals that are unable to stand or move about normally. These are sometimes referred to as "downer" animals. Any animal that fits this description may be withheld from slaughter for at least 24 h. An exception to this is the animal that has had a traumatic injury during transport to the stockyard. In this instance, emergency antemortem inspection (and salvage slaughter) is appropriate because there is a good explanation for the animal's condition. In all other instances, the 24 h period allows the inspector time to observe for clinical signs. If the animal's condition worsens during this time, it is condemned. If there is no change in the animal's health status, it may be slaughtered as a suspect. Samples may be taken to determine if there are any tissue residues of drugs that might be associated with the possible cause of the disabled condition.

Disabled animals, especially "downer" cows, present a complex problem, and some abattoirs no longer accept them. One reason is the possible greater risk of fecal contamination; i.e., they may be *Salmonella* carriers. Another concern is the associated animal welfare problem.

In the United Kingdom, it is unlawful to send disabled animals to the abattoir.

Pneumonia.   Dispositions of pneumonia cases vary. If the condition is advanced and generalized, the animal is usually condemned. In cases of lesser severity, the animal is sent to slaughter as a suspect.

Central Nervous System Damage.   There are several infections such as listeriosis that cause CNS damage and can be transmitted from food animals through the food chain to humans, in whom they produce similar lesions. Any animal with signs of current CNS infection is condemned.

Retained Fetal Membranes.   Animals that have recently given birth are fit for food. They are sent to slaughter as soon as the fetal membranes have passed. Parturient animals used to be held 10 d before slaughter, but the belief that these animals are unfit for food has been abandoned because parturition is a normal physiologic process.

Fever.   An animal is condemned at antemortem inspection if its body temperature is excessively high (cattle 40.6°C [105°F], swine 41.4°C [106°F]). The veterinarian must take the ambient temperature into consideration. Animals can become overheated on hot days or with crowding, for example. In these situations, the animals may be cooled with water, moved to shaded areas, or provided a more favorable shelter and reexamined later as suspects. Interestingly, minimum temperature guidelines have not been set; animals with subnormal body temperatures are **not** condemned solely on this basis.

Reactors.   Tuberculosis (TB) reactors (which are identified in the United States by a T brand on the left cheek) are classified as suspects at antemortem inspection. The ear tag number and all antemortem findings are recorded to facilitate tracing the animal to its source. Postmortem examinations are performed on any TB reactors that die in the holding pens. All information available about the animal is recorded and is extremely valuable to field investigation teams engaged in TB eradication. It is interesting that bovine and porcine **brucellosis reactors** are simply identified by their ear tag number and then sent to slaughter. In other words, unlike TB reactors, they are not classified as suspects. The law states that brucellosis reactor goats are not to be slaughtered in an abattoir. In the United States, bovine brucellosis reactors are further identified by a B brand on the left cheek.

Reportable (Notifiable) Diseases. Whenever signs of anthrax, hog cholera, CNS disease (possibly rabies), or vesicular disease are noticed, government (local, state, and national) livestock sanitary officials should be notified immediately, and they will recommend to the inspector a disposition on the animal. Some of these diseases are highly communicable; to avoid contamination of the entire area, the animals are held in isolation. The abattoir must not dispose of the animal before the appropriate government agencies are notified because examination of it is an important link in animal disease control, especially for suspected vesicular diseases.

Slaughter [20]

Humane Stunning Methods.[49,87,222] Four methods of stunning are permissible under the federal law.

Captive Bolt Guns. A captive bolt gun has a protruding bolt that may be of two types, either skull penetrating or nonpenetrating (mushroom head). The disadvantage of the skull-penetrating type is that the bolt passes through the hide and bone, thus contaminating the brain with bacteria, making it unacceptable for human food.

Rifle with Live Ammunition. Firearms have two disadvantages. First, safety, and second the brain and head meat cannot be used for food because, when the bullet hits the skull, it shatters and disperses fragments in many directions, thus contaminating tissues.

Electric Stunning. Electric shock is permissible only for pigs, sheep, and poultry in the United Kingdom. The electrical current and time are prescribed to stun the animal and produce a state of anesthesia. Death occurs from subsequent hemorrhage produced by severing major arteries and veins. An adverse effect is excessive current may cause petechial hemorrhages in various organs ("blood splash"). Because hemorrhage can indicate diseases, such as hog cholera, an inspector then has the added difficulty of determining the cause. "Blood splash" from this method also may be aesthetically unacceptable to consumers. In 1985, electric shock as a means of slaughter (as opposed to merely stunning) was approved. The procedure, referred to as *deep stunning* or *electrical slaughter,* is optional. (It is referred to as *stun kill* in the United Kingdom.) With this technique, which uses a third electrode placed on the back or foot in addition to the two normally applied to the head, the "splashing" that accompanies electrical stunning is reduced. The weight of blood lost during bleeding is comparable to that achieved using other stunning methods.

Carbon Dioxide. A carbon dioxide (CO_2) chamber is used in some larger abattoirs to produce anesthesia in sheep, calves, or swine. The animals are shackled and bled as they come out of the chamber. CO_2 stunning is not permitted in some countries.

Ritual Slaughter.[58,83,85,86,124] Ritual slaughter is one facet of the food-associated rules (dietary laws) observed by many religions. For instance, Hindus and Seventh Day Adventists have restrictions on eating meat, Jews and Moslems do not eat pork, and Sikhs seldom eat beef. Many Buddhists restrict the amount of meat in their diet. Many Catholics eat fish on Friday. Eating meat and other foods may be further restricted in relation to methods of preparation (e.g., using separate utensils) and eating (e.g., praying before eating, not touching food with the left hand, or not drinking milk and eating meat during the same meal), as well as specific days or periods of the year (e.g., Ramadan) considered of importance to the religion.

The practice of ritual slaughter is associated most often with Jews (Shechita) and Moslems (Halal). *Shechita* means "killing of animals for food," and when animals are killed in the prescribed manner, the meat is acceptable to eat (i.e., "kosher"). Cattle, goats, sheep, and poultry are slaughtered according to Shechita or Halal rules, which are similar because they are derived from a common origin. The Sikh practice of Jhatka involves decapitating small ruminants and birds with a single stroke. Poultry are also slaughtered in a manner to meet the religious requirements of some Confucians and Buddhists. Confucians do not eviscerate the bird, whereas Buddhists do not remove the head and feet from an eviscerated bird. The demand for meat and poultry derived from specific ritual slaughter varies in relation to the size of resident populations of the relevant religions. The details (i.e., allowances within religion) of ritual slaughter also vary around the world, usually as a result of interaction among the needs of church orthodoxy, the demands of state sanitary requirements, and pressure from animal welfare groups.

Traditional killing according to Shechita rules is performed by a designated Jew (Shochet) who makes a single cut across the throat of a fully conscious healthy animal, severing the jugular veins and carotid arteries. The head must be restrained sufficiently to prevent the animal from contributing to its own death. The animal becomes unconscious from blood loss and dies. The Shochet then makes an incision and examines the thoracic and abdominal cavities for adhesions. If the animal passes, only the forequarters are retained and marked as kosher unless the sciatic nerve and other unacceptable tissues are first removed from the hindquarters. The entire poultry carcass is acceptable.

Ritual slaughter practices conflict with animal welfare concerns, in particular practices such as shackling and hoisting fully conscious animals. This practice is not only inhumane but can be hazardous to employees especially when large cattle are slaughtered. Holding pens of several designs (e.g., rotary or casting) and other means of restraint have been developed that are acceptable for use in ritual slaughter. They vary greatly, however, in the amount of stress caused on the animal. A relatively humane system involves a center track, double-rail restrainer that can be modified to fit animals of various sizes from sheep to large calves. While animals move quietly along the restrainer track, slaughter, with or without stunning, and shackling can be done. In some countries, stunning (e.g., with captive bolt) after the throat is cut may be accepted by the religious community.

Postmortem Inspection[99,103,136,190,237]

General Considerations. Reliable evidence and careful reasoning are essential to making proper disposition of a carcass at postmortem inspection. Although the guidelines for making an appropriate decision are often not clear, veterinarians must understand the effects that observable lesions may have on the wholesomeness of meat. Diseases of public health importance have been well described, so guidelines related to them are clear, with the consumer being the first consideration. Nutritional value and wholesomeness are of prime concern although the aesthetic appearance is an important consideration. For example, wholesome meat may be pigmented, which may not appeal to consumers. Parenchymatous organs, such as the liver, spleen, or kidneys, must be examined for evidence of changes indicative of generalized disease, e.g., enlargement or diminution in size, multiple hemorrhages or abscessation, or changes in color or consistency.

Veterinarians must take care to prevent unnecessary wastage when making a disposition, and there must be good reasons for incising expensive cuts of meat. If an entire animal is condemned, no portion of it must be allowed to enter the human food chain. Figs. 4.6 and 4.7 illustrate the decision-making procedures in postmortem inspections.

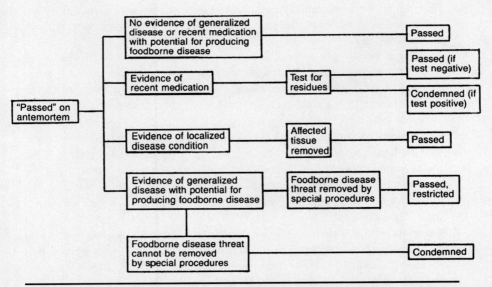

Fig. 4.6.   Postmortem inspection and dispositions for animals "passed" on antemortem inspection.

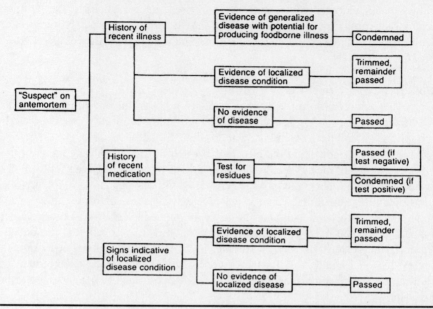

Fig. 4.7.   Postmortem inspection and dispositions for animals evaluated as "suspect" on antemortem inspection.

Basic Principles of Postmortem Disposition

1. **Normal versus abnormal tissue**. The first step is to separate normal from abnormal tissue. Minor lesions, such as small bruises, may be trimmed without further consideration.
2. **Localized versus generalized and acute versus chronic conditions**. Next, determine whether the condition is localized or generalized and if it is acute or chronic. It is essential to understand the pathologic process from which the lesion resulted, e.g., a well-encapsulated lesion is less hazardous than one that is active and spreading.
3. **Conditions affecting physiologic functions**. Although the primary lesion may be localized, it may in some instances have effects on other parts of the body. For example, a persistent urethral calculus will lead to uremia, and the animal must be condemned because of the generalized effect of the calculus. Another example is emaciation as a result of eating difficulty caused by excessively worn teeth.
4. **Conditions injurious to consumer health**. A condition may exist that can be hazardous to consumer health yet is difficult to recognize without postmortem lesions. For example, drug residues and contamination with salmonellae may pose threats.
5. **Consumer acceptability**. This determines whether or not a product passed

for food can be sold. For example, bovine and hog tongues may be sold easily whereas tongues from sheep and goats are not.

6. **Product wholesomeness**. Finally, disposition of the product is based on its wholesomeness. Consumer acceptance is related to what the public regards as wholesome. Consumers rightly expect that all animals are slaughtered under sanitary conditions and are inspected carefully.

Terminology. The following terms are used in this section.

1. *Condemned products.* These are unwholesome because of disease or severe contamination.
2. *Inedible products.* These are parts of the animal that are not usually eaten or are not expected in food. In some countries, edible portions of the thoracic and abdominal viscera (e.g., heart or liver) are considered, by definition, as meat. Therefore, all other **offal** would be inedible.
3. *Denaturant.* This term refers to any substance that changes the appearance, taste, or smell of a product and makes it unsuitable for food.
4. *Decharacterize.* This term refers to changing the physical appearance of a product, without further processing, in order to discourage its use as human food, e.g., staining condemned meat with dyes. Powdered charcoal can be used as an inexpensive alternative, but it can be washed off.
5. *Restricted products.* These are products that may not be sold without further processing, such as cooking or freezing.

Control of Condemned Products. Once a product has been condemned or deemed inedible at postmortem inspection, proper controls must be available to avoid subsequent mixing of these products with edible products inasmuch as they may appear similar. Unless the former are removed immediately, it may be difficult later to determine if mixing has occurred. Because inedible and condemned products are both considered unfit for human consumption, they can be disposed of in the same way. If the entire carcass, a cut, or boxed meat is to be retained for further inspection or analysis, it is placed in a lockable room or container (retained cage), the key being retained under the sole control of the inspector.

Methods Used to Destroy Inedible or Condemned Products.[128,178] There are four methods for destruction of inedible products.

1. **Hashing**. The hashing and crushing process involves grinding bones, meat, blood vessels, fat, etc., before rendering. This must be done in areas of the plant segregated from edible-product processing.
2. **Rendering**. Rendering (melting down) is practical only in high-volume plants (see Chapter 2). Hashing usually precedes rendering. Some rendered products (e.g., meat meal) are used in animal feeds.

3. **Incineration**. Incineration seldom is used because of cost and the problems of air pollution.

4. **Denaturing**. Application of a denaturing agent is a common method of disposing of these products. This is done by pouring or sprinkling the agent on the product. If the agent is applied to products in a barrel, it should be done in layers so not just the top is covered. Large cuts of meat and carcasses should be cut into or slashed before applying the agent. Cresylic disinfectants are commonly used denaturants because they are available in most plants for disinfecting trucks and premises.

Animal Food. In the United States, several conditions encountered in postmortem inspection require that the entire carcass be condemned for use as human food, although it may still be acceptable for animal food: for example, anasarca, emaciation, eosinophilic myositis, immaturity, nonseptic bruises, epithelioma, sarcosporidiosis, and unborn calves. The products must be separated from edible food products to avoid subsequent mixing. Decharacterizing with dyes provides a visible aid for separating products not further processed (i.e., not rendered). Some condemned livers can be used for animal foods. Conditions such as telangiectasis, "sawdust" liver, and livers with parasitic scars are acceptable. These must be slashed and decharacterized, however.

Postmortem Disposition of Infectious Conditions.   The conditions described herein are frequently encountered at postmortem inspection. The lymph nodes are key indicators of infection or neoplasia in the areas drained by them. If they are abnormally large, hemorrhagic, or abscessed, this should act as a red flag to the inspector.

Generalized Infectious Conditions.   There are three generalized conditions that may be encountered.

1. **Septicemia**. Septicemia is a generalized disease; the lesions vary but are often characterized by congestion, hyperemia, petechial hemorrhages, tissue edema, lymphadenopathy, and interference with normal blood clotting. Degenerative changes may be seen that are occurring in parenchymatous organs such as the liver, kidney, and spleen. Postmortem examination of these organs is essential. In this instance, the carcass is condemned because it is a potential health hazard to consumers.

2. **Pyemia**. Pyemia is generalized abscessation. The animal may have concurrent septicemia, but pyogenic organisms can be demonstrated. In addition to generalized abscesses, there may be hemorrhages involving the kidneys and/or lymph nodes. Lesions similar to septicemia may be noted in parenchymatous organs. The carcass must be condemned.

3. **Toxemia**. Toxemia results from systemic absorption of preformed toxins or

toxins that are produced originally at a local site of infection. The systemic changes, which are similar to those with pyemia and septicemia, result in carcass condemnation.

Pathologic Conditions of Bacterial and/or Viral Etiology.[217] There are several conditions of microbial etiology that require sound judgment when making a disposition at postmortem examination.

Although **bovine tuberculosis** is relatively infrequent in many areas, it remains important because it has not been eradicated and it poses a public health threat. TB is a granulomatous, inflammatory disease. The organism has a predilection for lymphoid tissues, and hence these are examined routinely at postmortem inspection. In cattle, the respiratory system is commonly involved. In the United States, granulomatous lesions found in TB reactor cattle are sent to the National Animal Disease Center, Ames, Iowa, for further investigation. In the abattoir, all lymph nodes from these animals are incised and examined. Infection with **mycobacteria in swine** is usually of the avian type and is associated with the gastrointestinal tract. Classic tubercles are not seen. It is important for farmers to segregate chickens from swine. Avian TB may be transmitted to hogs either by bird droppings or as a result of eating dead birds.

The carcass of a tuberculous animal can be passed for food without restrictions if the lesions are confined to a single organ system: for example, the mesenteric and cervical lymph nodes in swine. The gastrointestinal system is discarded and the remainder of the carcass and offal used for food. This might be regarded as an intermediate category. If the carcass is passed restricted (based on sites with lesions), it must be stamped appropriately and the meat cooked at a standardized time and temperature ($76.7°C$ [$170°F$] for 30 min). Cattle that are TB reactors and free of lesions are passed only for cooking. There are several circumstances in which cattle or swine carcasses must be condemned when lesions are noted at postmortem inspection: generalized lesions, antemortem pyrexia, or cachexia; lesions in muscle, bone, joint, or abdominal organ; extensive lesions in thoracic or abdominal cavity; and lesions that are multiple and active.

Arthritis often is of bacterial etiology. Enlarged joints may be congested, and there is hyperemia of an associated lymph node together with gross changes in synovial fluid. Arthritis and polyarthritis may be observed in animals at postmortem but are not reasons for condemnation unless associated systemic changes are present.

Although not all causes of **pneumonia** are infectious, a significant proportion are. The degree of lung involvement and the stage of the disease are important in determining carcass disposition. For example, a gray cast to the lung parenchyma indicates leucocytic infiltration, which may represent either a progressive disease process or convalescence. Therefore, the veterinarian must rely on associated signs when making an intelligent disposition. In this instance, antemortem inspection is important. Mediastinal lymph nodes and pleura are important indicators of the condition of the respiratory system. The gross appearance and extent of involvement of these structures aid in determining if a respiratory disease is localized or systemic.

Pericarditis is frequently caused by infectious organisms. The condition is encountered most often in cattle. The extent varies from a localized lesion to extensive systemic involvement. Disposition of the carcass depends on extent of involvement, that is, whether the condition is localized or generalized, acute or chronic, and active or inactive. Investigation to determine possible sites of secondary infection is important.

Actinobacillosis and **actinomycosis** are handled similarly to pericarditis. It must be determined if there is systemic involvement; if not, the affected part only need be excised.

Caseous lymphadenitis (**CLA**) is encountered in sheep. Depending on degree of involvement, the affected part may be excised or the entire carcass condemned.

Pathologic Conditions of Parasitic Etiology.[48,57,67,78,90,137,198,204]   Several parasitic conditions may be encountered in the abattoir, and they should be categorized according to the degree of involvement as (1) generalized, (2) local organ involvement with systemic manifestations, or (3) simply localized infection.

1. **Bovine taeniasis**. Bovine taeniasis ("cysticercosis" or "beef measles") is caused by the intermediate stage of the tapeworm *Taenia saginata.* Humans, the definitive host, contract the infection by eating meat of infected cattle. Special attention is given at postmortem to the heart and masseter muscles. In Europe, the diaphragm and esophagus also are routinely inspected. The cysts are destroyed by freezing bone-in carcasses at −10°C (15°F) or less for at least 10 d (or −18°C [0°F] for 3 d) or by freezing deboned meat at −10°C (15°F) or less for at least 20 d. The cysts are also destroyed when the meat is heated to 60°C (140°F), but this means of control cannot be regulated as effectively as freezing.

2. **Swine taeniasis**. Humans are the definitive host for the pork tapeworm and, more importantly, can also be the intermediate host. Disposition of carcasses affected with this disease is not the same as that for beef measles. If **any** lesion of taeniasis (*Cysticercus cellulosae*) is present, the carcass must be passed for cooking only. If excessive infestation is present, the carcass must be condemned.

3. **Swine ascariasis**. Normally, swine ascarids are localized in the intestine, but occasionally the worms occlude the common bile duct, causing obstructive jaundice, which requires condemnation of the carcass. Ascarids migrate through the liver during their development within the host and produce fibrous scars. The liver is edible, but the organ is usually condemned on aesthetic grounds.

4. **Tongue worms**. The parasite *Gongylonema pulchrum* inhabits the submucosa of the tongue of swine. Postmortem examination involves palpating the

tongue for the parasite, as well as abscesses. In some instances, large tongues must be split and each half palpated. If abnormalities are not detected, the tongue is passed for food after scalding.

5. **Trichinosis**. Trichinosis has been a major concern of consumers for many years. As a result of effective public education, its incidence has declined steadily in the United States. This parasite in swine is not detected readily at postmortem inspection. The requirement that all garbage for swine feeding be cooked and the application of newer testing methods have reduced the prevalence of trichinae in swine. In the United States, for control purposes, all pork is assumed to be infected. Proper preparation and cooking is relied on for effective control. For ready-to-eat products, one of four methods is utilized to destroy the trichinae: heating, freezing, salting and drying, or irradiation.

Pork must be heated to a temperature of 55°C (131°F) to kill trichinae; however, heating to 58.3°C (137°F) is required, which allows a safety margin. A combination of a lower temperature and an appropriate holding time may also be used.

The freezing temperature varies with the product and its salt content, but instant freezing to −40°C (−40°F) in the center of the meat is used to kill trichinae in some ready-to-eat products that will not be reheated.

Salting and drying are used primarily on products such as dry sausage. Drying by itself is not a reliable means of destroying trichinae, but in the presence of salt, the cysts become dehydrated by osmosis.

An irradiation dose level of 0.2–0.3 kGy prevents the maturation of encysted larvae in meat.[24]

The **pooled sample detection of trichinae** for the diagnosis of trichinosis has several essential components. First, sampling and analysis must be done rapidly before hogs are cut up and parts dispersed. Second, all incoming hogs must be identified so that trichina-positive animals can be traced to the farm of origin. The pooled sample technique involves pooling diaphragm samples from groups of 20 to 25 hogs, digesting the muscle, and examining the pooled samples microscopically for cysts. If the sample is positive, a second muscle sample is obtained from each hog and examined separately to identify the infected individuals.

Liver

1. **Abscess**. Approximately 10% of the cattle inspected in the United States have abscessed livers. These are condemned for human food, but some can be trimmed and used in animal food.
2. **Fascioliasis**. The liver fluke accounts for the condemnation of many livers. In the United States, any evidence of fluke infestation renders the liver unfit for human consumption. In other countries, however, flukes may be removed and the liver passed for food. Some flukes (e.g., *Dicrocelium* spp.) are transmitted by ingestion of raw liver. Because of the public health importance of flukes, infected livers should be condemned. In the United States, these livers usually

are condemned for aesthetic reasons because most liver is cooked before eating.

3. **Parasitic scars**. Scars caused by migrating parasites, other than flukes, are seen commonly in livers of sheep, swine, and cattle. The livers may be used for human food after the scars have been trimmed. The seasonal migration of first-instar larvae of *Hypoderma* spp. may also be a cause of the lesions in liver and other tissues.

Skin Conditions, Abscesses, and Eosinophilic Myositis. Many **skin conditions**, some of which are infectious, are considered as local lesions and are trimmed. If they are associated with systemic changes, the whole carcass is condemned.

Abscesses are usually localized and may be trimmed. Exudate from an abscess should not be washed off, as this disseminates the pus. The abscess and any exudate must be carefully excised to prevent carcass contamination.

Until recently, **eosinophilic myositis**, a condition seen usually in young, well-fattened cattle, was considered of uncertain etiology.[61,75,76,88,89,114,172,179,201] The lesions have been associated with the cysts of *Sarcocystis cruzi,* oocysts of which are shed in the feces of dogs. Other *Sarcocystis* spp. also may be involved. Lesions appear as yellowish green, spindle-shaped foci in the muscle fiber. *Sarcosporidiosis* is another name for the condition. Postmortem examination and disposition of affected meat is similar to that used for bovine cysticercosis.

Postmortem Disposition of Noninfectious Conditions

Emaciation and Asphyxia. Two noninfectious generalized conditions may be encountered at postmortem inspection: emaciation and asphyxiation (or suffocation).

Although **emaciation** is frequently suspected at antemortem inspection, it cannot be determined definitely until postmortem. It must be differentiated from cachexia and mere leanness. **Cachexia** is associated with chronic debilitating diseases whereas emaciation results from either inadequate caloric intake (starvation) or increased caloric demand, as a result of stresses such as cold weather; in other respects, the emaciated animal is essentially normal. In either emaciation or cachexia there is serous infiltration, or mucoid degeneration, of adipose tissue. A lean animal does not have this change and is normal and healthy with a minimum of fat stores. Emaciated animals are aesthetically unacceptable for food.

Asphyxiation (suffocation) is indicated when the tissues are engorged with blood. It results from improper sticking, after stunning, with subsequent inadequate bleeding. This condition is seen most commonly in hogs that drown in the scalding tank before hair removal. These animals are considered unsound and are, therefore, not accepted for food. Several lesions, such as petechial hemorrhages, may be masked by the retention of blood.

Pigmentary Conditions. Three pigmentary conditions may be encountered at postmortem: icterus, melanosis, and xanthosis. Pigments are produced either endogenously or exogenously.

1. **Icterus**. Icterus is the result of increased bilirubin in the body caused by a pathologic state in the hepatic or hemic systems. The affected animal is condemned. Icterus must be differentiated from similar-appearing conditions associated with diet or breed characteristics, such as seen in Jersey and Guernsey cattle. If the animal is icteric, tissues that should normally be white, such as the intima of large vessels, sclera, tendons, connective tissue, pleura, and joint surfaces, will have a distinct yellow color.
2. **Melanosis**. *Melanosis* refers to either an abnormal increase in, or an aberrant location of, melanin deposits. Melanin pigment is found normally in the skin, brain, tongue, and palate. If not excessive, melanin deposits found elsewhere can be trimmed, and the carcass passed for food. This condition must be differentiated from the neoplastic condition melanoma.
3. **Xanthosis**. Xanthosis, which develops from accumulation of waste pigments (lipochromes) in skeletal and cardiac muscles, is more common in older animals. Usually affected muscles can be trimmed and the remainder of the carcass used for food. In rare instances, xanthosis is so extensive that the carcass must be condemned.

Liver. There are many causes of *cirrhosis* of the liver, mostly noninfectious. These livers may be used for animal food but **not** human food. This condition is determined easily by incising the tough and fibrotic tissue.

Telangiectasis is a vascular dilation seen in bovine livers. Telangiectasis often is seen on the surface as red spots that extend into the parenchyma. With aging, the spots become somewhat larger, darker, and more indented. Disposition of a telangiectatic liver is similar to that of sawdust liver; i.e., if the foci are slight, they may be removed and the remaining liver used for food.

"Sawdust" liver is a term used in meat inspection for a liver, particularly a young well-fattened cow's liver, with characteristic small white or yellow necrotic spots on the surface. The etiology of this condition is uncertain, but it is thought to be related to telangiectasis or abscesses. However, many sawdust foci never become abscessed. At present, if the foci are slight, they may be removed and the remaining liver used for food.

Neoplasia. **Embryonal nephroma** is a well-circumscribed, benign tumor seen in swine, and less frequently in other species. Unless secondary changes are present, it is treated as a localized lesion, and the carcass is passed after trimming.

No matter how slight the extent, a carcass with **malignant lymphoma** must be condemned. Usually, involvement is generalized and the animal debilitated. Thus, recognition is not difficult. Involved lymph nodes have a claylike consistency with red streaks throughout.

Epithelioma, or squamous cell carcinoma (sometimes called "cancer eye"), is an eye tumor seen primarily in cattle. It must be differentiated from other ocular conditions such as traumatic injury or corneal dermoid. Animals with this condition may be consigned to slaughter subject to postmortem inspection when special attention is given to the parotid lymph nodes because this tumor usually metastasizes via this node to the atlantal and thoracic lymph nodes. Any extensions of this tumor beyond the eye is reason for carcass condemnation.

Miscellaneous Noninfectious Conditions. Several noninfectious conditions are encountered commonly at postmortem examination.

1. **Bruises**. Bruises are the result of traumatic injury with subsequent trapping of blood in and around muscle bundles and associated structures. Shortly after the injury, the regional lymph nodes begin to swell and become darkened. Bruised tissue may be trimmed out, with a wide margin, and the remaining normal tissue used for food.
2. **Pale, soft, exudative (PSE) pork**. PSE involves the muscle (see Chapter 2). The meat is considered wholesome and can be used as food if it is aesthetically acceptable. Occasionally, a slight odor is present, which can be detected after chilling the carcass for 24 h.
3. **Nephrosis**. In nephrosis, several systemic changes may occur depending on the duration and severity of the lesions. As severity increases, uremia becomes more pronounced. If the condition is advanced, there are secondary changes in the liver, spleen, etc., and the animal is unfit for food. If fat from a uremic animal is heated, it will emit a urinelike odor. Uremia may exist in the presence of only minimal gross lesions.
4. **Chemical residues**. Chemical residues are not obvious at postmortem examination. If an animal is suspected of containing residues (either from an evident injection site or suggestive history), the carcass is retained for subsequent laboratory analysis. Legal tolerance levels and withdrawal times are revised periodically, and it is the veterinarian's responsibility to be acquainted with current regulations.
5. **Sexual odor**. An offensive odor may be emitted when meat from adult male swine is cooked, whether they are boars, stags, or cryptorchids. Those that produce a strong sexual odor cannot be passed for food but may be incorporated in a comminuted product such as sausage (passed restricted). Boar meat often is sent directly to the rendering facility.
6. **Immaturity**. Meat from animals slaughtered at a very young age is usually pale in color and friable as a result of a higher water content than that found in meat from more mature animals.

Condemnation Basis of Common Noninfectious Conditions. Although the U.S. meat inspection program has done much to improve uniformity of inspection, the basis for dispositions may be obscure to those unfamiliar with the program. In 1968, the amendment to the WPPA stated that condemnations must be based on scientific facts or criteria. However, most commonly encountered noninfectious conditions are condemned for aesthetic reasons. Although the regulations do not so state, aesthetics is the basis for condemnation if a condition is offensive but not hazardous to human health.

Poultry Inspection[106,164,196]
Conditions Resulting in Carcass Condemnation

Avian Tuberculosis. Tuberculosis is almost always restricted to older fowl from areas with small farm flocks that are slaughtered in commercial abattoirs. *Mycobacterium avium* can infect mammals, including humans, swine, sheep, mink, and rabbits. Tuberculosis lesions in poultry are characterized by central caseation but are **not calcified**. Affected carcasses are condemned.

Leukosis (Lymphoid). Lymphoid tumors in chickens are usually caused by one of two viral agents: Marek's disease virus or the lymphoid leukosis virus group. The lesions may resemble those of tuberculosis, with tumors in the liver, spleen, and bone marrow.

With Marek's disease, changes such as thickening and enlargement of the nerves with loss of striations are usually detectable. Tumors in the ovary and testes may also be seen. Marek's disease also is manifested by changes in the feather follicles; only a few follicles or nearly the entire skin may be affected. Lymphoid leukosis is evidenced by an enlarged liver. There may also be bony changes. A carcass with evidence of leukosis is condemned.

Septicemia-Toxemia and Synovitis. Infectious diseases, other than tuberculosis and the leukosis complex, usually are classified as septicemia-toxemia or as synovitis. Any evidence of systemic involvement is reason for condemnation because there is a risk of transmitting infection to people eating the meat. Signs of systemic involvement are (1) dehydration, (2) color changes (of skin, liver, and other parts), (3) hemorrhages, (4) swelling of the liver (edges rounded rather than sharp), (5) swelling of the spleen, and (6) necrosis. Any one of these signs alone may not be sufficient for condemnation. In synovitis, the joints may contain thick, almost gelatinous, exudate tinged with blood. A purulent or caseous material may be found in some joints. Causes of the condition include species of *Mycoplasma, Staphylococcus, Salmonella,* and *Pasteurella.* Carcasses with synovitis are condemned.

Cadavers. A cadaver is a bird that died from causes other than normal slaughter procedures and is condemned.

Contamination. A carcass may be condemned for contamination if, for example, the skin was torn by the picking machine, which would allow nonpotable water from the scald vat or picking machine to get under the skin.

Airsacculitis.   This is an infection involving the air sacs. A severe case may be accompanied by pericarditis, perihepatitis, and peritonitis. The presence of these disease states results in condemnation of the carcass.

Carcass Temperature.   The temperature in a carcass that has reached the packing area must be 4.4°C (40°F) or less, one of the requirements designed to prevent microbial growth.

Labeling and Shipping of Meat and Meat Products

Essential Features of a Label.   For approval in the United States, the label must contain five pieces of information: product name, ingredients, firm's name and address, net weight, and inspection legend.

Product Names.   The product name must be the common name that would occur normally on the product or that fully describes it. This is important, for example, if buying luncheon meat. There is a difference between sliced ham and ham loaf!

Ingredients.   The ingredient statement must include **all** the ingredients included in the product. When more than one ingredient is used, the label must list them in order of decreasing amount.

Firm's Name and Address.   A complete address must be included on the label. It need not be the place where a particular product was produced; instead, it may be the name and address of the headquarters of a firm for whom the product is manufactured.

Net Weight.   The USDA regulations state explicitly the form in which net weight is indicated on a label. The requirements are as follows: (1) less than 1 lb, in ounces; (2) 1 lb but less than 4 lb, in ounces and also in pounds and a fraction or the decimal equivalent (e.g., 20 oz and also 1¼ lb **or** 20 oz and 1.25 lb); (3) 4 lb and over, in pounds and a fraction or the decimal equivalent (e.g., 5½ lb **or** 5.5 lb).

Inspection Legend.   The inspection legend is included in the label on all processed and packaged meat. A number in the legend is unique to the establishment at which the meat was inspected. The poultry inspection label is different in that the lettering is bolder than that used for meat and the letter *P* precedes the establishment number. The legend is circular for poultry and red meat. Only federally inspected meat may have this form. The legend for state-inspected meat and poultry must be an alternative form, such as an outline of the state.

Other Information as Required.   A warning statement, such as "Keep under Refrigeration," may be required on the label if the product has not been thermally processed to be commercially sterile. Such items are found in retail store self-service refrigerated display cases. Canned hams that are heated only to an internal temperature of 65.6°C (150°F) are not sterile, so the warning indicates the product must be refrigerated. Statements for heated smoked pork products (not canned) such as "cooked," "fully cooked," or "ready to eat" imply that the products have been heated to a temperature of 64.4°C (148°F) and will have a cooked appearance; that is, the meat has a cooked color and separates easily from the bone.

Labeling and Quality Standards. Improper labeling, whether intentional or accidental, can mislead consumers about product quantity or quality. The following are examples of misleading labeling.

Origin of the Product. It is important to indicate the origin of a product. Products, originally unique to a geographic area, but currently manufactured elsewhere, such as Italian, Swedish, or New England sausage must have the origin on the label. A product label with a term indicating a geographic significance is permissible provided the geographic term is qualified by a word such as *style, type,* or *brand* in the same size and style lettering as the geographic term.

Quality of the Product. A label should not misrepresent quality, as in instances in which imperfect bacon slices from ends and pieces are represented as a quality product.

Quantity of Product. If a product weighs between 1 and 4 lb, the label must represent the quantity with a dual declaration such as 20 oz (1 lb 4 oz). People are generally impressed with larger numbers, and thus some might be misled.

Nutritive Value of the Product. Generally, bacon contains relatively little red meat, which is the primary source of protein in the product. The statement "high in protein" is misleading, primarily because an established standard currently is not available to classify protein levels.

Color of the Label. Red lines in a transparent bacon wrapper make the label unacceptable because they make the product appear leaner than it actually is. Also, the transparent window must reveal the major portion of the length of a representative slice.

Proper Filling. Proper container filling is important to avoid misrepresentation. For example, if a jar contains Vienna-style sausages tightly packed against the wall, it may appear to contain more sausages than it actually does if the center is either empty or loosely filled.

Composition of the Product. Meat and poultry inspection regulations establish specific standards for each product produced in plants under federal inspection. For example, **ground beef** (or chopped beef) must consist of chopped fresh and/or frozen beef **without** the addition of beef fat and must not contain more than 30% fat. **Hamburger** must consist of chopped fresh and/or frozen beef **with or without** the addition of beef fat and must not contain more than 30% fat. The 30% fat for ground beef or hamburger established a standard of consumer expectancy.

Shipping.   Shipping, the final link in bringing food to retailers, is as important as processing if the consumer is to receive a wholesome product that has not become contaminated in transit to retailers. Before loading, trucks should be examined for the cleanliness of floors, walls, top hooks, and roof, and the drain holes blocked. If necessary, they should be washed and cleaned completely. If there are holes or protruding wood or metal, the truck should not be used as these conditions increase the probability of product contamination. The truck should be loaded in a manner that minimizes possibility of contamination. When carcasses are shipped, they should be hung from the overhead rail rather than laid on top of each other because hanging carcasses are contaminated less easily during transit and the air flow also helps to maintain temperature in refrigerated trucks. Carcass meat nearly always is boxed before shipment nowadays, however, which eliminates the risk of contamination.

Returned Products.   Some products are returned to the abattoir by dissatisfied purchasers. Before their reentry into the plant, an inspector must examine the products for wholesomeness. If the inspector cannot examine a product immediately, it should be placed temporarily in a retained cage maintained under refrigeration.

Acceptable Quality Level Program.   Because it is virtually impossible to inspect critically every carcass after it leaves the kill floor, the U.S. federal meat inspection program has developed the acceptable quality level (AQL) program in which carcasses are grouped in lots and the inspector selects sample carcasses at random and examines them very closely. Wholesomeness of a lot is determined by findings in this sample. For example, a few hairs on a carcass would be termed a minor defect, and the processor would have to reroute the whole lot of carcasses for subsequent trimming. A more serious defect, such as malignant lymphoma, would be termed critical, and all carcasses would be reinspected to determine disposition. A zero tolerance has been established for contamination of a carcass with feces, ingesta, or milk.

Egg Inspection[18,64,156,158,182,205,227,228]

Because of the numerous foodborne disease outbreaks associated with foods containing contaminated **egg products** during the 1940s–1960s, all egg products (whether liquid, dried, or frozen) prepared from shelled eggs must be pasteurized and are prepared in plants under government sanitary inspection.

For many years, gradings for size (weight) and quality (e.g., depth of air cell, which reflects duration and conditions of storage) were the only inspection criteria used for **shell eggs** sold for human consumption (see Chapter 2). Visible characteristics of the shell, such as cleanliness and absence of cracks, were the principal grading criteria related to food safety. Eggs with microbial or fungal growth or with developing embryos visible when candled were considered inedible. Although governmental agencies set the standards, the egg industry performed the grading. Industry-supported programs to eradicate *Salmonella pullorum* were initiated in the 1930s and required all **hatching eggs** to be obtained from tested flocks. Governments required health certification of source flocks for importation of hatching eggs. These requirements were the initial steps toward the health-related inspection of shell eggs that followed several decades later. The increase in the number of eggborne *S. enteritidis* outbreaks during the 1980s led to an increase in consumer concern regarding the safety of shell eggs. In the United States, public health and animal health agencies cooperated in tracing the source of eggs involved in foodborne salmonellosis outbreaks, and eggs from incriminated flocks were not allowed to be sold for human consumption. Disposal of these affected flocks was encouraged. The egg-producing industry has subsequently introduced a quality assurance program, including flock testing.

Benefits of Milk Inspection[1,93,183,224]

The principal objective of milk inspection is to safeguard human health from four major hazards: **milkborne infection, antibiotic residues, toxic chemicals,** and **hazardous radionuclides**. To meet this objective, there must be continuous surveillance of the product from the farm through retail sale. Although the principal objective of milk inspection is to protect human health from milkborne hazards, the consumers, distributors, and producers derive important additional benefits. Recognition of the safety of milk fosters consumer acceptance of this relatively inexpensive food, whereas distributors benefit economically from increased shelf life resulting from the high sanitary standards required for production and processing.

Milk Inspection in Australia
Historical Issues in Veterinary Public Health.[9,68,109] Tuberculosis of dairy
cattle had become widespread in Australia by the mid-to-late nineteenth century and was estimated to be present in 25% of slaughtered cattle. High stocking rates enhanced opportunity for transmission and infection in children, with enlarged calcified cervical lymph nodes from drinking raw milk a well-known clinical entity.

The principle of pasteurization was first applied to preserve beer and then to improve the keeping quality of milk in the 1880s, which increased shelf life during distribution and storage in homes, prior to use. Pasteurization of milk was also considered important in preventing infectious enteritis of infants. The prime concern of early legislation on milk quality was to apprehend and fine producers and distributors for adulterating milk. Because considerable quantities of raw milk were consumed, the government of Victoria ran education programs warning the public of the dangers of drinking raw milk.

Early in the twentieth century, legislation for production and sale of milk was gradually transferred from state health to state agricultural control. Lay inspectors were employed to examine the udders and lymph nodes of cattle and refer those suspected to have tuberculosis for veterinary examination. More intense supervision of industry practices led to improved hygiene in dairies, more effective culling of cows clinically affected with tuberculosis, and reduced prevalence of tuberculosis in swine that was attributed to decreased shedding of tubercle bacilli in cows' milk fed as swill.

Milk from unsanitary dairies was responsible for transmitting typhoid and streptococcal disease in early times, but this ceased as dairy hygiene improved gradually over the years. The last milkborne epidemic of typhoid occurred in 1943, with 440 cases attributed to a single human carrier bottling raw milk in a dairy. Raw cow's milk was undoubtedly the source of infection in some cases of human brucellosis.

Prior to World War II, testing for tuberculosis was done on a voluntary basis using Koch's old tuberculin; in some states agricultural departments had no legal power to compel farmers to have their cattle tested unless pigs found with tuberculosis at slaughter could be traced back to farms where cattle were raised. Where notified cases of extrapulmonary human tuberculosis were suspected to have occurred by drinking raw milk of identified herds, health authorities could order those cattle to be tested (by agricultural departments). It is of interest that one impetus for increasing testing of herds in Queensland was the arrival of the U.S. Armed Forces in Brisbane in 1942. Veterinarians on General MacArthur's staff refused to permit milk from nontested cattle to be passed for U.S. Army consumption. The veterinary profession used this opportunity to overcome the resistance of industry and consumers to make pasteurization of milk compulsory in southeast Queensland. The recognition and culling of infected cattle, the use of tuberculin testing, and later the outlawing of the commercial sale of raw milk significantly reduced the incidence of extrapulmonary tuberculosis (due to *Mycobacterium bovis)* in humans.

Postwar disease control programs had successfully eradicated tuberculosis from dairy herds by the late 1960s and gradually were extended to the beef herd through a national eradication policy (*c.* 1976), using more-sensitive and more-specific tuberculins, particularly purified protein tuberculin derived from *M. bovis* (PPD). This policy was based upon first obtaining data on prevalence of tuberculosis then regularly testing herds, slaughtering cattle reacting to tuberculin, tracing cattle found infected during slaughter to their herd of origin, complete mustering of herds for testing, and maintaining effective quarantine to prevent reinfection in clean herds and areas. In the terminal stages of eradication, tuberculin testing was complemented by a gamma interferon test to detect infection in anergic cattle. Australia has been declared provisionally free of bovine tuberculosis since 1992, although monitoring for presence of infection continues.

In New Zealand, a similar approach to eradication of bovine tuberculosis has been thwarted by a well-established and self-perpetuating infection in opossums, which are a feral reservoir of infection for cattle.

While anthrax had become widespread in the late nineteenth century, the milk of dairy cattle developing anthrax posed little health risk to consumers as anthrax bacilli reach the udder only shortly before death, when the cow is discernibly ill. In herds where anthrax is diagnosed today, monitoring the body temperature of all cows twice daily can identify any whose milk is at risk. Cows with elevated temperature must be milked by hand and the milk autoclaved and discarded.

The transition from hand to machine milking brought an increase in the prevalence of mastitis. The importance of detecting, controlling, and preventing udder infections has been succinctly summarized by Hughes:

> C.D. Wilson's 5-point program for controlling mastitis was commonly practised. New Zealand has been a leader in mastitis research from early times. By the 1930s a system for controlling mastitis based on disinfection of teats after milking and on the Whiteside test was widely practised in Australasia, whereby cows with the more significant reactions were milked last. Even by then, experimental and commercial vaccines had been tried extensively without benefit. Contributions to veterinary public health by Australian scientists include the pharmacodynamic studies of the therapeutic value of penicillin for mastitis, incorporation of dyes into intramammary antibiotic infusions to encourage farmers to discard milk containing antibiotics, modification of teat cups by installing baffle plates to prevent backflow of milk during milking, and development of herd health programs for the control of bovine mastitis. As part of health programs, dry cow intramammary therapy is now widely practised throughout Australasia.[109]

Having reduced the prevalence of bacterial infections, industry has been prohibited from using teat dipping prior to or during milking because of the potential for antimicrobial and detergent residues (including iodine) in milk. Current practice is to not disinfect the teats at all or to dry them with paper towels after spraying them with clean water. Teat dipping after milking is encouraged.

Quality Control of Dairy Products. Because of the unacceptably high prevalence of human infections with bovine tubercle bacilli in the early 1900s, milk laboratories were gradually established in all states to monitor milk quality and to conduct supporting research. By the 1940s each state had developed its own microbiological standards for quality and safety of milk. At that time continuous high-temperature short-time (HTST) pasteurization was in common use; the holding temperature was increased from 71.1°C (160°F) to 71.5°C (160.7°F) for 15 s upon the discovery that the lower temperature did not kill *Coxiella burnetii*. Today's standards are for milk to be heated to 72°C (161.6°F) for not less than 15 s and cooled immediately to 3.5°C (38.3°F). This regime also inactivates bovine leucosis virus in milk. The holder process is no longer used except by some cream processors.

Mechanical bottling of milk brought new challenges to industry, the most important development being the design of an inexpensive in-line procedure for sanitizing the final rinse water without overdosing it with chlorine.

With increased use of hot water in dairies to complement pasteurization procedures in factories, keeping quality of milk was enhanced by the installation on farms of refrigerated stainless steel vats for holding bulk milk prior to its transportation in bulk tankers. Samples of farm milk obtained at the time of collection were assayed and graded for thermoduric bacteria. To improve milk quality, farmers used iodophor disinfectants (since discontinued) for udders and milking equipment to reduce the count of thermoduric bacteria and to prevent the buildup of deposits that enhance bacterial contamination.

From the early 1980s industry increasingly relied upon the total plate count of bacteria in farm milk, adopting a standard of ≤50,000 organisms per ml of milk (after incubating at 30°C [86°F] for 3 d). This test provides the best assessment of bacteria in the sample, including psychrotrophic bacteria that multiply and can cause spoilage (tainting) of milk at refrigerated temperatures. The combined effects of improvements to milk quality have substantially increased its refrigerated shelf life to 10 d or longer after delivery to consumers.

Prior to World War II much of the cheese produced was of poor quality because of inadequate lactic acid production by cheese starter cultures that had originally been chosen to develop flavor in cream prior to churning it into butter. Quality improved with the use of pure cultures of *Streptococcus cremoris*. Considerable research effort was put into investigating failures in fermentation, identifying factors such as the role of bacteriophage lysing *S. cremoris,* the production of antimicrobial factors (nisin) by some bacteria, and the presence of adventitious antibacterial substances (e.g., penicillin and broad-spectrum antibiotics) in milk. Milk is now routinely tested for the presence of penicillin; the most common procedure being a disc assay test using *Sarcina lutea* that at pH 7.0 can detect penicillin at concentrations of 0.0015 IU/ml or greater (modified Delvotest).

Australia has been remarkably free of incidents of zoonotic disease transmitted via the medium of dairy products. In 1977 a serious epidemic of *S. bredeney* gastroenteritis (57 cases) occurred in infants ingesting infant formula containing contaminated milk powder. Contamination was traced to intermittent seeding of sterile product with *S. bredeney* harbored in insulation behind a crack in a steel drying cone. This had a cathartic effect upon the dairy industry, and many procedures were initiated to prevent further contamination (routine microbiological sampling of dried milk powder from all factories, education of factory workers about preventing contamination, and certification of export product to Australian microbiological standards). In the event of any contamination being detected, all products from the factory are detained, the source of infection ascertained, and the factory premises cleaned and disinfected thoroughly.

After several international epidemics of foodborne listeriosis were traced to pasteurized milk and unpasteurized cheese, the Australian dairy industry had become acutely aware that *Listeria* spp. multiply in food under some conditions of manufacture and storage. *Listeria* was later isolated from pasteurized milk in northern Queensland, and these findings led to authorities adopting comprehensive standards for *Listeria* clearance from dairy products, based on HACCP principles.

Current Trends in Industry. Approximately 6 billion l (1.6 billion gal) of whole milk are produced annually in Australia by a dairy industry that is distributed over a wide range of climates and ecological systems, resulting in marked fluctuations in seasonal supply. Marketing of fresh milk is controlled by state laws, while manufactured dairy products are controlled by federal law supplemented by state legislation. These laws work in tandem as milk may be switched to either market according to commercial opportunities.

For 20 yr the industry has been undergoing major restructuring, resulting in fewer but larger farms. It is now moving rapidly toward deregulation (1997) of government intervention, which is expected to reduce the cost of dairy products to consumers and enable the industry to compete more effectively in international markets. Decisions by producers to increase or decrease production can thus be made upon fluctuating market prices to suit their own enterprises. The current industry satisfies all domestic demand for milk products, exports about 90% of its manufactured milk powder to Japan and nearby regions, and is establishing a growing market for short shelf life dairy products in Southeast Asia.

Many producers utilize government herd recording services to improve productivity and to select bulls for artificial insemination programs. Queensland surveys show that those who use artificial insemination (over 75% of cows) and herd recording have 48% higher production per cow than those who don't. The information received provides basic data for genetic improvement of the national herd through an Australian Dairy Herd Improvement Scheme.

Occupational Health and Safety. With the eradication of bovine brucellosis from Australia in 1989, leptospirosis remains the only prevalent zoonosis with serious health implications for dairy workers. The design of herringbone dairies with pit access (and rotating dairies on raised platforms) markedly increases the exposure of dairyhands to the urine of infected cattle; many dairies have installed metal barriers behind the stalled cows to minimize splashing of urine and dispersal of microbes in aerosols.

Milk Inspection in Canada[36,37,132,133]

Milk inspection in Canada is a joint federal and provincial responsibility. The federal act governing milk quality, the Canada Product Standards Act, sets the minimum standards to which provincial regulations must conform. Standards set for milk are similar to those put forth in the Public Health Service milk ordinance in the United States.

Milk Marketing.   The Canadian dairy industry has been a supply-managed system since the early 1970s, with federal jurisdiction over trade between provinces and foreign trade. There are two markets for milk in Canada: the industrial market, which is federally regulated, and the fluid market, which is provincially regulated. The Canadian Milk Supply Management Committee, which has representation from producers and governments in all provinces except Newfoundland, sets a national production target that is adjusted periodically to reflect changes in demand. Each province allocates its share of the national quota to its producers according to its own policies. The target price for industrial milk is the level of return efficient milk producers should receive to cover cost of milk production and is based on an annual survey of 350 farms in Ontario, Quebec, and New Brunswick.

Each province has the responsibility for meeting the table milk and fresh cream requirements of its own residents. By convention (not by contract or written agreement), fluid milk products in one province are not shipped to another province, and hence there is no interprovincial competition in fluid milk products. The authority to market milk is delegated by provincial governments primarily through milk marketing boards financed and managed by producers. The degree of government control varies by province. Provinces establish price levels for fluid milk in a manner similar to that of the federal government for industrial milk. Provincial agencies may set fluid milk prices at the producer, wholesale, or retail levels and, in some instances, at all three levels.

The purpose of supply management is to provide a balance between the supply of raw milk and the demand for milk and milk products at the national level. The Canadian market is primarily supplied by Canadian production, except for a few cheeses and other products not available from domestic sources. Because of this, Canada has a wide range of measures to monitor and control imports of dairy products.

Responsibility for Quality Control of Milk.   Provincial departments of agriculture test milk from the farm for unacceptable bacteria levels (e.g., Standard Plate Count and Plate Loop Test), presence of added water, and inhibitors. On-site inspection of the farm is part of the provincial inspection process, with the frequency of inspection and testing differing from province to province. Each province may perform additional tests. For example, British Columbia tests for pesticide residues. In most instances, milk quality affects the price producers get for the milk through a penalty system. Provincial inspectors also monitor products for quality and safety in processing plants that do not ship out of province.

Processing plants for industrial milk (destined for powdered milk, butter, cheese, etc.), which ship out of province, are also federally registered and inspected under the mandate of Agriculture Canada's Food Inspection Directorate. In Ontario and Quebec, provincial inspectors conduct federal inspections, subject to federal audit. In other provinces, there is cooperation between federal and provincial inspectors, especially if there is a particular problem in a plant. Inspections in the plant, in addition to checking for coliforms, also check for toxins such as pesticides and herbicides.

Testing in retail outlets is undertaken nationally by Consumer and Corporate Affairs in Canada, in conjunction with Health and Welfare Canada and Agriculture Canada. Testing is done randomly to verify product composition. Some provinces also monitor product on the retail shelf.

Milk Inspection in the United Kingdom

Milk was recognized as an important source of *Mycobacterium bovis* for the human population in the early part of this century although it was not until 1925, when the Tuberculosis Order of 1925 became law, that any serious attempt was made to control the problem. The order made provision for the slaughter, with payment of compensation, of cattle giving tuberculous milk or affected with tuberculosis of the udder or tuberculous emaciation. This order had relatively little effect on the prevalence of tuberculosis in the cattle population because it was purely a public health measure. However, it did have an important side effect in that many local authorities set up a veterinary service to implement the order. These veterinarians subsequently joined the Ministry of Agriculture and Fisheries in 1938 when the Animal Health Division was formed. The program for eradicating *M. bovis* infection from cattle commenced in the United Kingdom during 1935, and it was essentially complete by the end of the 1960s although infection still persists in certain parts of the country, notably the Southwest. Pasteurization of milk was introduced at about the same time, principally to control the spread of *M. bovis* although it had the added advantage of killing pathogens such as *Brucella abortus* and the salmonellae. The legislation on milk hygiene has been updated recently by the Dairy Products (Hygiene) Regulations of 1994. Up to this time, the hygiene legislation only applied to cows' milk. The new regulations bring milk from goats, sheep, and buffaloes within the law for the first time. Cows' and buffaloes' milk must come from herds that are tuberculosis-free and brucellosis-free whereas milk-producing goats and sheep must only be brucellosis-free. Among other things, the regulations make it an offense to sell milk from animals that have signs of disease communicable to humans, enteritis with diarrhea and fever, or a vaginal discharge or a recognizable inflammation of the udder. It is also an offense to sell raw milk that contains added water, antibiotics, pesticides, or residues of substances having a pharmacologic or hormonal action. The regulations also set standards for the cellular and microbial content of the milk.

The production of cows' milk for human consumption was among the first sectors of the food industry to be legislated for on hygienic matters. The Dairy

Product (Hygiene) Regulations of 1994, which put into effect Directive 92/46/EEC, is the principal piece of legislation in England and Wales concerning milk hygiene. The regulations are enforced by the Dairy Hygiene Inspectorate of the Ministry of Agriculture, Fisheries and Food (MAFF), with checks on animal health carried out by part-time Local Veterinary Inspectors, usually private practitioners. Dairy farms (farms producing cow, sheep, goat, and buffalo milk) have to be registered as "production holdings" by the MAFF, and this is conditional on the attainment of structural standards, the provision of safe and adequate water supplies, and the maintenance of effective hygienic equipment and milking routines. Facilities suitable for isolating or separating diseased animals have to be provided. Dairy farms are inspected periodically by the MAFF and may be deregistered if any breaches of the regulations are not corrected within a specified time.

Premises where milk undergoes some form of heat treatment and further processing have to be registered as "dairy establishments" by the local authority. Milk for human consumption may be sold under a number of designations, e.g., raw, pasteurized, sterilized, and ultra-heat-treated (UHT). The regulations set down the conditions that have to be met in order for these designations to be used.

The bulk raw milk from dairy farms is tested at frequent intervals for the presence of antibacterial substances by the creamery. Any contaminated batch of milk is discarded, and further price penalties are imposed on the producer if there are repeated failures.

Milk Inspection in the United States[230]

In addition to economic regulations, there are sanitary regulations that have been (and in some areas still are) artificial trade barriers. Within the United States there are 20,000 state, county (or parish), and local (or municipal) health and sanitation authorities. Plants processing milk for fluid consumption within these authorities are inspected an average of 24 times/yr (in one state the average was 85), although the 1975 Public Health Service milk ordinance recommends once every quarter. This excessive number of inspections occurs because more than one agency is involved in an inspection program and has cost milk producers, processors, and distributors more than $1 million/yr in unnecessary expenditure.

Public Health Service Standards for Milk.   The Public Health Service (PHS) milk ordinance provides chemical, bacteriologic, and temperature standards as well as sanitation requirements for production and processing of Grade A raw and pasteurized milk and milk products. Some processors may offer monetary incentives for producers to meet more stringent standards to improve milk quality.

Grade A Raw Milk for Pasteurization.   Grade A raw milk for pasteurization must be cooled to 7.2°C (45°F) within 2 h after milking and maintained at that temperature or less until processed; milk from an individual producer cannot have more than 100,000 bacteria/ml or, after mixing with milk from other producers, not more than 300,000 bacteria/ml before pasteurization. There must be no detectable antibiotic residues.

Grade A Pasteurized Milk and Milk Products. Grade A milk after it has been pasteurized must be cooled to 7.2°C (45°F) or less and maintained at that temperature. There can be no more than 20,000 total bacteria/ml with no more than 10 coliform bacteria/ml. The phosphatase test must be negative.

Grade A Pasteurized Cultured Products. For Grade A pasteurized cultured products such as buttermilk or cottage cheese, there is no limit on the total bacterial count, but the coliform count must be less than 10/ml. The temperature and phosphatase requirements are the same as for pasteurized milk.

Responsibility for Quality Control of Milk. Primary responsibility for control of milk quality within a state rests with the state. A state agency, usually the health department, administers the program. In many states, actual inspection may be performed by personnel in a local agency, usually the health department or department of agriculture. This function has been delegated to the local agency by the state, which retains primary responsibility.

Fish and Shellfish Inspection[5,22,43]

Fish production and processing are regulated differently from any other food of animal origin. Organoleptic inspection of individual fish is not a part of most inspection programs. The majority of fish consumed by humans is caught in the wild, and commerce in seafood among all fishing nations is conducted in a complex international marketplace. In the United States, for example, over half the fish consumed comes from foreign sources. Imported fish must be inspected to protect consumers, and fish for export must be inspected to demonstrate product wholesomeness and safety to international markets. Fish inspection systems operate under standards that relate to all phases of production and processing: the microbial and chemical quality of harvest waters, the cleanliness of fishing vessels and processing plants, and the marketing of the final product.

Standards. Regulatory limits are set for microbial and chemical quality of seafood, the water from which it is harvested and, in some cases, for disease agents, particularly parasites. International standards are set by the Codex Alimentarius Commission (CAC). Agreement on an international standard is a very slow process. Currently the standards are limited to organic and inorganic chemical contaminants of fish and shellfish and cover less than a dozen contaminants, primarily heavy metals, DDT, and other pesticide residues. Acceptance of the standards is left up to individual member countries. The United States has accepted and taken action on approximately one-third of the current Codex recommendations. Some countries, including Denmark, Norway, Iceland, New Zealand, the United States, and Canada, have set limits on contaminants, including biohazards, not covered by the CAC.

Fish and Shellfish Inspection in the United States. Fish inspection in the United States comes under federal, state, and private (through industry compliance programs) jurisdictions. The federal government, through the Food and Drug Administration (FDA), the Environmental Protection Agency (EPA), and the National Marine Fisheries Service (NMFS) of the Department of Commerce, sets guidelines for surveillance and control. Environmental monitoring of water quality is a shared responsibility between federal and state governments, through the EPA and the National Shellfish Sanitation Program (NSSP), which monitors contaminants in shellfish-harvesting waters. The FDA is the agency with primary authority for seafood inspection. FDA inspection covers domestic fish plants, import control inspection, routine surveys for residues and chemicals, and product inspection for the NSSP. Sampling is an important feature of all regulatory programs and is focused on areas where the probability of a problem is highest. For example, in its product surveillance programs the FDA samples nearly five times as many imports as it does domestic fish products. The NMFS conducts a voluntary fee-for-service inspection program for the fish-processing industry. The goal is to establish product quality and safety, but the emphasis is on quality. The program is tailored to individual producers and may include vessel and plant sanitation inspection and product evaluation.

Because of regional differences in fish production and consumption, and the need to customize programs to address these differences, state regulatory programs play an important role in seafood safety. For example, *Vibrio vulnificus* is a problem in regions on warm waters and areas of the country where oysters are consumed raw. Geographically, the area producing the highest risk in the United States is in the coastal waters of the Gulf of Mexico. Red tides, which are associated with paralytic shellfish poisoning, are quite variable in occurrence, and more intensive surveillance is done by the states experiencing this phenomenon.

There are some industry-driven programs that have a regulatory component. The Mississippi Prime Program is a catfish promotional program that promotes consumer confidence in cultured catfish. A component of the program is compliance with the NMFS plant inspection and sanitation guidelines, with periodic and random reinspection.

Fish and Shellfish Inspection in Canada. The Canadian seafood inspection program is one of the most advanced in the world. It is an HACCP-based system that covers quality and safety inspection of fishing vessels, landing sites, and processing facilities. Standards of compliance have been established for a broad list of microbial and chemical contaminants, including agricultural chemicals and their derivatives. It is administered by the Department of Fisheries and

Oceans (DFO) under the authority of the Fish Inspection Act of 1970. Under this act and its regulations both fishing vessels and processing plants must maintain the same high standards of sanitation. Critical findings at the time of inspection can result in a failure of compliance or, if frequent, mandatory inspection, the cost of which is borne by the processor. Thus, financial resources are directed to problem areas. The same standards are applied to domestic and import plants and producers. The approximate cost of the service is 1 cent per pound of product. The Canadian program is directed at process control, thus preventing rather than detecting problems in the final product. This saves money for producers by avoiding detection of problems after value has been added to the product and has proven to be simpler and more efficient to conduct than a program aimed at the final product.

Fish and Shellfish Inspection Internationally.   Because of extensive international trade in both wild caught and farmed-raised seafood, and because of the growing importance of trade agreements between North America and Europe, international quality and safety standards are receiving increased attention. For example, the North American Free Trade Agreement (NAFTA) calls for the harmonization of safety and quality inspection procedures, which will most likely require significant changes in the American seafood inspection system. The same is true within the European Union (EU), which is driven by the relatively high standards set by Denmark and Norway. In order to maintain export trade in the more lucrative markets, other countries will have to come into compliance.

Bibliography

1. 3-A Sanitary Standards Committee. Sanitation in Dairy Equipment. In *3-A Sanitary Standards.* Des Moines, Iowa: International Association of Milk, Food and Environmental Sanitarians.

2. Adams, C.E. 1990. Use of HACCP in meat and poultry inspection. *Food Technol.* 44:169–170.

3. Agarwal, V.K. (ed.). 1992. *Analysis of Antibiotic/Drug Residues in Food Products of Animal Origin.* New York: Plenum Press.

4. Agriculture Canada. 1983. *Meat Hygiene Manual.* Ottawa, Ontario: Agriculture Canada, Food Protection and Inspection Branch.

5. Ahmed, F.E. 1991. *Seafood Safety.* Committee on the Evaluation of the Safety of Fishery Products. Washington, D.C.: National Academy Press.

6. Amstutz, H.E. (ed.). 1980. *Bovine Medicine and Surgery.* 2d ed. 2 vols. Santa Barbara, Calif.: American Veterinary Publications.

7. Andriessen, E.H. 1987. *Meat Inspection and Veterinary Public Health in Australia,* Sydney: Rigby, Chatswood.

8. ———. 1992. *Risk Management.* Proc. Aust. Vet. Publ. Health, AQIS. Canberra.

9. Anonymous. 1944. *Conference on Milk.* Proc. Aust. Assoc. Sci. Work. Queensland, Brisbane.

10. ———. 1984. *Livestock Feed Additives. Their Recommended Claims, Use Levels, and Limitations.* Dept. Prim. Ind., Aust. Bur. Anim. Health, Aust. Gov. Publ. Serv. Canberra, 16 pp.

11. ——. 1994. *Risk Assessment Models of the Animal and Plant Risk Assessment Network.* Animal and Plant Health Directorate, Food Production and Inspection Branch, Agriculture and Agri-Food Canada. Ottawa.

12. Archer, D.L. 1989. Food counseling for persons infected with HIV: Strategy for defensive living. *Public Health Rep.* 104:196–198.

13. Association of Medical Milk Commissions. 1972. *Methods and Standards for the Production of Certified Milk.* Association of Medical Milk Commissions, Inc., 1824 N Hillhurst Ave., Los Angeles, Calif. 90027.

14. Bauman, H. 1990. HACCP: Concept, development, and application. *Food Technol.* 44:156–158.

15. Bennett, R. 1990. A negative residue test that wasn't. *Dairy Herd Manage.* 27:53–55.

16. Berends, B.R., J.M.A. Snijders, and J.G. van Logtestijn. 1993. Efficacy of current EC meat inspection procedures and some proposed revisions with respect to microbiological safety: A critical review. *Vet. Rec.* 133:411–415.

17. van den Berg, M.G. 1986. Quality assurance for raw milk in the Netherlands. *Neth. Milk Dairy J.* 40:69–84.

18. Bergquist, D.H. 1979. Sanitary processing of egg products. *J. Food Prot.* 42:591–595.

19. Berkowitz, D.B. 1990. Immunoassays in meat inspection: Uses and criteria. *Am. Chem. Soc. Symp. Ser.* 1990:15–20.

20. Blackmore, D.K., and M.W. Delany. 1988. *Slaughter of Stock: A Practical Review and Guide.* Publ. 118. Palmerston North, New Zealand: Massey University.

21. Blood, D.C., J.A. Henderson, and O.M. Radostits. 1989. *Veterinary Medicine.* 7th ed. Philadelphia: Lea and Febiger.

22. Bonnell, A.D. 1994. *Quality Assurance in Seafood Processing: A Guide.* New York: Chapman and Hall.

23. Boyle, D.L., J.N. Sofos, and G.R. Schmidt. 1990. Growth of *Listeria monocytogenes* inoculated in waste fluids collected from a slaughterhouse. *J. Food Prot.* 53:102–104, 118.

24. Brake, R.J., K.D. Murrell, E.E. Ray, et al. 1985. Destruction of *Trichinella spiralis* by low-dose irradiation of infected pork. *J. Food Safety* 7:127–143.

25. British Veterinary Association. 1991. A policy for a unified meat service. *Vet. Rec.* 129:122–123.

26. Bryan, F.L. 1981. Hazard analysis critical control point approach: Epidemiologic rationale and application to foodservice operations. *J. Environ. Health* 44:7–14.

27. ——(chrmn.). 1988. *Procedures to Investigate Foodborne Illness.* 4th ed. Ames, Iowa: International Association of Milk, Food, and Environmental Sanitarians, Inc.

28. ——. 1990. Hazard analysis critical control point (HACCP) concept. *Dairy Food Environ. Sanit.* 10:416–418.

29. ——. 1991. Teaching HACCP techniques to food processors and regulatory officials. *Dairy Food Environ. Sanit.* 11:562–568.

30. ——. 1992. *Hazard Analysis Critical Control Point Evaluations.* Geneva: World Health Organization.

31. Buchanan, R.L. 1990. HACCP: A re-emerging approach to food safety. *Trends Food Sci. Technol.* 1:104–106.

32. Burger, J., K. Staine, and M. Gochfeld. 1993. Fishing in contaminated waters: Knowledge and risk perception of hazards by fishermen in New York City. *J. Toxicol. Environ. Health* 39:95–105.

33. Buttery, P.J., D.B. Lindsay, and N.B. Haynes (eds.). 1986. *Control and Manipulation of Animal Growth.* London: Butterworths.

34. Carlsson, A., and L. Bjorck. 1992. Liquid chromatography verification of tetra-

cycline residues in milk and influence of milk fat lipolysis on the detection of antibiotic residues by microbial assays and the Charm II test. *J. Food Prot.* 55:374–378.

35. Canada Agricultural Products Act. 1988. *Chapter 27.* Ottawa, Ontario.

36. Canada Agricultural Products Standards Act. 1979. *Dairy Products Regulations. Canada Gazette Part II.* Vol. 113. Ottawa, Ontario, 4260–4314.

37. Canada Dairy Commission. 1989. *The Canadian Dairy Industry.* Ottawa, Ontario.

38. Chagonda, L.S., and J. Ndikuwera. 1989. Antibiotic residues in milk supplies in Zimbabwe. *J. Food Prot.* 52:731–732.

39. Charm, S.E. 1978. *The Fundamentals of Food Engineering.* 3d ed. Westport, Conn.: AVI Publishing.

40. Clesceri, L.S. (ed.). 1989. *Standard Methods for the Examination of Water and Wastewater.* 17th ed. Washington, D.C.: American Public Health Association.

41. Collins-Thompson, D.L., D.S. Wood, and I.Q. Thomson. 1988. Detection of antibiotic residues in consumer milk supplies in North America using the Charm II procedure. *J. Food Prot.* 51:632–633, 650.

42. Conacher, H.B.S., and J. Mes. 1993. Assessment of human exposure to chemical contaminants in foods. *Food Addit. Contam.* 10:5–15.

43. Connell, J.J. 1980. *Control of Fish Quality.* 2d ed. Farnham, Surrey, England: Fishing News Books.

44. Corlett, D.A., Jr. 1991. Regulatory verification of industrial HACCP systems. *Food Technol.* 45:144–146.

45. ———. 1992. Importance of the hazard analysis and critical control point system in food safety evaluation and planning. *Am. Chem. Soc. Symp. Ser.* 1992:120–130.

46. Corrigan, P.J., and P. Seneviratna. 1990. Occurrence of organochloride residues in Australian meat. *Aust. Vet. J.* 67:56–58.

47. Crawford, L.M., and D.A. Franco (eds.). 1994. *Animal Drugs and Human Health.* Lancaster, Pa.: Technomic Publishing.

48. Dada, B.J.O. 1980. Taeniasis, cysticercosis and echinococcosis/hydatidosis in Nigeria. II—Prevalence of bovine and porcine cysticercosis, and hydatid disease in slaughtered food animals based on retrospective analysis of abattoir records. *J. Helminthol.* 54:287–291.

49. Daly, C.C., E. Kallweit, and F. Ellendorf. 1988. Cortical function in cattle during slaughter: Conventional captive bolt stunning followed by exsanguination compared with shechita slaughter. *Vet. Rec.* 122:325–329.

50. Davies, E.B. 1993. Law enforcement: Carrot or stick? *Vet. Rec.* 132:465.

51. Dawson, P.S. 1992. Control of *Salmonella* in poultry in Great Britain. *Int. J. Food Microbiol.* 15:215–217.

52. Delahaut, P.H., M. Dubois, I. Pri-Bar, et al. 1991. Development of a specific radioimmunoassay for the detection of clenbuterol residues in treated cattle. *Food Addit. Contam.* 8:43–53.

53. Delany, M.F., J.U. Bell, and S.F. Sundlof. 1988. Concentrations of contaminants in muscle of the American alligator in Florida. *J. Wildl. Dis.* 24:62–66.

54. Dixon, B.R. 1992. Prevalence and control of toxoplasmosis—a Canadian perspective. *Food Control* 3:68–75.

55. Dubbert, W.H. 1988. Assessment of *Salmonella* contamination in poultry—past, present, and future. *Poult. Sci.* 67:944–949.

56. Dubey, J.P. 1988. Long-term persistence of *Toxoplasma gondii* in tissues of pigs inoculated with *T. gondii* oocysts and effect of freezing on viability of tissue cysts in pork. *Am. J. Vet. Res.* 49:910–913.

57. Dunn, A.M. 1978. *Veterinary Helminthology.* 2d ed. London: William Heinemann Medical Books.

58. Dunn, C.S. 1990. Stress reactions of cattle undergoing ritual slaughter using two methods of restraint. *Vet. Rec.* 126:522–525.

59. Elliott, C.T., W.J. McCaughey, and H.D. Shortt. 1993. Residues of the beta-agonist clenbuterol in tissues of medicated farm animals. *Food Addit. Contam.* 10:231–244.

60. Elliott, C.T., J.D.G. McEvoy, W.J. McCaughey, et al. 1993. Effective laboratory monitoring for the abuse of the beta-agonist clenbuterol in cattle. *Analyst* 118:447–448.

61. Ely, R.W., and J.C. Fox. 1989. Elevated IgG antibody to *Sarcocystis cruzi* associated with eosinophilic myositis in cattle. *J. Vet. Diagn. Invest.* 1:53–56.

62. Engel, R.E. 1977. Nitrites, nitrosamines, and meat. *J. Am. Vet. Med. Assoc.* 171:1157–1160.

63. Eriksen, P.J. 1978. *Slaughterhouse and Slaughterslab Design and Construction.* FAO Anim. Prod. Health Pap. 9. Rome: Food and Agriculture Organization.

64. Ernst, R.A. 1987. Microbiological monitoring of hatchery and hatching egg sanitation. *Worlds Poult. Sci. J.* 43:56–63.

65. Evanson, D.J. 1991. Pesticides and food safety. *Dairy Food Environ. Sanit.* 11:196–199.

66. Farb, P., and G. Armelagos. 1980. *Consuming Passions: The Anthropology of Eating.* Boston: Houghton Mifflin.

67. Fayer, R., and J.P. Dubey. 1985. Methods for controlling transmission of protozoan parasites from meat to man. *Food Technol.* 39:57–60.

68. Fenner, F. 1990. *History of Microbiology in Australia.* Australia Society of Microbiology. Canberra: Brolga Press, 610 pp.

69. Filice, G.A., and C. Pomeroy. 1991. Preventing secondary infections among HIV-positive persons. *Public Health Rep.* 106:503–517.

70. Food and Agriculture Organization. 1979–1986. *Manuals of Food Quality Control.* FAO Food Nutr. Pap. 14/1–7. Rome: FAO.

71. ———. 1991. *Guidelines for Slaughtering, Meat Cutting and Further Processing.* FAO Anim. Prod. Health Pap. 91. Rome: FAO.

72. Franco, D.A., J. Webb, and C.E. Taylor. 1990. Antibiotic and sulfonamide residues in meat: Implications for human health. *J. Food Prot.* 53:178–185.

73. Frank, R., K.I. Stonefield, and H. Luyken. Monitoring wood shaving litter and animal products for polychlorophenol residues, Ontario, Canada, 1978–1986. *Bull. Environ. Contam. Toxicol.* 40:468–474.

74. Fraser, C.M. (ed.). 1991. *Merck Veterinary Manual.* 7th ed. Rahway, N.J.: Merck.

75. Gajadhar, A.A., and W.C. Marquardt. 1992. Ultrastructural and transmission evidence of *Sarcocystis cruzi* associated with eosinophilic myositis in cattle. *Can. J. Vet. Res.* 56:41–46.

76. Gajadhar, A.A., W.D.G. Yates, and J.R. Allen. 1987. Association of eosinophilic myositis with an unusual species of *Sarcocystis* in a beef cow. *Can. J. Vet. Res.* 51:373–378.

77. Gast, R.K., and C.W. Beard. 1993. Research to understand and control *Salmonella enteritidis* in chickens and eggs. *Poult. Sci.* 72:1157–1163.

78. Gemmell, M.A. 1990. Australasian contributions to an understanding of the epidemiology and control of hydatid disease caused by *Echinococcus granulosus*—past, present and future. *Int. J. Parasitol.* 20:431–456.

79. Goodhand, R.H. 1983. The future role of meat inspection in the field of meat hygiene. *J. R. Soc. Health* 103:11–15.

80. Grace, F.H. 1986. Microbial ecology of meat and meat products. In *Advances in Meat Research,* Vol. 2. Westport, Conn.: AVI Publishing, 1–47.

81. Gracey, J.F., and D.S. Collins. 1992. *Meat Hygiene.* 9th ed. London: Bailliere Tindall.

82. Graham-Rack, B., and R. Binstead. 1973. *Hygiene in Food Manufacturing and Handling.* 2d ed. London: Food Trade Press.

83. Grandin, T. 1980. Problems with kosher slaughter. *Int. J. Stud. Anim. Prob.* 1:375–390.

84. ———. 1983. *Livestock Handling from Farm to Slaughter.* Dept. Prim. Ind., Aust. Bur. Anim. Health, Aust. Gov. Publ. Serv. Canberra, 130 pp.

85. ———. 1987. New humane slaughter system installed at Utica Veal. *Off. Proc. Ann. Meet. Livest. Conserv. Inst.* 1987:13–16.

86. ———. 1989. Improving kosher slaughter. *Humane Society News* Spring:9–10.

87. ———. 1994. Euthanasia and slaughter of livestock. *J. Am. Vet. Med. Assoc.* 204:1354–1360.

88. Granstrom, D.E., R.K. Ridley, Y. Baoan, et al. 1989. Type-I hypersensitivity as a component of eosinophilic myositis (muscular sarcocystosis) in cattle. *Am. J. Vet. Res.* 50:571–574.

89. Granstrom, D.E., R.K. Ridley, Y. Baoan, et al. 1990. Immunodominant proteins of *Sarcocystis cruzi* bradyzoites isolated from cattle affected or nonaffected with eosinophilic myositis. *Am. J. Vet. Res.* 51:1151–1155.

90. Griffiths, H.J. 1978. *A Handbook of Veterinary Parasitology.* Minneapolis: University of Minnesota Press.

91. Guthrie, R.K. (ed.). 1988. *Food Sanitation.* 3d ed. Westport, Conn.: AVI Publishing.

92. Haagsma, N., A. Ruiter, and P.B. Czedik-Eysenberg (eds.). 1990. *Proceedings of EuroResidue Conference on Residues of Veterinary Drugs in Food.* Netherlands: Faculty of Veterinary Medicine, University of Utrecht.

93. Hansen, T.J. 1990. Affinity column cleanup and direct fluorescence measurement of aflatoxin M_1 in raw milk. *J. Food Prot.* 53:75–77.

94. Harrison, R. (chrmn.). 1985. *Report on the Welfare of Livestock When Slaughtered by Religious Methods.* Ref. Book 262. Farm Animal Welfare Council. London: Her Majesty's Stationery Office.

95. Hartung, M., and K. Gerigk. 1991. *Yersinia* in effluents from the food-processing industry. *Rev. Sci. Technol. Off. Int. Epizoot.* 10:799–811.

96. Hathaway, S.C. 1993. *Risk Analysis and Meat Hygiene.* Vol. 12. Revue Scientifque et Technique, Office International des Epizooties, 5–12.

97. Hathaway, S.C., and A.I. McKenzie. 1989. Impact of ovine meat inspection systems on processing and production costs. *Vet. Rec.* 204: 189–193.

98. ———. 1991. Meat inspection in New Zealand: Prospects for change. *N. Z. Vet. J.* 39:1–7

99. ———. 1991. Postmortem meat inspection programs: Separating science and tradition. *J. Food Prot.* 54:471–475.

100. Hedman, S. 1994. Recommendations for successful ISO 9000 registration. *Cereal Foods World* 39:389–392.

101. Heise, S. 1991. Pesticide residues in cattle and environmental samples from the northern Ivory Coast. *Toxicol. Environ. Chem.* 33:85–91.

102. Henken, A.M., K. Frankena, J.O. Goelema, et al. 1992. Multivariate epidemiological approach to salmonellosis in broiler breeder flocks. *Poult. Sci.* 71:838–843.

103. Herenda, D.C., and D.A. Franco. 1991. *Food Animal Pathology and Meat Hygiene.* St. Louis, Mo.: Mosby Year Book.

104. Herrick, J.B. 1993. Milk and dairy beef quality assurance program: A food safety issue. *J. Am. Vet. Med. Assoc.* 203:1389.

105. Hobbs, B.C., and D. Roberts. 1993. *Food Poisoning and Food Hygiene.* 6th ed. London: Edward Arnold.

106. Hofstad, M.S., B.W. Calnek, C.F. Helmboldt, et al. (eds.). 1984. *Diseases of Poultry.* 8th ed. Ames: Iowa State University Press.

107. Howard, J.L. (ed.). 1992. *Current Veterinary Therapy: Food Animal Practice.* Philadelphia: W.B. Saunders.

108. Hughes, B.J., J.H. Forsell, S.D. Sleight, et al. 1985. Assessment of pentachlorophenol toxicity in newborn calves: Clinicopathology and tissue residues. *J. Anim. Sci.* 61:1587–1603.

109. Hughes, K.L. 1991. History of veterinary public health in Australasia. *Rev. Sci. Technol. Off. Int. Epizoot.* 10:1019–1040.

110. Hutchinson, D.N. 1992. Foodborne botulism. *Br. Med. J.* 305:264–265.

111. Jaglan, P.S., F.S. Yein, R.E. Hornish, et al. 1992. Depletion of intramuscularly injected ceftiofur from the milk of dairy cattle. *J. Dairy Sci.* 75:1870–1876.

112. Jensen, R., and D.R. Mackey. 1979. *Diseases of Feedlot Cattle.* 3d ed. Philadelphia: Lea and Febiger.

113. Jensen, R., and B.L. Swift. 1982. *Diseases of Sheep.* 2d ed. Philadelphia: Lea and Febiger.

114. Jensen, R., A.F. Alexander, R.R. Dahlgren, et al. 1986. Eosinophilic myositis and muscular sarcocystosis in the carcasses of slaughtered cattle and lambs. *Am. J. Vet. Res.* 47:587–593.

115. Jones, G.M., and E.H. Seymour. 1988. Cowside antibiotic residue testing. *J. Dairy Sci.* 71:1691–1699.

116. Jones, T.C., and R.D. Hunt. 1983. *Veterinary Pathology.* 5th ed. Philadelphia: Lea and Febiger.

117. Jubb, K.V.F., and P.C. Kennedy. 1985. *Pathology of Domestic Animals.* 3d ed. 2 vols. New York: Academic Press.

118. Jukes, D.J. 1988. The structure of food law enforcement in the United Kingdom. *Br. Food J.* 90:239–249.

119. Kerschen, B. 1991. Quality assurance and residue avoidance: Monfort's approach. *Proc. Annu. Conv. Am. Assoc. Bov. Pract.* 23:133–134.

120. Kindred, T. 1993. Risk assessment. *Proc. World Conf. Meat Poult. Insp.* Sponsored by the United States Department of Agriculture, Food Safety Inspection Service. College Station, Tex., 1–9.

121. Kindred, T.P., and W.T. Hubbert. 1993. Residue prevention strategies in the United States. *J. Am. Vet. Med. Assoc.* 202:46–49.

122. Kindred, T.P., W.T. Hubbert, and J.C. Prucha. 1993. Residue prevention in meat, milk, and poultry: A challenge of the 21st century. *Proc. 11th Int. Symp. WAVFH,* 377–380.

123. Klaasen, C.D., M.O. Amdur, and J. Doull (eds.). 1986. *Casarett and Doull's Toxicology: The Basic Science of Poisons.* New York: Macmillan.

124. Koorts, R. 1992. The development of a restraining system to accommodate the Jewish method of slaughter (Shechita). *Proc. Meat Sci. Technol. Serv. Meat Ind,* vol. 7. Irene, Republic of South Africa: Irene Animal Production Institute, 41–51.

125. Kotula, A.W., J.P. Dubey, A.K. Sharar, et al. 1991. Effect of freezing on infectivity of *Toxoplasma gondii* tissue cysts in pork. *J. Food Prot.* 54:687–690.

126. Kramer, A. 1980. *Food and the Consumer.* Rev. ed. Westport, Conn.: AVI Publishing.

127. Kramer, A., and B.A. Twigg. 1973. *Quality Control for the Food Industry.* 3d ed. Vol. 1, *Fundamentals* (1970); Vol. 2, *Applications.* Westport, Conn.: AVI Publishing.

128. Krenk, P. 1991. An overview of rendering structure and procedures in the European Community. *Curr. Top. Vet. Med. Anim. Sci.* 55:161–167.

129. Laing, A.D.M.G. 1970. The history of meat hygiene and inspection in New Zealand up to the formation of the Meat Division in 1963. *N. Z. Vet. J.* 18:241–243.

130. Lasta, J.A., R. Rodriguez, M. Zanelli, et al. 1992. Bacterial count from bovine carcasses as an indicator of hygiene at slaughtering places: A proposal for sampling. *J. Food Prot.* 55:271–278.

131. Lawrie, R.A. 1985. *Meat Science.* 4th ed. New York: Pergamon Press.

132. Laws of Prince Edward Island. 1987. *Dairy Industry Act.* Charlottetown, Prince

Edward Island.

133. ——. 1988. *Dairy Industry Act. Regulations.* Charlottetown, Prince Edward Island.

134. van der Leek, M.L., G.A. Donovan, R.L. Saltman, et al. 1991. Effect of an insecticide controlled-release bolus on a milk antibiotic residue test. *J. Dairy Sci.* 74:433–435.

135. Leman, A.D. (ed.). 1986. *Diseases of Swine.* 6th ed. Ames: Iowa State University Press.

136. Levie, A. 1979. *The Meat Handbook.* 4th ed. Westport, Conn.: AVI Publishing.

137. Levine, N.D. 1980. *Nematode Parasites of Domestic Animals and of Man.* 2d ed. Minneapolis: Burgess Publishing.

138. Libby, J.A. (ed.). 1975. *Meat Hygiene.* 4th ed. Philadelphia: Lea and Febiger.

139. Longree, K., and G. Armbruster. 1987. *Quantity Food Sanitation.* 4th ed. New York: John Wiley and Sons.

140. Longree, K., and G.G. Blaker. 1982. *Sanitary Techniques in Food Service.* 2d ed. New York: Macmillan.

141. MacNeil, J.D., G.O. Korsrud, J.O. Boison, et al. 1991. Performance of five screening tests for the detection of penicillin G residues in experimentally injected calves. *J. Food Prot.* 54:37–40.

142. Mann, I. 1960. *Meat Handling in Underdeveloped Countries.* Rome: Food and Agriculture Organization.

143. Mason, J., and E. Ebel. 1992. USDA task force favors voluntary SE control plan. *Feedstuffs* 64(26):4–5, 22.

144. Matyas, Z. (chrmn.). 1988. *Salmonellosis Control: The Role of Animal and Product Hygiene.* WHO Tech. Rep. Ser. 774. Geneva: World Health Organization.

145. Mawhinney, H., and S. Oakenfull. 1990. Pesticide and antibiotic residues in pig products—current monitoring methodology and results. *Proc. Aust. Soc. Anim. Prod.* 18:94–96.

146. McCapes, R.H., B.I. Osburn, and H. Riemann. 1991. Safety of foods of animal origin: Model for elimination of *Salmonella* contamination of turkey meat. *J. Am. Vet. Med. Assoc.* 199:875–880.

147. McDougall, K.W. 1990. Organochlorine residues in pasture and livestock. *Proc. Aust. Soc. Anim. Prod.* 18:19–27.

148. McEwen, S.A., W.D. Black, and A.H. Meek. 1991. Antibiotic residue prevention methods, farm management, and occurrence of antibiotic residues in milk. *J. Dairy Sci.* 74:2128–2137.

149. McEwen, S.A., A.H. Meek, and W.D. Black. 1991. A dairy farm survey of antibiotic treatment practices, residue control methods and associations with inhibitors in milk. *J. Food Prot.* 54:454–459.

150. McIntyre, C.R. 1991. Hazard analysis critical control point (HACCP) identification. *Dairy Food Environ. Sanit.* 11:357–358.

151. McMahon J., S. Kahn, R. Batey, et al. 1987. Revised post-mortem inspection procedures for cattle and pigs slaughtered at Australian abattoirs. *Aust. Vet. J.* 64(6):183–187.

152. McNab, W.B., C.M. Forsberg, and R.C. Clarke. 1991. Application of an automated hydrophobic grid membrane filter interpreter system at a poultry abattoir. *J. Food Prot.* 54:619–622.

153. Mead, G.C. 1990. Food poisoning salmonellas in the poultry-meat industry. *Br. Food J.* 92:32–36.

154. Medina, M.B., R.A. Barford, M.S. Palumbo, et al. 1992. Evaluation of commercial immunochemical assays for detection of sulfamethazine in milk. *J. Food Prot.* 55:284–290.

155. Microbiology and Food Safety Committee of the National Food Processors As-

sociation. 1992. HACCP and total quality management—winning concepts for the 90's: A review. *J. Food Prot.* 55:459–462.

156. Moats, W.A. 1982. A staining procedure for detecting cracked eggs. *Poult. Sci.* 61:1007–1008.

157. Mock, D.E., and D.C. Cress. 1986. Pesticides, meat animal production and residues in food. *Dairy Food Sanit.* 6:234–239.

158. Monsey, J.B., and J.B. Jones. 1979. A simple, enzymatic test for monitoring the efficient thermal pasteurization of chicken egg-white. *J. Food Technol.* 14:381–388.

159. Mossel, D.A.A. 1988. Impact of foodborne pathogens on today's world, and prospects for management. *Anim. Human Health* 1:13–23.

160. Mossel, D.A.A., and D.M. Drake. 1990. Processing food for safety and reassuring the consumer. *Food Technol.* 44:63–67.

161. Mossel, D.A.A., and P. van Netten. 1991. Microbiological reference values for foods: A European perspective. *J. Assoc. Off. Anal. Chem.* 74:420–432.

162. Mossel, D.A.A., and C.B. Struijk. 1992. The contribution of microbial ecology to management and monitoring of the safety, quality and acceptability (SQA) of foods. *J. Appl. Bact. Symp. Suppl.* 73:1S–22S.

163. National Research Council. Commission on Life Sciences Food and Nutrition Board. 1985. *Meat and Poultry Inspection. The Scientific Basis of the Nation's Program.* Washington, D.C.: National Academy Press.

164. ———. 1987. *Poultry Inspection. The Basis for a Risk-Assessment Approach.* Washington, D.C.: National Academy Press.

165. Newman, S., and A. McKenzie. 1991. Organisation of veterinary public health in Australasia and the Pacific Islands. *Rev. Sci. Technol. Off. Int. Epizoot.* 10:1159–1184.

166. Nicholson, M.D. 1980. The value of traditional meat inspection procedures in Australia—the view from industry. *Aust. Adv. Vet. Sci, Hobart*, Aust. Vet. Assoc., 40–43.

167. Noble, A. 1990. The relation between organochlorine residues in animal feeds and residues in tissues, milk and eggs: A review. *Aust. J. Exp. Agric.* 30:145–154.

168. Nurmi. E., L. Nuotio, and C. Schneitz. The competitive exclusion concept: Development and future. *Int. J. Food Microbiol.* 15:237–240.

169. Oosterom, J. 1991. Epidemiological studies and proposed preventive measures in the fight against human salmonellosis. *Int. J. Food Microbiol.* 12:41–51.

170. Palmer, S.R., and B. Rowe. 1983. Investigation of outbreaks of salmonella in hospitals. *Br. Med. J.* 287(6396):891–893.

171. Pan American Health Organization. 1982. *Sanitary Control of Food.* Sci. Publ. 421. Washington, D.C.: World Health Organization.

172. Panciera, R.J., S.A. Ewing, E.M. Johnson, et al. 1993. Eosinophilic mediastinitis, myositis, pleuritis, and pneumonia of cattle associated with migration of first-instar larvae of *Hypoderma lineatum. J. Vet. Diagn. Invest.* 5:226–231.

173. Persson, U., and S. Jendteg. 1992. The economic impact of poultry-borne salmonellosis: How much should be spent on prophylaxis? *Int. J. Food Microbiol.* 15:207–213.

174. Peterson, M.S., and D.K. Tressler (eds.). 1963/1965. *Food Technology the World Over: Vol. 1, Europe, Canada and the United States, Australia: Vol. 2, South America, Africa and the Middle East, Asia.* Westport, Conn.: AVI Publishing.

175. Place, A.R. 1990. Animal drug residues in milk: The problem and the response. *Dairy Food Environ. Sanit.* 10:662–664.

176. Potter, N.N. 1986. *Food Science.* 4th ed. Westport, Conn.: AVI Publishing.

177. Price, J.F., and B.S. Schweigert (eds.). 1987. *The Science of Meat and Meat Products.* 3d ed. San Franscisco: W.H. Freeman.

178. Prokop, W.H. 1985. Rendering systems for processing animal by-product materials. *J. Am. Oil Chem. Soc.* 62:805–811.

179. Pullen, M.M., and G.R. Ruth. 1981. Eosinophilic myositis in a slaughtered heifer. *J. Am. Vet. Med. Assoc.* 178:140.

180. Pylypiw, H.M., Jr., and L. Hankin. 1991. Herbicides in pooled raw milk in Connecticut. *J. Food Prot.* 54:136–137.

181. Reinert, R.E., B.A. Knuth, M.A. Kamrin, et al. 1991. Risk assessment, risk management, and fish consumption advisories in the United States. *Fisheries* 16:5–12.

182. Ricard, D. 1982. A comparative analysis of the egg grading industry in Ontario and Quebec. *Can. Farm Econ.* 17(3):1–14.

183. Richardson, G.H. (ed.). 1985. *Standard Methods for the Examination of Dairy Products.* 15th ed. Washington, D.C.: American Public Health Association.

184. Roberts, T., and D. Smallwood. 1991. Data needs to address economic issues in food safety. *Am. J. Agric. Econ.* 73:933–942.

185. Ryan, J.J., E.E. Wildman, A.H. Duthie, et al. 1986. Detection of penicillin, cephapirin, and cloxacillin in commingled raw milk by the Spot Test. *J. Dairy Sci.* 69:1510–1517.

186. Ryder, C.J. 1990. U.K. food legislation. *Lancet* 336:1559–1562.

187. Saini, P.K., and D.W. Webert. 1991. Application of acute phase reactants during antemortem and postmortem meat inspection. *J. Am. Vet. Med. Assoc.* 198:1898–1901.

188. Samarajeewa, U., C.I. Wei, T.S. Huang, et al. 1991. Application of immunoassay in the food industry. *Crit. Rev. Food Sci. Nutr.* 29:403–434.

189. Sastry, M.S. 1983. Monitoring of pesticide residues in animal feeds and animal products. *Pesticides* 17(12):36–38.

190. Schell, O. 1985. *Modern Meat.* New York: Random House.

191. Selmer-Olsen, A. 1985. Guidelines for bacterial counts on carcasses at Cato Ridge abattoir. *J. S. Afr. Vet. Assoc.* 56:99–100.

192. Senyk, G.F., J.H. Davidson, J.M. Brown, et al. 1990. Comparison of rapid tests used to detect antibiotic residues in milk. *J. Food Prot.* 53:158–164.

193. Sharp, J.C.M. 1991. Foodborne infections and poultry. *J. R. Soc. Health* 111:35–37.

194. Shiffman, M. A. (chrmn.). 1974. *Foodborne Disease: Methods of Sampling and Examination in Surveillance Programmes.* WHO Tech. Rep. Ser. 543. Geneva: World Health Organization.

195. Silliker, J.H. (chrmn.). 1988. *Microorganisms in Foods 4. Application of the Hazard Analysis Critical Control Point (HACCP) System to Ensure Microbiological Safety and Quality.* Oxford: Blackwell Scientific.

196. Silverside, D., and M. Jones. 1992. *Small-Scale Poultry Processing.* FAO Anim. Prod. Health Pap. 98. Rome: Food and Agriculture Organization.

197. Sinclair, U. 1906. *The Jungle.* Reprint. New York: New American Library, 1961.

198. Sloss, M.W., and R.L. Kemp. 1978. *Veterinary Clinical Parasitology.* 5th ed. Ames: Iowa State University Press.

199. Smeltzer, T.I. 1981. Ecology of Salmonella in the abattoir environment. *Adv. Vet. Publ. Health,* Aust. Coll. Vet. Sci. 1: 43–48.

200. ———. 1985. *Salmonella* contamination of beef in the abattoir environment. *Proc. Int. Symp. Salmonella,* 262–271.

201. Smith, H.J., K.E. Snowdon, and G.G. Finley. 1991. Eosinophilic myositis in Canadian cattle. *Can. J. Vet. Res.* 55:94–95.

202. Smith. L.L.W., and L.J. Minor (eds.). 1974. *Food Service Science.* Westport, Conn.: AVI Publishing.

203. Snijders, J.M.A., J.G. van Logtestijn, and B.R. Berends. 1993. Integrated quality control and HACCP as prerequisites for a new meat inspection system. *Proc. 11th Int. Symp. WAVFH,* 123–127.

204. Soulsby, E.J.L. 1982. *Helminths, Arthropods and Protozoa of Domesticated An-*

imals. 7th ed. Baltimore: Williams and Wilkins.

205. Stadelman, W.J., and O.J. Cotterill (eds.). 1990. *Egg Science and Technology.* 3d ed. Binghamton, N.Y.: Food Products Press.

206. Stallones, R.A. (chrmn). 1980. *The Effects on Human Health of Subtherapeutic Use of Antimicrobials in Animal Feeds.* Washington, D.C.: National Academy of Sciences.

207. Stoloff, L., H.P. van Egmond, and D.L. Park. 1991. Rationales for the establishment of limits and regulations for mycotoxins. *Food Addit. Contam.* 8:213–221.

208. Strauch, D. (ed.). 1987. *Animal Production and Environmental Health.* World Anim. Sci., B6. Amsterdam: Elsevier.

209. Sundlof, S.F., J.E. Riviere, and A.L. Craigmill. 1991. *Food Animal Residue Avoidance Databank Trade Name File.* Gainesville: Institute of Food and Agricultural Sciences, University of Florida.

210. Swann, M.M. 1969. *Joint Committee on the Use of Antibiotics in Animal Husbandry and Veterinary Medicine: Report.* London: Her Majesty's Stationery Office.

211. Szczawinska, M.E., D.W. Thayer, and J.G. Phillips. 1991. Fate of unirradiated *Salmonella* in irradiated mechanically deboned chicken meat. *Int. J. Food Microbiol.* 14:313–324.

212. TerHune, T.N., and D.W. Upson. 1989. Factors affecting the accuracy of the live animal swab test for detecting urine oxytetracycline and predicting oxytetracycline residues in calves. *J. Am. Vet. Med. Assoc.* 194:918–921.

213. Thompson, D.R., I.D. Wolf, K.L. Nordsiden, et al. 1979. Home canning of food: Risks resulting from errors in processing. *J. Food Sci.* 44:226–233.

214. Thompson, P., P.A. Salsbury, C. Adams, et al. 1990. U.S. food legislation. *Lancet* 336:1557–1559.

215. Thomson, R.G. 1984. *General Veterinary Pathology.* 2d ed. Philadelphia: W.B. Saunders.

216. Thorner, M.E. 1973. *Convenience and Fast Food Handbook.* Westport, Conn.: AVI Publishing.

217. Timony, J.F. 1988. *Hagan and Bruner's Microbiology and Infectious Diseases of Domestic Animals.* 8th ed. Ithaca, N.Y.: Cornell University Press.

218. Todd, J. 1987. The Pasteur Institute of Australia: Success and failure. In *Louis Pasteur and the Pasteur Institute in Australia,* ed. J. Chaussivert and M. Blackman. University of New South Wales.

219. Tompkin, R.B. 1990. The use of HACCP in the production of meat and poultry products. *J. Food Prot.* 53:795–803.

220. Townsend, W.E., and L.C. Blankenship. 1989. Methods for detecting processing temperatures of previously cooked meat and poultry products—a review. *J. Food Prot.* 52:128–135.

221. Trichilo, C.L. 1987. EPA pesticide contaminant concerns for residues in food and feed. *Cereal Foods World* 32:806, 808–810.

222. Universities Federation for Animal Welfare. 1987. *Humane Slaughter of Animals for Food.* Proceedings of a Symposium. South Mimms, England: UFAW.

223. United Nations Conference on Trade and Development/General Agreement on Tariffs and Trade (UNCTAD/GATT). 1991. *Quality Control for the Food Industry: An Introductory Handbook.* Geneva: International Trade Centre UNCTAD/GATT.

224. United Nations. Food and Agriculture Organization. 1970. *Joint FAO/WHO Expert Committee on Milk Hygiene, Third Report.* FAO Agric. Stud. 83. Rome: FAO.

225. ———. 1979–1986. *Manual of Food Quality Control.* FAO Food Nutr. Pap. 14/1–14/9. Rome: FAO.

226. United Nations. World Health Organization. 1989. *Evaluation of Programmes to Ensure Food Safety: Guiding Principles.* Geneva: WHO.

227. U.S. Department of Agriculture. 1983. Egg products inspection act. *Agric. Handb. U.S. Dept. Agric.* 556(Rev.):127–142.

228. ———. 1983. *Egg-Grading Manual. Agric. Handb. U.S. Dept. Agric.* 75(Rev.).

229. U.S. Department of Health and Human Services. 1988. *Dollars and Sense.* Proc. Symp. Anim. Drug Use. DHHS Publ. (FDA) 88-6045. Washington, D.C.: U.S. Government Printing Office.

230. U.S. Department of Health and Human Services. Public Health Service. 1991. *Grade "A" Pasteurized Milk Ordinance: Recommendations to the United States Public Health Service, Food and Drug Administration.* PHS Publ. 229. Washington, D.C.: U.S. Government Printing Office.

231. Vanderzant, C., and D.F. Splittstoesser (eds.). 1992. *Compendium of Methods for the Microbiological Examination of Foods.* 3d ed. Washington, D.C.: American Public Health Association.

232. Vazquez-Moreno, L., M.C. Bermudez, A.A. Langure, et al. 1990. Antibiotic residues and drug resistant bacteria in beef and chicken tissues. *J. Food Sci.* 55:632–634, 657.

233. Veary, C.M. 1993. The application of the HACCP principle in a large rural poultry abattoir in the Republic of South Africa. *Proc. 11th Int. Symp. WAVFH,* 68–71.

234. Vickery, J.R. 1990. Meat. In *Food Science and Technology in Australia: A Review of Research Since 1990.* Melbourne: CSIRO, 162 pp.

235. Vorster, S.M. 1992. Foodborne pathogens in the South African meat industry. *Proc. Meat Sci. Tech. Serv. Meat Ind.,* vol 7. Irene, Republic of South Africa: Irene Animal Production Institute, 91–95.

236. Wierup. M., H. Wahlstrom, and B. Engstrom. 1992. Experience of a 10-year use of competitive exclusion treatment as part of the *Salmonella* control programme in Sweden. *Int. J. Food Microbiol.* 15:287–291.

237. Wiggins, G.S., and A. Wilson. 1976. *Color Atlas of Meat and Poultry Inspection.* New York: Van Nostrand Reinhold.

238. Wilkinson, P.J., S.P. Dart, and C.J. Hadlington. 1991. Cook-chill, cook-freeze, cook-hold, *sous vide:* Risks for hospital patients? *J. Hosp. Infect.* 18(Suppl. A):222–229.

239. Willeberg, P. 1979. Epidemiological applications of Danish swine slaughter inspection data. *Proc. 2d Int. Symp. Vet. Epidemiol. Econ.,* 161–167.

240. Willeberg, P., M.-A. Gerbola, B.K. Petersen, et al. 1984. The Danish pig health scheme: Nation-wide computer-based abattoir surveillance and follow-up at the herd level. *Prev. Vet. Med.* 3:79–91.

241. Wilson, D.J., C.E. Franti, and B.B. Norman. 1991. Antibiotic and sulfonamide agents in bob veal calf muscle, liver, and kidney. *Am. J. Vet. Res.* 52:1383–1387.

242. Wilson, L.L., and J.R. Dietrich. 1993. Assuring a residue-free food supply: Special-fed veal. *J. Am. Vet. Med. Assoc.* 202:1730–1733.

243. World Health Organization. 1993. *Guidelines for Cholera Control.* Geneva: WHO.

244. Wythes, J.R. 1981. Transport and handling of livestock from farm to slaughter. *Adv. Vet. Publ. Health,* Aust. Coll. Vet. Sci. 1: 22–35.

245. Zomer, E., S. Saul, and S.E. Charm. 1992. HPLC receptorgram: A method for confirmation and identification of antimicrobial drugs by using liquid chromatography with microbial receptor assay. I. Sulfonamides in milk. *J. AOAC Int.* 75:987–993.

246. U.S. Department of Agriculture. *Statistical Summary, Federal Meat and Poultry Inspection,* fiscal years 1986–90.

247. World Health Organization. 1988. Salmonellosis control: The role of animal and product hygiene. WHO Technical Report Series 774. Geneva.

Index

Printed in the United States
87881LV00004B/27/A